Sustainable Textiles: Production, Processing, Manufacturing & Chemistry

Series Editor

Subramanian Senthilkannan Muthu, Head of Sustainability, SgT and API, Kowloon, Hong Kong

This series aims to address all issues related to sustainability through the lifecycles of textiles from manufacturing to consumer behavior through sustainable disposal. Potential topics include but are not limited to: Environmental Footprints of Textile manufacturing; Environmental Life Cycle Assessment of Textile production; Environmental impact models of Textiles and Clothing Supply Chain; Clothing Supply Chain Sustainability; Carbon, energy and water footprints of textile products and in the clothing manufacturing chain; Functional life and reusability of textile products; Biodegradable textile products and the assessment of biodegradability; Waste management in textile industry; Pollution abatement in textile sector; Recycled textile materials and the evaluation of recycling; Consumer behavior in Sustainable Textiles; Eco-design in Clothing & Apparels; Sustainable polymers & fibers in Textiles; Sustainable waste water treatments in Textile manufacturing; Sustainable Textile Chemicals in Textile manufacturing. Innovative fibres, processes, methods and technologies for Sustainable textiles; Development of sustainable, eco-friendly textile products and processes; Environmental standards for textile industry; Modelling of environmental impacts of textile products; Green Chemistry, clean technology and their applications to textiles and clothing sector; Eco-production of Apparels, Energy and Water Efficient textiles. Sustainable Smart textiles & polymers, Sustainable Nano fibers and Textiles; Sustainable Innovations in Textile Chemistry & Manufacturing; Circular Economy, Advances in Sustainable Textiles Manufacturing; Sustainable Luxury & Craftsmanship; Zero Waste Textiles.

More information about this series at https://link.springer.com/bookseries/16490

Subramanian Senthilkannan Muthu · Ali Khadir
Editors

Advanced Oxidation Processes in Dye-Containing Wastewater

Volume 1

Springer

Editors
Subramanian Senthilkannan Muthu
SgT Group and API
Hong Kong, Kowloon, Hong Kong

Ali Khadir
Western University
London Ontario, Canada

ISSN 2662-7108 ISSN 2662-7116 (electronic)
Sustainable Textiles: Production, Processing, Manufacturing & Chemistry
ISBN 978-981-19-0986-3 ISBN 978-981-19-0987-0 (eBook)
https://doi.org/10.1007/978-981-19-0987-0

This Springer imprint is published by the registered company Springer Nature Singapore Pte Ltd.
The registered company address is: 152 Beach Road, #21-01/04 Gateway East, Singapore 189721, Singapore

Contents

About the Editors

Dr. Subramanian Senthilkannan Muthu currently works for SgT Group as Head of Sustainability and is based out of Hong Kong. He earned his PhD from The Hong Kong Polytechnic University and is a renowned expert in the areas of Environmental Sustainability in Textiles & Clothing Supply Chain, Product Life Cycle Assessment (LCA), and Product Carbon Footprint Assessment (PCF) in various industrial sectors. He has 5 years of industrial experience in textile manufacturing, research and development, and textile testing and over a decade's experience in life cycle assessment (LCA), carbon and ecological footprints assessment of various consumer products. He has published more than 100 research publications, written numerous book chapters, and authored/edited over 100 books in the areas of Carbon Footprint, Recycling, Environmental Assessment and Environmental Sustainability.

Dr. Ali Khadir is an environmental engineer and a member of the Young Researcher and Elite Club, Islamic Azad University of Shahre Rey Branch, Tehran, Iran. He has published several articles and book chapters in reputed international publishers, including Elsevier, Springer, Taylor & Francis, and Wiley. His articles have been published in journals with IF of greater than four, including Journal of Environmental Chemical Engineering and International Journal of Biological Macromolecules. He also has been the reviewer of journals and international conferences. His research interests center on emerging pollutants, dyes, and pharmaceuticals in aquatic media, advanced water and wastewater remediation techniques and technology.

Fundamental of Advanced Oxidation Processes

Pallavi Jain, Prashant Singh, and Madhur Babu Singh

Abstract Today's most concerning situation is wastewater releasing from printing, dyeing, and textile industries. It contains reactive dye residuals, harmful chemicals, and many inorganic and organic pollutants. Pre-treatment of these pollutants is very important before release. For the treatment of these pollutants, the use of biotechnology with highly concentrated organic and inorganic chemicals is more difficult. So, there is a need for an alternative for it, which has the ability to degrade this non-biodegradable pollutant. The application of advanced oxidation processes (AOPs) is a sustainable solution. In AOPs, there is the formation of radical specifically hydroxyl radical ($^{\bullet}OH$) which is an excellent oxidant. Nowadays, the use of AOPs is started in many places for the treatment of wastewater. In this chapter, there is detailed information on all AOPs.

Keywords Textile industries · Pre-treatment · Organic and inorganic · Advanced oxidation process · Hydroxyl radical

1 Introduction

Water is the prime natural resource on our planet. Without water, life cannot be possible beyond a few days and the lack of access to adequate water leads to the spread of many diseases. As we know that the ocean contains most of the water on the planet and is highly prone to getting polluted [1]. Major sources of water contamination are agriculture, construction, industrialization, and marine dumping. The studies in last 10 years show that industrialization, tourism, and fishing occupation are mostly affecting surface water and pose too many diseases [2]. "In 2008, the European Union

P. Jain (✉) · M. B. Singh
Department of Chemistry, SRM Institute of Science and Technology, Delhi-NCR Campus, Modinagar, Ghaziabad, Uttar Pradesh 201204, India
e-mail: palli24@gmail.com

P. Singh
Department of Chemistry, Atma Ram Sanatan Dharma College, University of Delhi, New Delhi, India

S. S. Muthu and A. Khadir (eds.), *Advanced Oxidation Processes in Dye-Containing Wastewater*, Sustainable Textiles: Production, Processing, Manufacturing & Chemistry,
https://doi.org/10.1007/978-981-19-0987-0_1

developed the Marine Strategy Framework Directive (2008/56/EC to scan the marine environment and related activities." This was made to achieve "Good Environmental Status (GES)" by 2020. The ugly unregulated trend of releasing effluents in the water bodies comprising organic dyes, plastic, toxic chemicals, and agricultural waste destroying the quality of water [3, 4]. Presently, the convention method for treating water has drawbacks and challenges with satisfactory results. The adsorption process has the drawback that it generates a huge amount of toxic sludge, convert into another form of pollutant, and remain in the ecosystem [5, 6].

With the increasing civilization, we have to focus on the techniques that are able to meet up the growing demand for clean water [7, 8]. Hence, advanced oxidation method for the treatment of water was developed. This method is one of the most efficient methods used for the oxidation of biological pollutants that are not conceived by the convention method due to their chemical stability. The mechanism of degradation involves the production of reactive and unstable products like ozone, hydroxyl, and hydrogen peroxide radicals that react with the pollutant into decomposed product and other inorganic salts [9]. In the reaction, hydroxyl radical ($^{\bullet}OH$) undergoes electron transfer or abstraction of hydrogen and forms a double bond with the pollutant [10, 11]. Advanced oxidation technology involves the four different methods for producing $^{\bullet}OH$ using different reacting system [12].

2 Advanced Oxidation Process (AOPs)

It is a chemical treatment that is used to remove the organic and inorganic compounds from the water and wastewater by the oxidation reactions with the $^{\bullet}OH$. It is basically called the in situ production of the highly reactive $^{\bullet}OH$. These are the reactive species that can oxidize the compounds present in the water matrix. $^{\bullet}OH$ reacts as it can produce, oxidize, and separate the inorganic molecules [13, 14]. Aromatic, volatile organic chemicals, petroleum components, pesticides, sewage, and other harmful and non-degradable compounds are removed from wastewater using this method [15]. Figures 1 and 2 show the steps involved in AOPs and main features of $^{\bullet}OH$, respectively.

3 Ozone (O_3)-Based AOPs

Ozone is utilized as both an oxidant and a disinfectant in water treatment. O_3-based AOPs basically attack the functional groups that are rich in electrons like double bonds, amides, and aromatic compounds because they form $^{\bullet}OH$ [16] in the aqueous solution and the reaction is second-order reaction which is quite slow having the rate constant 70 M^{-1} s^{-1}.

O_3 treatment is found to be a more effective advanced oxidation method as it effectively oxidizes a wide range of pollutants and is used to decolorize bleaching

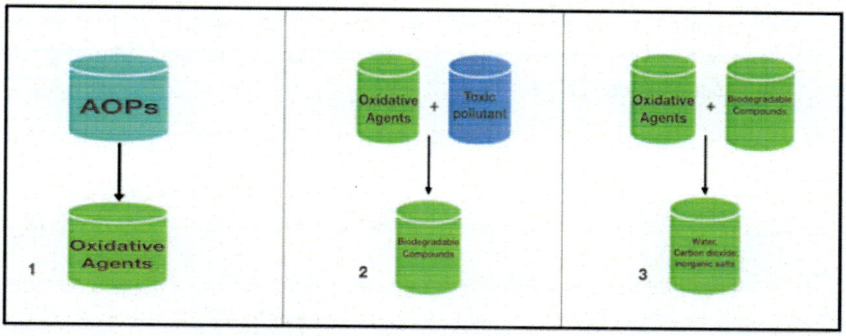

Fig. 1 Steps involved in advanced oxidation process

Fig. 2 Features of hydroxyl radical

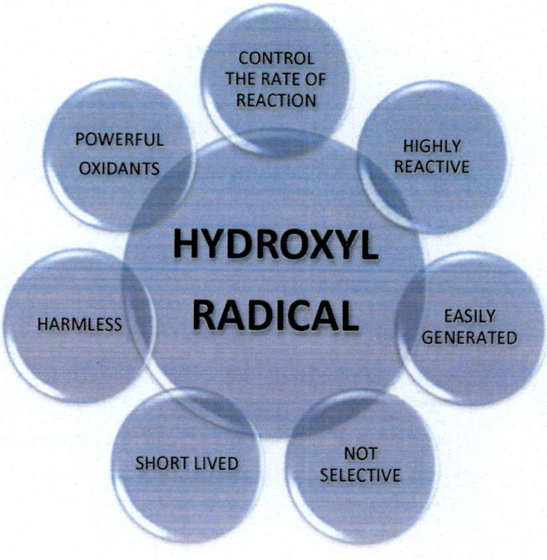

water in the textile, pulp, and paper industries. It is widely used for the treatment of pharmaceutical in water by combining O_3 as oxidant. The O_3 gas is completely unstable in an aqueous medium, so it directly attacks the substrate or generates $^{\bullet}OH$ that attacks the substrate [17]. The generation of O_3 involves different methods such as corona discharge (production of ozone by passing dust-free oxygen through the high-energy electrical field between two electrodes separated by the dielectric material), photocatalytic method, or conversion of oxygen molecule into O_3. Most economic and environmentally preferable methods are corona discharge and UV radiation method. In corona discharge method temperature control is important because at high temperature decomposition of O_3 get started [18]. The spontaneous degradation of O_3 occurs above 35 °C, and therefore must be equipped with a cooling system.

Equations 1 and 2 explain when the oxygen gas passes through the electrode it gets split into oxygen atom, i.e., reactive radical which combines with oxygen molecule to generate ozone [19, 20].

$$O_2 + e^- \rightarrow 2O^- \quad (1)$$

$$O_2 + O \rightarrow O_3 \quad (2)$$

The UV radiation method involves photo-dissociation of oxygen molecule which gives unstable O_2 atom at 140–190 nm wavelength of UV radiation. However, the O_3 gas gets decomposed above the wavelength of 180 nm. Therefore, this method is not widely used in the industrial areas because of the poor yield of O_3 gas.

3.1 *Ozonation at Elevated pH*

Oxidation performed by ozone can be achieved by two following processes:

(a) Indirect oxidation followed by generation of ($^{\bullet}OH$).
(b) Direct oxidation with dissolved O_3.

pH of the medium is an important factor on which the degradation of a compound depends. Generally, when pH <4 then direct ozonation wins.

$$3O_3 + OH^- + H^+ \rightarrow 2OH + 4O_2 \quad (3)$$

When pH > 9 then indirect oxidation wins the race. According to rule, rate of degradation in ozonation increases with increase in pH. Hence, elevated pH promotes indirect oxidation via radical generation

$$O_3 + OH^- \rightarrow O_2 + HO_2^- \quad (4)$$

3.2 *O_3/H_2O_2*

The peroxone process involves the reaction of O_3 with peroxide (H_2O_2) and forms the $^{\bullet}OH$ precursors which act as an oxidizing radical. In essence, O_3 and H_2O_2 (in ionized form, hydroperoxide H_2O^-) had a fast reaction, leading in the creation of

$^{\bullet}OH$ radicals (Eqs. 5–7). This process is more efficacious as H_2O_2 boosts the O_3 decomposition rate, leading to a significant amount of highly reactive $^{\bullet}OH$ radicals.

$$H_2O_2 + 2O_3 \rightarrow 2^{\bullet}OH + 2O_3 \tag{5}$$

$$H_2O_2 \leftrightarrow H^+ + H_2O^- \tag{6}$$

$$H_2O^- + O_3 \rightarrow {}^{\bullet}OH + {}^{\bullet}O_2 + O_2 \tag{7}$$

Additional mechanisms that occur under optimum conditions, such as an H_2O_2/O_3 ratio of 0.5 and a pH of 7.7, are utilized to degrade H_2O_2 by the radicals. This technique is employed to reduce micropollutants from industrial water, potable water, and groundwater, and is also used to lower the bromate formation [21].

3.3 O_3/Catalysts

Ozonation performed by the use of catalysts increases the rate of ozone decomposition followed by generation of hydroxyl radical. d-block elements are generally used as catalysts in the homogeneous catalytic oxidation method for degradation of pollutant (organic). In this process, metal ions mainly alter the reaction rate, the ability of ozone employment, and ozone oxidation selectivity.

There are two following approaches through which homogenous catalytic ozonation will be achieved:

(a) Generation of free radical by decomposition of O_3 through metal ion.
(b) Complex formation within organic molecule and catalyst through complex oxidation.

$$Fe^{2+} + O_3 \rightarrow FeO^{2+} + O^{\bullet} \tag{8}$$

$$FeO^{2+} + H_2O \rightarrow Fe^{3+} + OH^{\bullet} + OH^- \tag{9}$$

$$FeO^{2+} + Fe^{2+} + 2H^+ \rightarrow 2Fe^{3+} + H_2O \tag{10}$$

Heterogeneous catalytic ozonation method is an effective process for ozonation of poorly biodegradable organic by using solid catalyst. Frequently used catalyst in this process are oxide of metals such as manganese dioxide, iron(III)oxide, copper iron oxide, and some porous material like zeolite and activate carbon. Potential of these catalytic methods mainly relay on pH of the solution, targeted pollutant, and type of catalyst. Heterogeneous catalytic ozonation can be achieved via two steps: (a) hydroxyl method and (b) interfacial reaction method.

4 AOPs Based on UV Irradiation

Combining UV with H_2O_2 is the most commonly used UV-based AOP. Persulfate and chlorine are among the other radical promoters that generate sulfate radical, $^{\bullet}OH$ and $^{\bullet}Cl$ radical, respectively, and are investigated. This type of process is called advance oxidation. The sources used for UV irradiation are low-pressure and medium-pressure mercury lamps [22, 23].

4.1 *UV/H_2O_2*

UV radiation has been intensively investigated for its potential to decrease contaminants in water and wastewater. The radiation is commonly used to kill microorganisms and breakdown aquatic organic compounds [24]. Various researches have shown that this technique is effective in removing drugs and pharma products from various kinds of surface water [25]. This method is limited to photosensitive chemicals containing water and the water having low COD value [26]. UV/H_2O_2 is a potential alternative to eliminate photoactive organic compounds having less reactivity to $^{\bullet}OH$ and O_3.

When UV radiation is coupled with H_2O_2, the photolytic cleavage of H_2O_2 results in the formation of two •OH (Eq. 6), which oxidizes another organic moiety further [27]. A series of radical formation reactions is as follows (Eqs. 11–15):

$$H_2O_2 + h\upsilon \rightarrow 2^{\bullet}OH \tag{11}$$

$$H_2O_2 + 2^{\bullet}OH \rightarrow {}^{\bullet}OOH + H_2O \tag{12}$$

$$H_2O_2 + {}^{\bullet}OOH \rightarrow {}^{\bullet}OH + O_2 + H_2O \tag{13}$$

$$2^{\bullet}OOH \rightarrow H_2O_2 + O_2 \tag{14}$$

$$2^{\bullet}OH + 2^{\bullet}OOH \rightarrow 2H_2O_2 + O_2 \tag{15}$$

Though the photolytic cleavage of H_2O_2 produces $^{\bullet}OH$ radicals, excessive quantities of H_2O_2 may have a shielding effect on the $^{\bullet}OH$ radicals, reducing the efficacy of the oxidation reaction [28, 29]. As a result, the H_2O_2 preliminary concentration should be strictly managed to enhance the removal process' effectiveness.

4.2 UV/O$_3$

In UV/O$_3$ processes, dissolved O$_3$ is cleaved due to the UV radiation followed by the reaction of water with atomic O$_2$ that forms the thermally excited H$_2$O$_2$ which decompose into the two $^{\bullet}$OH radicals.

$$2H_2O_2 + O_3 \rightarrow 2^{\bullet}OH + O_2 \rightarrow 2H_2O_2 + O_2 \tag{16}$$

The cage recombination of generated H$_2$O$_2$ decomposed to $^{\bullet}$OH in free $^{\bullet}$OH 0.1 quantum yields [30].

4.3 UV/Peroxydisulfate (UV/PDS)

UV/PDS AOP is more effective than UV/H$_2$O$_2$ under identical situations. The degradation rates of ATZ, TCA, and TCS rise when the oxidant dosage is increased while the concentrations of these three chemicals are decreased. The presence of organic compounds scavenging the sulfate radical and $^{\bullet}$OH reduces the removal rate of the targeted compounds.

4.4 UV/Cl$_2$

In the UV/Cl$_2$ AOPs, UV-activated chloride forms the free radicals, i.e., Cl$^{\bullet}$ and $^{\bullet}$OH radicals that oxidize the target compounds. Here Cl$^{\bullet}$ is more favorable than the $^{\bullet}$OH as it selectively reacts with the electron-rich contaminants and impurities. This type of AOP is favoring to those that have lower pH value water like reverse osmosis permeate.

5 Physical AOPs

5.1 Microwave (MW) AOP

The utilization of extremely energetic waves or radiation under the microwave region (300 MHz–300 GHz) has been examined for effective water contaminant oxidation process. The microwaves are utilized together with catalysts and oxidants (mostly H$_2$O$_2$) to help in the devastation of the organic pollutants. These microwaves are capable of enhancing the rate of the reactions and an inner molecule vibration that helps to induce particular heating of various contaminants. Furthermore, microwaves can produce UV radiations through a discharge lamp which is electrodeless for united UV/MW reactors. Regrettably, a larger amount of applied microwave energy is transferred into heat. Due to the small electrical productivity, various kinds of cooling appliances must be operated for preventing the purified water from excess heating.

5.2 Electron Beam AOP

The implementation of the ionizing radiation origin by an electron beam source (0.01–10 meV) has been investigated since 1980 for wastewater treatment. The electrons that are accelerated go through the surface of the water; as a result, an electronically excited species is formed in water that includes a range of free radicals and ionic species. The highest penetration strength of the fast-moving electrons and the incident electron energy is directly proportional to each other (around 7 mm) [31]. The electron beam process indicates a high degree of oxidizing strength and a slight intrusion via water surrounding substances and electrical capability is under the practicability range.

5.3 Ultrasound AOP

Water sonication through the ultrasound process results in the development and collapse of mini-bubbles through sound propagation contraction and unusual faction. These resulting bubbles break down viciously after getting a significant resonance dimension (size) and create transitory high temperature (>500 K), pressure (>1000 bar), and excessively reactive free radicals. Due to the thermal decay and diverse radical chemical reactions, the devastation of aqueous contaminants takes place. The cavitation through ultrasound shows less obstruction from the aqueous matrix and low heat flow in contrast to UV radiation. The sonochemical methods are well known for oxidizing the various water contaminants in laboratory-scale experimentation [32]. Although the function of ultrasound is very much energy in depth and produces in an extremely less electrical order of AOPs as compared to other techniques [33]. Hence, the combination of ultrasound with UV radiation (known as sonophotolysis), catalysts (TiO_2) or oxidants (mainly O_3 and H_2O_2) or both (known as sonophotocatalysis) achieves great consideration [34]. These fused routes can result in additional benefits. Although a large enhancement of energy effectiveness is gained frequently because of the large competence of the united additional methods or processes (UV/H_2O_2 in ultrasound/UV/H_2O_2) [35].

5.4 Plasma

Aqueous-phase electrically liberating reactors are used as AOPs for wastewater treatment. In this process, a powerful electric field is applied inside the aqueous medium (known as electrohydraulic discharge) or in between the gas phase (non-thermal plasma) and water which commence both physical and chemical processes [36]. Along with the immediate oxidation of pollutant species present in water, numerous oxidizing active radicals, revelation waves, and UV radiations are produced at the time of discharge that can stimulate further oxidation.

6 Catalytic AOPs

6.1 Fenton AOP

Fenton process is preferred most because of its ability to degrade pollutants completely and converting them into harmless products like H_2O, CO_2, and many more inorganic salts. When ferrous ion Fe^{2+} and H_2O_2 were mixed by maintaining the condition acidic, it leads to the production of $^{\bullet}OH$, and this process is known as the Fenton reaction. For obtaining optimum catalytic activity of iron, the pH must be 3.0 because at higher pH there is a precipitation of ferric oxyhydroxide [37]. Surplus addition of H_2O_2 results in the reduction of Fe^{3+} to Fe^{2+}. For achieving increased trace organic chemical removal activity, iron oxide (FeO) can be substituted by transition group metals [38]. Acidic condition is maintained in the Fenton process to avoid precipitation of iron. That's why the replacement for iron-free Fenton methods is recently investigated as summarized [39]. Due to its low operational cost and simple residual iron separation by a magnet, the process is more feasible and economical. As a result, the Fenton process has been established in a number of full-scale applications [40, 41].

6.2 Photo-Fenton AOP

In the photo-Fenton process, H_2O_2 and UV radiation combined with Fe^{2+} and Fe^{3+} led to an increase in the production of $^{\bullet}OH$ in comparison with the Fenton method or photolysis, and as a result the rate of organic pollutant degradation has increased. When it comes to the photo-Fenton reaction, Fe^{3+} builds up in the solution, and if all of the Fe^{2+} ions are consumed, the reaction will stop.

$$Fe(OH)^{2+} + h\upsilon \rightarrow Fe^{2+} + {}^{\bullet}OH \quad (17)$$

Equation 17 represents the regeneration of ferrous ion by photo-reduction of ferric ion Fe^{3+} which occurs in photo-Fenton reaction. The reaction of H_2O_2 with this newly produced ferrous ion leads to the generation of $^{\bullet}OH$ and ferric ion and like the same manner cycle continues [42].

Organic pollutant shows better degradation by Fenton reaction in conventional radiation zone of the visible and near-UV range. Some pollutants which were effectively degraded by the photo-Fenton method are ethylene glycol, herbicides, anisole, and 4-chlorophenol [43]. For getting $^{\bullet}OH$ which is used in organic pollutant degradation, direct photolysis of H_2O_2 is done (Eq. 11).

For getting optimum performance in the photo-Fenton process, pH must be at 3.0, hydroxy-Fe^{3+} must be soluble, and $Fe(OH)^{2+}$ are more photoactive. In comparison with Fenton process, photo-Fenton process has a better result. For an economical approach, sometime sunlight is used instead of UV radiation. This approach of using sunlight results in a lower rate of pollutant degradation [44]. Acidic condition (pH 3) is found better because at this pH there is a conversion of carbonate and bicarbonate family to carbonic acid, and these have a low rate of reactivity with $^{\bullet}OH$ [45].

6.3 *Photocatalysis AOP*

The photocatalytic process has received great attention as it can deal with a wide range of organic pollutants at ambient temperature and pressure [46, 47]. In this method, excitation and transfer of electron from the valence band to the empty conduction band, and the excitation of electron takes place only when catalyst absorbs enough energy to cross the bandgap. The process produces hollow-electron (h^+/e^-) pairs that react with O_2 and H_2O molecule which further produce highly reactive superoxide anion and $^{\bullet}OH$ that oxidize the substrate with an oxidation reaction. The energy absorbed by the photocatalyst in form of photons is equal or higher than the bandgap energy for the excitation of electrons and hence the generation of h^+/e^- [48] (Fig. 3).

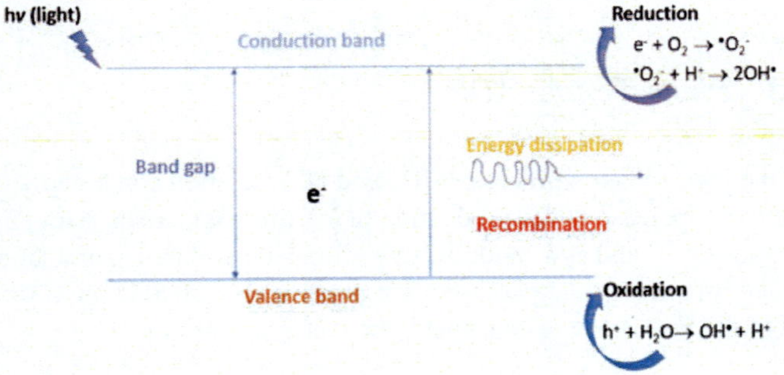

Fig. 3 Mechanism of photocatalysis AOP [49]

The mostly used photocatalyst are TiO_2 and ZnO_2 as they are chemically stable and have good photosensitivity but in this mechanism must have efficient reducing and oxidizing power [50]. However, the large bandgap (3.2 eV) of these catalyst makes them unfavorable for the application. For improving their efficiency and stability, reduction in h^+/e^- pairs regeneration can be done via TiO_2 and ZnO_2 doping (Eqs. 18–21). In doping, the electron of conduction band or holes in valance band are confined in defect sites, reducing h^+/e^- pairs and improving interfacial charge transfer. Later it reduces the bandgap and enables the absorption which occurs in visible light [51].

$$TiO_2 + h\upsilon \rightarrow TiO_2(e^-) + TiO_2(h^+) \tag{18}$$

$$TiO_2(e^-) + O_2 \rightarrow TiO_2 + O_2^- \tag{19}$$

$$TiO_2(h^+) + H_2O \rightarrow TiO_2 + {}^{\bullet}OH + H \tag{20}$$

$$TiO_2(h^+) + OH^- \rightarrow TiO_2 + {}^{\bullet}OH \tag{21}$$

7 Electrochemical AOPs (EAOPs)

Electrochemical oxidation process is the oxidation process of pollutants through ($^{\bullet}$OH) generation with no secondary sludge produced. The mechanism of degradation of organic matter is reaction of highly reactive radical with substrate through addition, substitution, or electron transfer reaction that completely decompose the pollutant into organic matter [52, 53]. The generation of ($^{\bullet}$OH) is pH and temperature independent. However, the efficiency of treating water is pH and temperature dependent, it also depends on the quantity of carbonate ions [54].

7.1 Boron-Doped Diamond Electrodes (BDD)

In electrochemical advanced oxidation process, most favorable and researched electrode is boron-doped diamond electrode. By using chemical vapor deposition (CVD) method mostly BDD electrodes are made. CVD method is widely used in this because of its low cost. Owing to low-point charge carrier activation energy (0.37 eV) boron is chosen as dopant for diamond electrode. Substitution of boron atom in the lattice of diamond for carbon atom results in a p-type of semiconductor in which the extra electron is consumed for bonding which results in the formation of holes in semiconductor. When level of doping is low (~10^{17} atoms cm^{-3}) semiconductor property is shown by diamond having conduction performed by hopping method. When level of doping is high (~10^{20} atoms cm^{-3}) semi-metallic property is shown by diamond because of its impurity bands having low energy which permit electron conduction having metallic resistivities <0.1 Ω cm generally achieved [55, 56].

7.2 Doped TiO_2 Electrode

For water processing conductive Magneli phase suboxide of doped TiO_2 and TiO_2 emerged as efficient electrode materials. Electrical conductivity ~10^{-9} Ω^{-1} cm^{-1} of stoichiometric TiO_2 shows that it is an insulator. At temperature above 900 °C and within the H_2 atmosphere, electronic property of TiO_2 can be altered drastically with the help of generating O_2 deficiencies in their lattice structure or by doping of group five elements which results in the formation of Ti^{4+} to Ti^{3+} and behavior to n-type semiconductor [57].

7.3 Lead Oxide (PbO_2) Electrodes

Packed bed reactors consisting oxidized Pb pellets have been used in earlier studies of compound oxidation at PbO_2. On exploration of PbO_2 and doped PbO_2 anodes at a various substrate (e.g., Ta, Ti), many studies show the formation of hydroxyl radical on PbO_2 electrodes. But till now how $^{\bullet}OH$ is formed at PbO_2 electrode is not well explored. Research focused on PbO_2 electrodes, lead-acid battery, and ($^{\bullet}OH$) generated in a hydrated PbO_2 gel layer, which is formed on the outer surface of the electrode. In this method, there is an assumption that an equilibrium exists between the hydrated gel layer of PbO_2 and PbO_2 crystal. A photon and electrically conductive linear polymer chain are formed by gel layer.

7.4 Doped SnO_2 Electrodes

For getting optimum conductivity of SnO_2 electrode, it has to be doped which make it to work as efficient EAOP electrode. The dopant which is used to increase the conductivity of electrode and provide potential for O_2 evolution versus SHE is Sb. But using Sb as a dopant is quite harmful because of its toxicity. That's why researchers move to its alternative (B, F, P, Ar). After many investigations and study, researchers conclude that the use of doped SnO_2 electrode is in laboratory but it is not used in a large scale or commercially because of its short life service. The two theories which were proposed for the deactivation of these electrodes (i) on the outer surface of the anode nonconductive $Sn(OH)_2$ layer is formed and (ii) there is a delamination in

SnO_2 film because of formation of layer of the underline Ti substrate. For improving the doped SnO_2 electrode, layering of Ti support has to be minimized by placing of IrO_2 inter-layer between Ti and SnO_2-Sb_2O_5 coating [58, 59].

8 Advantages and Limitations of Conventional AOPs

As already discussed, the chemical stability of AOPs proved it as one of the most efficient methods for removal of contaminants that are not envisioned by conventional methods. Still, it has some limitations which are illustrated in Table 1.

Table 1 Advantages and limitations of conventional AOPs

AOP	Advantages	Limitations	References
O_3	Effectively high oxidation performance, less time needed, no residue left, O_3 left residue can be easily decomposed	Not pocket friendly, requires lots of energy, pre-treatment of compound is required	[60, 61, 20]
O_3/UV	Better result with good efficiency, hydroxyl radical generation is more, better than UV and O_3 as compared alone	Requires more money, lots of energy needed, limits to the mass transfer, muck production	[62–66]
H_2O_2/UV	Disinfectant, simple process	Light penetration can be interfered by turbidity, generate $^{\bullet}OH$ less efficiently	[67–69]
O_3/H_2O_2/UV	Effective degradation of polyphenols and aromatics, non-selectively with complete solution species	Expensive, COD removal not complete, sludge production	[70, 66]
Fenton-based process	Economic, eco-friendly, rate of reactions is fast, production of powerful $^{\bullet}OH$, can degrade variety of pollutants	No full-scale application, small of sludge production, acidic environment	[71–75]
Electrochemical-based process	Eco-friendly, able to automate, worked as single-electron oxidizing agent $^{\bullet}OH$ effectively	Reliability toward electrode, expensive in terms of both money and energy	[76–79]

9 Future Perspectives

In above-discussed AOPs, many issues are neglected such as while treating underground wastewater many pollutants are present but targeted portion is organic pollutants not inorganic which decrease the efficiency of process and water is also not pollutant free. So, there is a need to work on treating inorganic pollutants more. Nowadays, use of AOPs is limited and it is used only in labs or in a small scale. Work is needed to be done to implement the AOPs on a large scale which helps us to treat wastewater and pollutants released from textile industries in a large scale.

10 Conclusion

Old methods used for treatment of pollutants released from industries were insufficient and inappropriate because those methods were not cost-effective as well not green process. So, there is a need of an alternative for treatment of these organic pollutants. AOP emerged as boom alternative for the treatment of organic pollutant for industry wastes. AOP falls under green process for degradation of waste released from an industry and for textile wastewater treatment. There are different AOPs which have been reported and give ease for selecting the more appropriate methods for degrading pollutants. Different AOP had been tested for removal of organic pollutants and inorganic pollutants from wastewater and it gives a way to find a more suitable and efficient method. According to literature, the photochemical AOPs and Fenton AOP are found to be more potent for reduction of pollutant from textile wastewater. In comparison of UV and $^{\bullet}OH$ or $SO_4^{\bullet-}$ with UV photolysis or per-sulfate oxidation, it was found that the combination of UV and radicals has more potency in removal of pollutants.

References

1. Renou S, Givaudan JG, Poulain S, Dirassouyan F, Moulin P (2008) Landfill leachate treatment: review and opportunity. J Hazard Mater 150:468–493. https://doi.org/10.1016/j.jhazmat.2007.09.077
2. Adnan A, Mavinic DS, Koch FA (2003) Pilot-scale study of phosphorus recovery through struvite crystallization-Examining the process feasibility. J Environ Manag 37:315–324. https://doi.org/10.1139/s03-040
3. Gültekin I, Ince NH (2007) Synthetic endocrine disruptors in the environment and water remediation by advanced oxidation processes. J Environ Manage 85(4):816–32. https://doi.org/10.1016/j.jenvman.2007.07.020
4. Tsai WT, Lee MK, Chang JH, Su TY, Chang YM (2009) Characterization of bio-oil from induction-heating pyrolysis of food-processing sewage sludges using chromatographic analysis. Bioresour Technol 100(9):2650–54. https://doi.org/10.1016/j.biortech.2008.11.023
5. Kasprzyk-Hordern B, Dinsdale RM, Guwy AJ (2009) The removal of pharmaceuticals, personal care products, endocrine disruptors and illicit drugs during wastewater treatment and its impact

on the quality of receiving waters. Water Res 43(2):363–80. https://doi.org/10.1016/j.watres.2008.10.047
6. Maletz S, Floehr T, Beier S, Klümper C, Brouwer A, Behnisch P, Higley E, Giesy JP, Hecker M, Gebhardt W, Linnemann V (2013) In vitro characterization of the effectiveness of enhanced sewage treatment processes to eliminate endocrine activity of hospital effluents. Water Res 47(4):1545–57. https://doi.org/10.1016/j.watres.2012.12.008
7. Brown AM, Fisher S, Kathryn Iovine M (2009) Osteoblast maturation occurs in overlapping proximal-distal compartments during fin regeneration in zebrafish. Developmental dynamics: an official publication of the American Association of Anatomists. 238(11):2922–2928. https://doi.org/10.1002/dvdy.22114
8. Mendez-Arriaga F, Esplugas S, Gimenez J (2010) Degradation of the emerging contaminant ibuprofen in water by photo-Fenton. Water Res 44(2):589–95. https://doi.org/10.1016/j.watres.2009.07.009
9. Kurniawan TA, Lo WH, Chan GYS (2006) Physicochemical treatments for removal of recalcitrant contaminants from landfill leachate. J Hazard Mater 129:80–100. https://doi.org/10.1016/j.jhazmat.2005.08.010
10. Sharma N, Singh NK, Singh OP, Pandey V, Verma PK (2011) Oxidative stress and antioxidant status during transition period in dairy cows. Asian-Australasian J Animal Sci 24(4):479–84. https://doi.org/10.5713/ajas.2011.10220
11. Oller I, Malato S, Sáhez-pérez JA (2011) Combination of advanced oxidation processes and biological treatments for wastewater decontamination: a review. Sci Total Environ 409:4141–4166. https://doi.org/10.1016/j.scitotenv.2010.08.061
12. Pulgarin C (2015) Effect of advanced oxidation processes on the micropollutants and the effluent organic matter contained in municipal wastewater previously treated by three different secondary methods. Water Res 84:295–306. https://doi.org/10.1016/j.watres.2015.07.030
13. Bartolomeu M, Neves MGPMS, Faustino MAF, Almeida A (2018) Wastewater chemical contaminants: remediation by advanced oxidation processes. Photochem Photobiol Sci 17:1573–1598. https://doi.org/10.1039/C8PP00249E
14. Chaplin BP (2014) Critical review of electrochemical advanced oxidation processes for water treatment applications. Environ Sci Process Impacts 16:1182–1203. https://doi.org/10.1039/c3em00679d
15. Murray CA, Parsons SA (2004) Advanced oxidation processes: flowsheet options for bulk natural organic matter removal. Water Sci Technol 4:113–119. https://doi.org/10.2166/ws.2004.0068
16. Stock NL, Peller J, Vinodgopal K, Kamat PV (2000) Sonolysis of 2,4-dichlorophenoxyacetic acid in aqueous solutions. Evidence for •OH-radical-mediated degradation. Environ Sci Technol 34:1747–1750. https://doi.org/10.1021/jp003478y
17. Schutte AL, Vlok JH, Van Wyk BE (1995) Fire-survival strategy—a character of taxonomic, ecological and evolutionary importance in fynbos legumes. Plant Syst Evol 195(3):243–59. https://doi.org/10.1007/BF00989299
18. Rosenfeldt EJ, Linden KG, Canonica S, von Gunten U (2006) Comparison of the efficiency of OH radical formation during ozonation and the advanced oxidation processes O_3/H_2O_2 and UV/H_2O_2. Water Res 40:3695–3704. https://doi.org/10.1016/j.watres.2006.09.008
19. Bokare AD, Choi W (2014) Review of iron-free Fenton-like systems for activating H_2O_2 in advanced oxidation processes. J Hazard Mater 275:121–35. https://doi.org/10.1016/j.jhazmat.2014.04.054
20. Ikehata K, El-Din MG (2004) Degradation of recalcitrant surfactants in wastewater by ozonation and advanced oxidation processes: a review. Ozone Sci Eng 26:327–343. https://doi.org/10.1080/01919510490482160
21. Qin W, Song Y, Dai Y, Qiu G, Ren M, Zeng P (2015) Treatment of berberine hydrochloride pharmaceutical wastewater by $O_3/UV/H_2O_2$ advanced oxidation process. Environ Earth Sci 73:4939–4946. https://doi.org/10.1007/s12665-015-4192-2
22. Ahn Y, Lee D, Kwon M, Choi IH, Nam SN, Kang JW (2017) Characteristics and fate of natural organic matter during UV oxidation processes. Chemosphere 184:960–968. https://doi.org/10.1016/j.chemosphere.2017.06.079.

23. Ao X, Liu W (2016) Degradation of sulfamethoxazole by medium pressure UV and oxidants: peroxymonosulfate, persulfate, and hydrogen peroxide. Chem Eng J 313:629–637. https://doi.org/10.1016/j.cej.2016.12.089
24. Bolton JR, Valladares JE, Zanin JP, Cooper WJ, Nickelsen MG, Kajdi DC, Kurucz CNW (1998) Figures-of-Merit for advanced oxidation technologies: a comparison of homogeneous UV/H_2O_2, heterogeneous UV/TiO_2 and electron beam processes. J Adv Oxid Technol 3:174–181. https://doi.org/10.1515/jaots-1998-0211
25. Yuan F, Hu C, Hu X, Wei D, Chen Y, Qu J (2011) Photodegradation and toxicity changes of antibiotics in UV and UV/H_2O_2 process. J Hazard Mater 185:1256–1263. https://doi.org/10.1016/j.jhazmat.2010.10.040
26. Homem V, Santos L (2011) Degradation and removal methods of antibiotics from aqueous matrices—a review. J Environ Manag 92:2304–2347. https://doi.org/10.1016/j.jenvman.2011.05.023
27. Jain P, Raghav S, Kumar D (2021) Advanced oxidation technologies for the treatment of wastewater. In: Inamuddin, Ahamed MI, Boddula R, Rangreez TA (eds) Applied water science: remediation technologies, vol 2. Wiley, pp 469–484. https://doi.org/10.1002/9781119725282.ch14
28. Peñalver JJL, Polo MS, Pacheco CVG, Utrilla JR (2010) Photodegradation of tetracyclines in aqueous solution by using UV and UV/H_2O_2 oxidation processes. J Chem Technol Biotechnol 85:1325–1333. https://doi.org/10.1002/jctb.2435
29. Wols BA, Caris, CHMH, Harmsen DJH, Beerendonk EF (2013) Degradation of 40 selected pharmaceuticals by UV/H_2O_2. Water Res 47:5876–5888. https://doi.org/10.1016/j.watres.2013.07.008
30. Paucar NE, Kim I, Tanaka H, Sato C, Paucar NE (2019) Effect of O_3 Dose on the O_3/UV treatment process for the removal of pharmaceuticals and personal care products in secondary effluent. Chem Eng 3:53. https://doi.org/10.3390/chemengineering3020053
31. Nickelsen MG, Cooper WJ, Lin K, Kurucz CN, Waite TD (1994) High energy electron beam generation of oxidants for the treatment of benzene and toluene in the presence of radical scavengers. Water Res 28(5):1227–37. https://doi.org/10.1016/0043-1354(94)90211-9
32. Wang S, Wu X, Wang Y, Li Q, Tao M (2008) Removal of organic matter and ammonia nitrogen from landfill leachate by ultrasound. Ultrason Sonochem 15:933–937. https://doi.org/10.1016/j.ultsonch.2008.04.006
33. Makino K, Mossoba MM, Riesz P (1983) Chemical effects of ultrasound on aqueous solutions, formation of hydroxyl radicals and hydrogen atoms. J Phys Chem 87:1369–1377. https://doi.org/10.1021/j100231a020
34. Bousiakou L, Kazi M, Lianos P (2013) Wastewater treatment technologies in the degradation of hormones and pharmaceuticals with focus on TiO_2 technologies. Pharmakeftiki 25:37–48
35. Mahamuni NN, Adewuyi YG (2010) Advanced oxidation processes (AOPs) involving ultrasound for waste water treatment: a review with emphasis on cost estimation. Ultrason Sonochem 17(6):990–1003. https://doi.org/10.1016/j.ultsonch.2009.09.005
36. Vazquez HA, Jefferson B, Judd SJ (2004) Membrane bioreactors vs conventional biological treatment of landfill leachate: a brief review. J Chem Technol Biotechnol 79:1043–1049. https://doi.org/10.1002/jctb.1072
37. Santos LVDS, Meireles AM, Lange LC (2015) Degradation of antibiotics norfloxacin by Fenton, UV and UV/H_2O_2. J Environ Manag 154:8–12. https://doi.org/10.1016/j.jenvman.2015.02.021
38. Zhang H, Choi HJ, Huang CP (2006) Treatment of landfill leachate by Fenton's reagent in a continuous stirred tank reactor. J Hazard Mater 136:618–623. https://doi.org/10.1016/j.jhazmat.2005.12.040
39. Pradhan AA, Gogate PR (2010) Degradation of p-nitrophenol using acoustic cavitation and Fenton chemistry. J Hazard Mater 173:517–522. https://doi.org/10.1016/j.jhazmat.2009.08.115
40. Rivas FJ, Beltran F, Gimeno O, Carvalho F (2003) Fenton-like oxidation of landfill leachate. J Environ Sci Health 38:371–379. https://doi.org/10.1081/ESE-120016901

41 Neyens E, Baeyens J (2003) A review of classic Fenton's peroxidation as an advanced oxidation technique. J Hazard Mater 98:33–50. https://doi.org/10.1016/s0304-3894(02)00282-0
42. Levec J, Pintar A (2007) Catalytic wet air oxidation processes: a review. Catal Today 124:172–184. https://doi.org/10.1016/j.cattod.2007.03.035
43. Sedlak DL, Andren AW (1991) Oxidation of chlorobenzene with Fenton's reagent. Environ Sci Technol 25:777–782. https://doi.org/10.1021/es00016a024
44. Perez JAS, Roman SIM, Carra I, Cabrera RA, Casas LJL, Malato S (2013) Economic evaluation of a combined photo-Fenton/MBR process using pesticides as model pollutant. Factors affecting costs. J Hazard Mater 244–245:195–203. https://doi.org/10.1016/j.jhazmat.2012.11.015
45. Muthuvel I, Swaminathan M (2007) Photoassisted fenton mineralization of acid violet 7 by heterogeneous Fe(III)-Al_2O_3 catalyst. Catal Commun 8:981–986. https://doi.org/10.1016/j.catcom.2006.10.015
46. Daneshvar E, Zarrinmehr MJ, Hashtjin AM, Farhadian O, Bhatnagar A (2018) Versatile applications of freshwater and marine water microalgae in dairy wastewater treatment, lipid extraction and tetracycline biosorption. Bioresour Technol 268:523–30. https://doi.org/10.1016/j.biortech.2018.08.032
47. Karami-Mohajeri S, Abdollahi M (2011) Toxic influence of organophosphate, carbamate, and organochlorine pesticides on cellular metabolism of lipids, proteins, and carbohydrates: a systematic review. Hum Exp Toxicol 30(9):1119–40. https://doi.org/10.1177/0960327110388959
48. Dong S, Feng J, Fan M, Pi Y, Hu L, Han X, Liu M, Sun J, Sun J (2015) Recent developments in heterogeneous photocatalytic water treatment using visible light-responsive photocatalysts: a review. RSC Adv 5:14610–14630. https://doi.org/10.1039/C4RA13734E
49. Mishra NS, Reddy R, Kuila A, Rani A, Mukherjee P, Nawaz A, Pichiah S (2017) A review on advanced oxidation processes for effective water treatment. Curr World Environ 12:470–490. https://doi.org/10.12944/CWE.12.3.02
50. Khan MM, Ansari SA, Pradhan D, Ansari MO, Lee J, Cho MH (2014) Band gap engineered TiO_2 nanoparticles for visible light induced photoelectrochemical and photocatalytic studies. J Mater Chem A 2(3):637–44. https://doi.org/10.1039/C3TA14052K
51. Yue L, Wang K, Guo J, Yang J, Luo X, Lian J, Wang L (2014) Enhanced electrochemical oxidation of dye wastewater with Fe_2O_3 supported catalyst. J Ind Eng Chem 20:725–731. https://doi.org/10.1016/j.jiec.2013.06.001
52. Li D, Haneda H (2003) Morphologies of zinc oxide particles and their effects on photocatalysis. Chemosphere 51(2):129–37. https://doi.org/10.1016/S0045-6535(02)00787-7
53. Vassilev P, van Santen RA, Koper MT (2005) Ab initio studies of a water layer at transition metal surfaces. J Chem Phys 122(5):54701. https://doi.org/10.1063/1.1834489
54. Bagastyo AY, Batstone DJ, Kristiana I, Gernjak W, Joll C, Radjenovic J (2012) Electrochemical oxidation of reverse osmosis concentrate on boron-doped diamond anodes at circumneutral and acidic pH. Water Res 46:6104–6112. https://doi.org/10.1016/j.watres.2012.08.038
55. Zhao G, Lu X, Zhou Y, Gu Q (2013) Simultaneous humic acid removal and bromate control by O_3 and UV/O_3 processes. Chem Eng J 232:74–80. https://doi.org/10.1039/C8EW00520F
56. Bergmann MEH, Iourtchouk T, Rollin J (2011) The occurrence of bromate and perbromate on BDD anodes during electrolysis of aqueous systems containing bromide: first systematic experimental studies. J Appl Electrochem 41:1109–1123. https://doi.org/10.1007/s10800-011-0329-5
57. Bernard C, Colin JR, Anne LDD (1997) Estimation of the hazard of landfills through toxicity testing of leachates, comparison of physico-chemical characteristics of landfill leachates with their toxicity determined with a battery of tests. Chemosphere 35:2783–2796. https://doi.org/10.1016/S0045-6535(97)00332-9
58. Lutze HV, Bakkour R, Kerlin N, Sonntag VC, Schmidt TC (2014) Formation of bromate in sulfate radical based oxidation: mechanistic aspects and suppression by dissolved organic matter. Water Res 53:370–377. https://doi.org/10.1016/j.watres.2014.01.001

59. Abbas AA, Jingsong G, Ping LZ, Ya PY, Al-Rekabi WS (2009) Review on landfill leachate treatment. Am J Appl Sci 6:672–684. https://doi.org/10.3844/ajassp.2009.672.684
60. Amr SS, Aziz HA (2012) New treatment of stabilized leachate by ozone/Fenton in the advanced oxidation process. Waste Manage 32(9):1693–98. https://doi.org/10.1016/j.wasman.2012.04.009
61. Fanchiang JM, Tseng DH (2009) Degradation of anthraquinone dye CI Reactive Blue 19 in aqueous solution by ozonation. Chemosphere 77(2):214–21. https://doi.org/10.1016/j.chemosphere.2009.07.038
62. Guittonneau S, De Laat J, Duguet JP, Bonnel C, Dore M (1990) Oxidation of parachloronitrobenzene in dilute aqueous solution by O_3+ UV and H_2O_2+ UV: a comparative study. 73–94. https://doi.org/10.1080/01919519008552456
63. Ruppert L, Becker KH (2000) A product study of the OH radical-initiated oxidation of isoprene: formation of C5-unsaturated diols. Atmos Environ 34(10):1529–42. https://doi.org/10.1016/S1352-2310(99)00408-2
64. Oleskowicz-Popiel P, Lisiecki P, Holm-Nielsen JB, Thomsen AB, Thomsen MH (2008) Ethanol production from maize silage as lignocellulosic biomass in anaerobically digested and wet-oxidized manure. Bioresour Technol 99(13):5327–34. https://doi.org/10.1016/j.biortech.2007.11.029
65. Lucas MS, Peres JA, Puma GL (2010) Treatment of winery wastewater by ozone-based advanced oxidation processes (O_3, O_3/UV and $O_3/UV/H_2O_2$) in a pilot-scale bubble column reactor and process economics. Sep Purif Technol 72(3):235–41. https://doi.org/10.1016/j.seppur.2010.01.016
66. Rao YF, Chu W (2010) Degradation of linuron by UV, ozonation, and UV/O_3 processes—Effect of anions and reaction mechanism. J Hazard Mater 180(1):514–23. https://doi.org/10.1016/j.jhazmat.2010.04.063
67. Elmorsi TM, Riyad YM, Mohamed ZH, Abd El Bary HM (2010) Decolorization of Mordant red 73 azo dye in water using H_2O_2/UV and photo-Fenton treatment. J Hazard Mater 174(1):352–58. https://doi.org/10.1016/j.jhazmat.2009.09.057
68. Hu AL, Liu YH, Deng HH, Hong GL, Liu AL, Lin XH, Xia XH, Chen W (2014) Fluorescent hydrogen peroxide sensor based on cupric oxide nanoparticles and its application for glucose and l-lactate detection. Biosens Bioelectron 61:374–78. https://doi.org/10.1016/j.bios.2014.05.048
69. Karci A, Arslan-Alaton I, Olmez-Hanci T, Bekbölet M (2012) Transformation of 2, 4-dichlorophenol by H_2O_2/UV-C, Fenton and photo-Fenton processes: oxidation products and toxicity evolution. J Photochem Photobiol A 230(1):65–73. https://doi.org/10.1016/j.jphotochem.2012.01.003
70. Monteagudo JM, Carmona M, Duran A (2005) Photo-Fenton-assisted ozonation of p-Coumaric acid in aqueous solution. Chemosphere 60(8):1103–10. https://doi.org/10.1016/j.chemosphere.2004.12.063
71. Zhang H, Choi HJ, Huang CP (2005) Optimization of Fenton process for the treatment of landfill leachate. J Hazard Mater 125(1):166–74. https://doi.org/10.1016/j.jhazmat.2005.05.025
72. Lee SH, Oh JY, Park YC (2006) Degradation of phenol with Fenton-like treatment by using heterogeneous catalyst (modified iron oxide) and hydrogen peroxide. Bull Korean Chem Soc 27(4):489–94
73. Pignatello JJ, Oliveros E, MacKay A (2006) Advanced oxidation processes for organic contaminant destruction based on the Fenton reaction and related chemistry. Critical Rev Environ Sci Technol 36(1):1–84. https://doi.org/10.1080/10643380500326564
74. Catrinescu C, Arsene D, Apopei P, Teodosiu C (2012) Degradation of 4-chlorophenol from wastewater through heterogeneous Fenton and photo-Fenton process, catalyzed by Al–Fe PILC. Appl Clay Sci 58:96–101. https://doi.org/10.1016/j.clay.2012.01.019
75. Giroto JA, Teixeira AC, Nascimento CA, Guardani R (2008) Photo-Fenton removal of water-soluble polymers. Chem. Eng. Process 47(12):2361–69. https://doi.org/10.1016/j.cep.2008.01.014

76. Huang L, Logan BE (2008) Electricity generation and treatment of paper recycling wastewater using a microbial fuel cell. Appl Microbiol Biotechnol 80(2):349–55. https://doi.org/10.1007/s00253-008-1546-7
77. Anotai J, Lu MC, Chewpreecha P (2006) Kinetics of aniline degradation by Fenton and electro-Fenton processes. Water Res 40(9):1841–1847. https://doi.org/10.1016/j.watres.2006.02.033
78. Su CC, Chang AT, Bellotindos LM, Lu MC (2012)Degradation of acetaminophen by Fenton and electro-Fenton processes in aerator reactor. Sep Purif Technol 99:8–13. https://doi.org/10.1016/j.seppur.2012.07.004
79. Nidheesh PV, Gandhimathi R (2012) Trends in electro-Fenton process for water and wastewater treatment: an overview. Desalination 299:1–15. https://doi.org/10.1016/j.desal.2012.05.011

Fenton Processes in Dye Removal

Helhe Daiany Cabral Silva Pimentel, Lívia Fernandes da Silva, Anna Karla dos Santos Pereira, Grasiele Soares Cavallini, and Douglas Henrique Pereira

Abstract Dye wastewater comprises contaminants that are difficult to degrade. When released into water bodies without proper treatment, it can cause a variety of problems for human health and aquatic ecosystems. Among several alternatives for treating wastewater contaminated by dyes, the Fenton process is an option with many advantages. The Fenton process involves the formation of the hydroxyl ($HO^•$) and hydroperoxyl ($HO_2^•$) radicals through the oxidation of Fe^{2+} ($Fe^{2+} + H_2O_2 \rightarrow Fe^{3+} + HO^- + HO^•$). It is an advanced oxidative process that involves easily accessible reagents, is low-cost, and the conditions, such as pH range, temperature, and concentration ratio, are well established in the literature. The Fenton process can proceed in a homogeneous or heterogeneous manner, and through both pathways, total mineralization through oxidation can be achieved. Methods that cannot achieve total mineralization can be coupled with the Fenton process to achieve greater efficiency, and in some cases, reduced consumption of the hydrogen peroxide oxidizer. The Fenton process has high applicability in environmental fields, mostly because its mechanism involves the total destruction of the contaminant. However, the degradation process often results in the formation of new toxic by-products, which means that the end product remains toxic.

1 Introduction

A large volume of wastewater is currently generated worldwide due to rapid industrial, economic, and population growth [5, 63]. These large volumes of effluent-contaminated water bodies, especially freshwater, generating complex effluents, and

H. D. C. S. Pimentel · L. F. da Silva · G. S. Cavallini · D. H. Pereira (✉)
Chemistry Collegiate, Federal University of Tocantins, Campus Gurupi – Badejós, P.O. Box 66, 77 402-970 Gurupi, Tocantins, Brazil
e-mail: doug@uft.edu.br

A. K. dos Santos Pereira
Institute of Chemistry, University of Campinas – UNICAMP, PO Box 6154, 13083-970 Campinas, SP, Brazil

S. S. Muthu and A. Khadir (eds.), *Advanced Oxidation Processes in Dye-Containing Wastewater*, Sustainable Textiles: Production, Processing, Manufacturing & Chemistry,
https://doi.org/10.1007/978-981-19-0987-0_2

have become a critical problem worldwide in recent decades [25, 59, 63]. Organic contaminants have high toxicity, low biodegradability, refractory characteristics [27, 79], and among organic contaminants highlight color agents such as dyes [55].

Dyes are synthetic or natural substances that are used for various applications, such as in the textile, paint, tanning, paper, cellulose, food, beverage, pharmaceutical, and wood industries [33, 55, 63, 66, 78]. Dyes can be classified as disperse, reactive, direct, acid, or basic, according to their chemical structure: azo, anthraquinone, indigo, phthalocyanines, sulfur, nitrated, and nitrosated [41, 78]. Dyes have chromophore groups responsible for color and auxochromes responsible for solubility [11, 63]. It is estimated that there are more than 100,000 dyes currently available [66].

When untreated discharge enters waterways, the dyes in the wastewater can cause numerous problems, such as high chemical and biological oxygen demands [59], limited access to sunlight, which affects the photosynthesis capacity of plants [55], contamination of food chains [47, 53], carcinogenicity, toxicity, and mutagenicity in aquatic organisms [3, 10, 16, 28, 37]. Prolonged exposure in humans can cause tumors, respiratory problems, allergies, and congenital diseases [70].

In recent decades, several treatments for wastewater containing dyes have been proposed, such as biological treatments with microorganisms and activated sludge [23, 44], and physical treatments such as adsorption [54, 69], ion exchange [46], and chemical treatments [43]. Biological treatments have the disadvantages of substantial operational cost, potential for microorganism inactivation, and sludge generation [70]. Physical treatment has the disadvantages of transferring the contaminant to another material and the need to constantly search for low-cost and easily obtainable materials [20].

Owing to the disadvantages of biological and physical treatments, chemical treatments are desirable, especially advanced oxidative processes (AOPs) [57]. AOPs can be homogeneous or heterogeneous, and the main AOPs are the Fenton, electrochemical, photochemical, and sonochemical processes the processes that use ozone [57].

Among the AOPs, the Fenton process is a simple, low-cost technique capable of degrading soluble and insoluble dyes [66]. The Fenton process involves the oxidation of Fe^{2+} ions, which generates $HO^{\bullet}$ via the catalytic decomposition of hydrogen peroxide [70].

$$Fe^{2+} + H_2O_2 \rightarrow Fe^{3+} + HO^- + HO^{\bullet} \quad (1)$$

The radicals formed by the Fenton process can repeatedly attack organic compounds in a non-selective manner, degrading their chemical structures, potentially causing total mineralization [29, 36]. Thus, the Fenton process is highly applicable for the treatment of dye-containing effluents. There are many publications that demonstrate the applicability of the Fenton process to treat effluents with dyes. A Science Direct [67] search for the phrase "*Fenton and Dyes*" produces a large number of results corresponding to research in several related areas (Fig. 1). These results confirm the increase in annual citations and highlight the importance of dye studies.

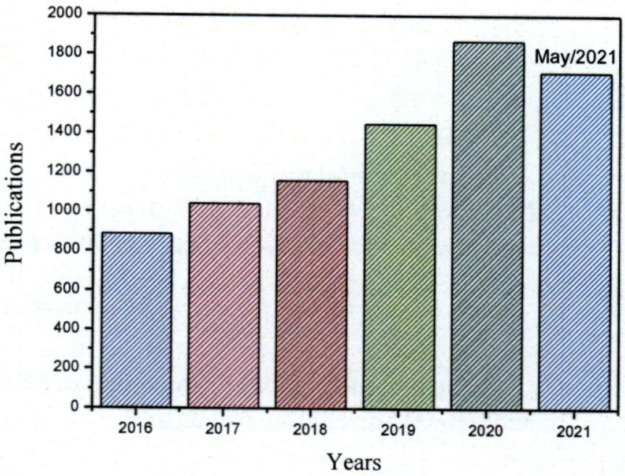

Fig. 1 Publications relating to the search words "*Fenton* and *Dyes*" from 2016 to May 2021

Therefore, a review was conducted to examine the Fenton process since its inception, addressing the various types of Fenton processes and their application for the removal and degradation of dyes, as well as the toxicity of the process when used for this purpose.

2 Fenton Process

The process known as "Fenton" was discovered by Henry John Horstman Fenton in 1876, when he observed the strong oxidizing power of a mixture of hydrogen peroxide (H_2O_2), a low concentration of Fe^{2+} ions, and tartaric acid ($C_4H_6O_6$) [34, 35, 81]. Henry JH Fenton observed that a violet-colored solution was formed by mixing an oxidizer (H_2O_2) with tartaric acid, sodium hydroxide, and ferrous salt; however, if a ferric salt was used instead of a ferrous salt, the same color was not observed [34, 35, 81].

In his research, Fenton demonstrated that iron can also be removed by precipitation and that other oxidants could be used to replace H_2O_2; however, the results were disappointing [35]. Fenton did not propose the mechanisms of radical formation, and the reactions that occur in the process have been studied and debated in the literature for years [17, 40, 81].

In 1930, mechanisms for the Fenton process were proposed by Haber and Weiss [42] through kinetic studies. The main stages (reactions (2)–(5)) are described below:

$$Fe^{2+} + H_2O_2 \rightleftharpoons Fe^{3+} + HO^- + HO^\bullet \tag{2}$$

$$HO^\bullet + H_2O_2 \rightleftharpoons H_2O + HO_2\bullet \tag{3}$$

$$HO_2^{\bullet} + H_2O_2 \rightleftarrows O_2 + H_2O + HO^{\bullet} \quad (4)$$

$$Fe^{2+} + HO^{\bullet} \rightleftharpoons Fe^{3+} + HO^{-} \quad (5)$$

In Bray and Gorin [15], through a series of kinetic studies, demonstrated the formation of an intermediate composed of tetravalent iron (FeO^{2+}). Their results demonstrate the reduction of Fe^{3+} in the presence of H_2O according to reaction (6):

$$2Fe^{3+} + H_2O \rightleftharpoons Fe^{2+} + FeO^{2+} + 2H^{+} \quad (6)$$

In the presence of hydrogen peroxide, Fe^{2+} ions can react to form FeO^{2+}, water, and oxygen (reactions (7) and (8)).

$$Fe^{2+} + H_2O_2 \rightarrow FeO^{2+} + H_2O \quad (7)$$

$$FeO^{2+} + H_2O_2 \rightarrow Fe^{2+} + H_2O + O_2 \quad (8)$$

Barb et al. [7–9] investigated the iron system with hydrogen peroxide and their results conflicted somewhat with those of Haber and Weiss. A new chain mechanism was proposed with the formation of $^{\bullet}OH$ and $^{\bullet}OOH$ radicals, oxygen, water, and iron ions (reactions (9)–(13)):

$$Fe^{2+} + H_2O_2 \rightarrow Fe^{3+} + HO^{-} + HO^{\bullet} \quad (9)$$

$$H_2O_2 + HO\bullet \rightarrow H_2O + HO_2\bullet \quad (10)$$

$$Fe^{2+} + HO^{\bullet} \rightarrow Fe^{3+} + HO^{-} \quad (11)$$

$$Fe^{2+} + HO_2^{\bullet} \rightarrow Fe^{3+} + HO_2^{-} \quad (12)$$

$$Fe^{3+} + HO_2^{\bullet} \rightarrow Fe^{2+} + O_2 + H^{+} \quad (13)$$

The results of studies by Barb et al. [7–9] demonstrate that reactions (9), (10), and (11) are similar to those reported by Haber and Weiss, who state that the formation of oxygen (O_2) occurs via a reaction between the Fe^{3+} ion with the radical $HO_2^{\bullet}$ (reaction (13)), and that the ferrous ions are regenerated.

According to the authors, when there is a high ratio of ferric ions to both ferrous ions and hydrogen peroxide, it can lead to the formation of $Fe(OH)^{+++}$ or FeO^{++} species from the $HO^{\bullet}$ according to reaction (14):

$$Fe^{3+} + HO^{\bullet} \rightarrow Fe(OH)^{3+} \rightleftharpoons FeO^{2+} + H^{+} \quad (14)$$

Many other reports in the literature suggest potential mechanisms [56, 57, 60, 84]. Briefly, the equations that describe the classical Fenton process can be expressed as follows:

$$Fe^{2+} + H_2O_2 \rightarrow Fe^{3+} + HO^- + HO^\bullet \quad (15)$$

$$H_2O_2 + HO^\bullet \rightarrow H_2O + HO_2^\bullet \quad (16)$$

$$Fe^{2+} + HO^\bullet \rightarrow Fe^{3+} + HO^- \quad (17)$$

$$Fe^{2+} + HO_2^\bullet \rightarrow Fe^{3+} + HO_2^- \quad (18)$$

$$Fe^{3+} + HO_2^\bullet \rightarrow Fe^{2+} + O_2 + H^+ \quad (19)$$

$$Fe^{3+} + H_2O_2 \rightarrow FeOOH^{2+} + H^+ \quad (20)$$

$$FeOOH^{2+} \rightarrow Fe^{2+} + HO_2^\bullet \quad (21)$$

$$HO_2^\bullet + HO_2^\bullet \rightarrow H_2O_2 \quad (22)$$

All studies agree that the first step of the Fenton process involves the catalytic decomposition of H_2O_2 by Fe^{2+} (reaction (15)), and the termination steps proceed according to reactions (17) and (22) [56, 60]. Iron complexes and intermediates with a high oxidation number can also be formed under specific conditions or when irradiated with light [13, 19, 64].

Fenton reactions have a broad application spectrum, as they can take place in aqueous media at a wide variety of temperatures under constant pressure, and can form $HO^\bullet$ and $HO_2^\bullet$ radicals that are non-selective [64]. The formed radicals have a high oxidizing power, 2.8 V and 1.42 V (SHE) for $HO^\bullet$ and $HO_2^\bullet$, respectively [22, 38, 80]. The determining factors are the easily obtained reagents (Fe^{2+}, Fe^{3+}, and H_2O_2), the pH dependence, and the organic constituents of the medium [64, 84].

The Fenton process can be heterogeneous or homogeneous; it can be modified by changing the reagents or manipulating external factors such as electromagnetic radiation, electrons, or a combination of different processes, thus ensuring greater applicability and a broader application scope [19, 52, 64, 84].

Figure 2 shows the diverse range of disciplines that make use of the Fenton process, as well as the percentages of articles searchable using the query "*Fenton*" for each subject area [67]. The fields in which the process is most used are as follows: (i) environmental science (19%); (ii) biochemistry, genetics, and molecular biology (16%); (iii) chemical engineering (15%); (iv) medicine and dentistry (12%); and (v) chemistry (10%) (Fig. 2).

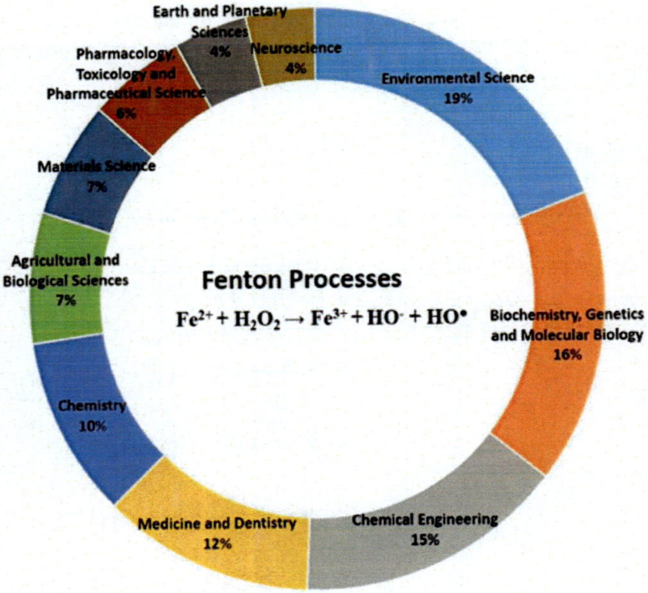

Fig. 2 Main application fields of the Fenton process

In environmental science, chemical engineering, and chemistry, the Fenton process can be applied to degrade a wide variety of contaminants, such as pesticides [61], medicines [83], and dyes [19]. Thus, different methods for removing dyes from effluents have been reported in the literature.

3 Applications of the Fenton Process for Dye Degradation: Factors Affecting Process Efficiency

Studies on dye degradation by the Fenton process gained visibility in the 1990s. In a study by Kuo [48], the Fenton process is described as a new treatment method for dye wastewater, with high efficiency of chemical oxygen demand (COD) reduction and decolorization for different types of dyes (disperse, reactive, direct, acid, and basic dyes).

The factors most often reported to affect the efficiency of dye degradation in the Fenton process are pH, temperature, agitation gradient, concentration (ratio) of reagents, and the presence of salts and inhibitors [14, 30, 48].

3.1 pH

The Fenton process proceeds optimally between pH 3 and 4 [4]. Above pH 4, the availability of free iron species declines due to the formation of ferrous complexes or the precipitation of ferric oxyhydroxides [49, 85], which are less reactive with H_2O_2 and therefore reduce the rate of oxidative degradation. The reaction rate and degradation efficiency of the Fenton process are also affected between pH 1 and 3 because of the formation of iron(III)-hydroperoxy complexes [39]. In addition, H_2O_2 can react with excess H^+ ions to form the ion ($H_3O_2^+$) with Fe^{2+} to form the $HO^\bullet$, which could react with H^+ to form water [32, 73].

Bouasla et al. [14] evaluated the degradation of methyl violet dye by the Fenton process and reported that the use of phosphoric acid for acidification of the medium was not suitable, as phosphate anions have the capacity to complex Fe^{2+} and Fe^{3+} ($FeH_2PO_4^+$ and $FeH_2PO_4^{2+}$, respectively).

3.2 Temperature

The effect of temperature was evaluated by Emami et al. [30], who observed that an increase in temperature (20–60 °C) contributed positively to the increase in dye discoloration and reaction speed as a function of the increase in the rate of $HO^\bullet$ generation. According to Bouasla et al. [14], the ideal temperature range for the Fenton process is between 30 and 40 °C; above this temperature, H_2O_2 decomposes, thus reducing the degradation rate [14].

3.3 Agitation Gradient

When evaluating the effect of agitation during the Fenton process for the degradation of Reactive Blue 19 dye, Emami et al. [30] observed an increase in the reaction speed with an increase in rotation from 1000 to 2000 rpm ($k = 0.29$ and $0.44\ min^{-1}$, respectively). Bouasla et al. [14] reported that the ideal agitation during the Fenton process is 240 rpm, because the oxygen input improves the efficiency and decolorization kinetics through the generation of $HO^\bullet$ (reactions (23)–(26)); however, above this value, atmospheric CO_2 is dissolved and the $HO^\bullet$ is consumed, reducing the efficiency of the process (reactions (27)–(34)).

$$R^\bullet + O_2 \rightarrow ROO^\bullet \tag{23}$$

$$ROO^\bullet + RH \rightarrow ROOH + R^\bullet \tag{24}$$

$$ROO^{\bullet} + H_2O \rightarrow ROH + HOO^{\bullet} \quad (25)$$

$$HOO^{\bullet} + H_2O_2 \rightarrow HO^{\bullet} + H_2O + O_2 \quad (26)$$

$$CO_2 + H_2O \rightarrow HCO_3^- + H^+ \quad (27)$$

$$HCO_3^- \rightarrow CO_3^{2-} + H^+ \quad (28)$$

$$Fe^{2+} + HCO_3^- \rightarrow FeHCO_3^+ \quad (29)$$

$$Fe^{2+} + CO_3^{2-} \rightarrow FeCO_3 \quad (30)$$

$$Fe^{2+} + 2CO_3^{2-} \rightarrow Fe(CO_3)_2^{2-} \quad (31)$$

$$Fe^{2+} + CO_3^{2-} \rightarrow HO^- + Fe(CO_3)_2(OH^-) \quad (32)$$

$$HO^{\bullet} + HCO_3^- \rightarrow H_2O + CO_3^{\bullet -} \quad (33)$$

$$HO^{\bullet} + CO_3^{2-} \rightarrow HO^- + CO_3^{\bullet -} \quad (34)$$

3.4 Reagent Concentration

In their evaluation of the contribution of Fe^{2+} concentration to the degradation of azo dye Orange G (OG) by the Fenton process, Sun et al. [74] observed that an Fe^{2+} concentration of 5.0×10^{-6} mol L^{-1} reflected a degradation efficiency of approximately 40%, while a concentration of 3.5×10^{-5} mol L^{-1} Fe^{2+} resulted in greater than 90% degradation ([OG] = 6.63×10^{-5} mol L^{-1}; [H_2O_2] = 1.0×10^{-2} mol L^{-1}; pH = 4.0; at 20 °C; 60 min), the authors explained that an insufficient Fe^{2+} concentration limits the formation of $HO^{\bullet}$ and that the H_2O_2:Fe^{2+} ratio directly affects the production of $HO^{\bullet}$. The H_2O_2:Fe^{2+} ratio has great significance in the Fenton process for the degradation efficiency of the dye, and an increase in the H_2O_2 concentration tends to contribute to the formation of $HO^{\bullet}$, and consequently an increase in discoloration. However, according to Sun et al. [73], above a critical concentration of H_2O_2, the reaction rate reduces due to the formation of $HOO^{\bullet}$, according to reaction (35):

$$H_2O_2 + HO^{\bullet} \rightarrow H_2O + HOO^{\bullet} \quad (35)$$

Fan et al. [32] observed a reduction in the degradation efficiency of crystal violet in the Fenton process with a H_2O_2:Fe^{2+} molar ratio of 1000:1. According to Souza et al., there is an optimal H_2O_2:Fe^{2+} molar ratio. However, there is no consensus, and values ranging from 1:1 to 400:1 (H_2O_2:Fe^{2+}) have been reported in the literature.

Notably, the use of Fe^{2+} or Fe^{3+} ions in the Fenton process can result in similar degradation efficiencies for dyes, but the Fe^{3+} ions react more slowly with H_2O_2, reducing the reaction speed [57].

Silva and Baltrusaitis [71] theorized that the consumption of H_2O_2 in the Fenton process depends on the concentrations of the pollutants and the concentration of dissolved organic matter (DOM) in the effluent. Shanmugam et al. [68] observed a decrease in the decolorization efficiency of acid blue dye 113 from 40 to 30% in a study comparing synthetic and real effluents. In this study, the lower rate of degradation of the dye by the Fenton process was attributed to the presence of organic impurities in the real effluent, which reinforces the importance of combination treatments.

3.5 *Presence of Salts and Inhibitors*

Salts that come from dyes in textile effluents can also interfere with the efficiency of the Fenton process. Emami et al. [30] evaluated different concentrations of sodium sulfate and sodium chloride in a Reactive Blue 19 dye solution and observed that an increase in chloride and sulfate ions impaired dye discoloration in the Fenton process. The negative effect was attributed to the reactions between chloride and sulfate ions with $HO^•$, producing chlorine (1.36 V), sulfate radicals, and peroxydisulfate ions (2.01 V), which have lower oxidation potentials than $HO^•$ (2.8 V).

Another factor that can affect the efficiency of the Fenton process is the presence of inhibitors. Ashraf et al. [4] evaluated the degradation of the dye Rhodamine B using the Fenton process in the presence of surfactants and observed that anionic surfactants can completely inhibit the degradation of the dye, and that cationic surfactants inhibited degradation better than non-ionic surfactants.

4 The Heterogeneous Fenton Process

Although the homogeneous Fenton process is efficient for dye degradation, the formation of iron complexes can increase sludge production, which is disadvantageous, and the alternative for sludge reduction is the use of the heterogeneous Fenton process, which uses solid catalysts [51]. When using the heterogeneous Fenton process, supports are required for iron ions [12], iron oxides, other metal oxides, and metal–organic frameworks (MOFs) [51, 71].

Thomas et al. [76] reported that the most accepted mechanism to describe the true heterogeneous Fenton process was published by Lin and Gurol [50], who described that the interaction between the surface of goethite (≡FeIII–OH) and H_2O_2 promotes the formation of a complex. By transferring the charge from the ligand to the metal, a new transition state complex is formed (≡FeII•O_2H), which dissociates into a $HO_2^{\bullet}$ and ≡FeII and generates $HO^{\bullet}$ in the presence of H_2O_2.

The heterogeneous Fenton process can also occur through the leaching of iron from solid catalysts. According to Cai et al. [18], the heterogeneous Fenton process can occur via the corrosion of metallic iron (Fe^0) to Fe^{2+}, which reacts with H_2O_2 in an acidic medium to form $HO^{\bullet}$ and Fe^{3+}, and subsequently Fe^0 reduces Fe^{3+} to Fe^{2+}, promoting a cyclic process. However, the reduction kinetics from Fe^{3+} to Fe^{2+} are inherently low. Thus, the use of electrical energy or ultraviolet or visible radiation can be used to accelerate the process [76].

5 Fenton Process Combined with Biological Degradation

Studies on the Fenton process for the degradation of azo dyes are very common in scientific literature because this class of dyes represents 60–70% of all textile dyes [62, 72, 78]. Furthermore, most azo dyes are resistant to bacterial activity, and direct biological treatment is not effective [45, 74].

The BOD:COD ratio of textile effluents is estimated to be <0.1 [31, 75], that is, of low biodegradability due to biorecalcitrance of dyes [21]. However, because of their lower cost, biological treatments are the most commonly used processes for the treatment of these effluents, even with limited efficiency for the complete degradation of the dye [1, 24, 68].

Xie et al. [82] evaluated the degradation of Reactive Black 5 and Remazol Brilliant Blue R dyes using the Fenton process and biological degradation. They described increased discoloration and higher dissolved organic carbon reduction rates using the Fenton process. They also reported that the biological degradation of dyes resulted in increased antiestrogenic activity, which can interfere with reproduction and development in organisms. This was not observed when using the Fenton process.

The use of hybrid treatment processes that integrate the Fenton process with biological processes to increase the efficiency of contaminant degradation is described for textile effluents. Bae et al. [6] described good results for research aimed at treating real textile effluent using activated sludge with pure oxygen followed by the Fenton process. The authors reported that the initial biological process reduced the organic load of the effluent, which contributed to Fenton reagent being more available for the degradation of recalcitrant compounds in the second stage of the reaction. The presence of ferric ions also promoted the coagulation of the effluent, increasing the efficiency of the treatment. According to Arslan-Alaton et al. [2], in addition to contributing to the complete discoloration of dyes and partial reduction of organic carbon, the Fenton process does not inhibit anaerobic, anoxic, and aerobic processes.

Shanmugam et al. [68] evaluated the pre-treatment with the Fenton process followed by biological degradation using a defined bacterial consortium and obtained

40% removal of the dye in the Fenton process and another 45% during biological treatment of a synthetic effluent containing the azo dye acid blue 113. The authors obtained a COD reduction of 93.7% in real effluent, which led them to conclude that the method is viable for the treatment of industrial effluents on a large scale.

6 Fenton Process-Related Ecotoxicity

In the textile industry, 10–15% of the dye is lost during the dyeing step and this figure doubles in the case of direct and reactive dyes [58]. In water bodies, these contaminants interfere with the passage of solar radiation, inhibiting photosynthesis, and can also cause harm to aquatic organisms due to their toxicity [26].

According to Rizzo [65], studies that report a reduction in the toxicity of dyes after exploiting the Fenton reaction prevail in the literature. However, the use of AOPs can generate degradation by-products with higher toxicity than the dye, mostly due to the formation of toxic aromatic intermediates [21] such as aromatic amines [86].

Vieira et al. [77] observed an increase in the toxicity of methylene blue dye after its degradation by the Fenton process, using the test organism *Girardia tigrina*. In this study, the authors warned of the toxicity of residual H_2O_2.

7 Conclusion

The Fenton process is one of the most used AOPs for the treatment of dye-containing effluent and has a growing number of applications owing to its cost-effectiveness. The great applicability of the Fenton process in dye removal can be justified by its good yield, easy applicability, and well-established reaction conditions.

The Fenton process has been described by numerous mechanisms since its discovery by Fenton in 1876, and even today, the mechanisms are still studied. Its versatility allows the realization of combined processes. It is most often coupled with biological processes, as a pre- or post-treatment, to assist in degrading the dye more effectively. Furthermore, the process can be easily optimized through already established changes, such as an external source of light irradiation, electrons, or through the change of some reagent, which makes the process well adaptable to real operating conditions.

In conclusion, the Fenton process is a consolidated chemical process with demonstrated utility for the treatment of dye wastewater; however, as with any process based on the chemical transformation of contaminants, more research is needed to mitigate the problem of toxic by-products.

References

1. Alves de Lima RO, Bazo AP, Salvadori DMF et al (2007) Mutagenic and carcinogenic potential of a textile azo dye processing plant effluent that impacts a drinking water source. Mutat Res Genet Toxicol Environ Mutagen 626:53–60. https://doi.org/10.1016/j.mrgentox.2006.08.002
2. Arslan-Alaton I, Gursoy BH, Schmidt J-E (2008) Advanced oxidation of acid and reactive dyes: effect of Fenton treatment on aerobic, anoxic and anaerobic processes. Dyes Pigm 78:117–130. https://doi.org/10.1016/j.dyepig.2007.11.001
3. Asghar A, Ramzan N, Jamal B ul et al (2020) Low frequency ultrasonic-assisted Fenton oxidation of textile wastewater: process optimization and electrical energy evaluation. Water Environ J 34:523–535. https://doi.org/10.1111/wej.12482
4. Ashraf U, Chat OA, Dar AA (2014) An inhibitory effect of self-assembled soft systems on Fenton driven degradation of xanthene dye Rhodamine B. Chemosphere 99:199–206. https://doi.org/10.1016/j.chemosphere.2013.10.074
5. Badmus KO, Irakoze N, Adeniyi OR, Petrik L (2020) Synergistic advance Fenton oxidation and hydrodynamic cavitation treatment of persistent organic dyes in textile wastewater. J Environ Chem Eng 8:103521. https://doi.org/10.1016/j.jece.2019.103521
6. Bae W, Won H, Hwang B et al (2015) Characterization of refractory matters in dyeing wastewater during a full-scale Fenton process following pure-oxygen activated sludge treatment. J Hazard Mater 287:421–428. https://doi.org/10.1016/j.jhazmat.2015.01.052
7. Barb WG, Baxendale JH, George P, Hargrave KR (1949) Reactions of ferrous and ferric ions with hydrogen peroxide. Nature 163:692–694
8. Barb WG, Baxendale JH, George P, Hargrave KR (1951) Reactions of ferrous and ferric ions with hydrogen peroxide. Part I. The ferrous ion reaction. Trans Faraday Soc 47:462–500
9. Barb WG, Baxendale JH, George P, Hargrave KR (1951) Reactions of ferrous and ferric ions with hydrogen peroxide. Part II. The ferric ion reaction. Trans Faraday Soc 47:591–616
10. Barbusiński K (2005) The modified fenton process for decolorization of dye wastewater. J Environ Stud 14:281–285
11. Benkhaya S, M' rabet S, El Harfi A, (2020) A review on classifications, recent synthesis and applications of textile dyes. Inorg Chem Commun 115:107891. https://doi.org/10.1016/j.inoche.2020.107891
12. Blanco M, Martinez A, Marcaide A et al (2014) Heterogeneous fenton catalyst for the efficient removal of azo dyes in water. AIDS Patient Care STDs 05:490–499. https://doi.org/10.4236/ajac.2014.58058
13. Bokare AD, Choi W (2014) Review of iron-free Fenton-like systems for activating H_2O_2 in advanced oxidation processes. J Hazard Mater 275:121–135. https://doi.org/10.1016/j.jhazmat.2014.04.054
14. Bouasla C, Samar ME-H, Ismail F (2010) Degradation of methyl violet 6B dye by the Fenton process. Desalination 254:35–41. https://doi.org/10.1016/j.desal.2009.12.017
15. Bray WC, Gorin MH (1932) Ferryl ion, a compound of tetravalent iron. J Am Chem Soc 54(5):2124–2125. https://doi.org/10.1021/ja01344a505
16. Brillas E, Martínez-Huitle CA (2015) Decontamination of wastewaters containing synthetic organic dyes by electrochemical methods. An updated review. Appl Catal B 166–167:603–643. https://doi.org/10.1016/j.apcatb.2014.11.016
17. Buda F, Ensing B, Gribnau MCM, Baerends EJ (2001) DFT study of the active intermediate in the Fenton reaction. Chem Eur J 7:13. https://doi.org/10.1002/1521-3765(20010702)7:13%3c2775::aid-chem2775%3e3.0.co;2-6
18. Cai M, Su J, Zhu Y et al (2016) Decolorization of azo dyes Orange G using hydrodynamic cavitation coupled with heterogeneous Fenton process. Ultrason Sonochem 28:302–310. https://doi.org/10.1016/j.ultsonch.2015.08.001
19. Carlos TD, Bezerra LB, Vieira MM, Sarmento RA, Pereira DH, Cavallini GS (2021) Fenton-type process using peracetic acid: efficiency reaction elucidations and ecotoxicity. J Hazard Mater 403:123949. https://doi.org/10.1016/j.jhazmat.2020.123949

20. Chang M-W, Chern J-M (2010) Decolorization of peach red azo dye, HF6 by Fenton reaction: initial rate analysis. J Taiwan Inst Chem Eng 41:221–228. https://doi.org/10.1016/j.jtice.2009.08.0
21. Chang S-H, Chuang S-H, Li H-C et al (2009) Comparative study on the degradation of I.C. Remazol Brilliant Blue R and I.C. Acid Black 1 by Fenton oxidation and Fe^0/air process and toxicity evaluation. J Hazard Mater 166:1279–1288. https://doi.org/10.1016/j.jhazmat.2008.12.042
22. Chu L, Wang J, Dong J, Liu H, Sun X (2012) Treatment of coking wastewater by an advanced Fenton oxidation process using iron powder and hydrogen peroxide. Chemosphere 86:409–414. https://doi.org/10.1016/j.chemosphere.2011.09.007
23. Cui D, Li G, Zhao M, Han S (2014) Decolourization of azo dyes by a newly isolatedKlebsiel-lasp. strain Y3, and effects of various factors on biodegradation. Biotechnol Biotechnol Equip 28(3):478–486. https://doi.org/10.1080/13102818.2014.926053
24. de Aragão UG, Freeman HS, Warren SH et al (2005) The contribution of azo dyes to the mutagenic activity of the Cristais River. Chemosphere 60:55–64. https://doi.org/10.1016/j.chemosphere.2004.11.100
25. De Gisi S, Notarnicola M (2017) Industrial wastewater treatment. Encycl Sustain Technol Elsevier 23–42. https://doi.org/10.1016/B978-0-12-409548-910167-8
26. de Luna LAV, da Silva THG, Nogueira RFP et al (2014) Aquatic toxicity of dyes before and after photo-Fenton treatment. J Hazard Mater 276:332–338. https://doi.org/10.1016/j.jhazmat.2014.05.047
27. Dhangar K, Kumar M (2020) Tricks and tracks in removal of emerging contaminants from the wastewater through hybrid treatment systems: a review. Sci Total Environ 738:140320. https://doi.org/10.1016/j.scitotenv.2020.140320
28. dos Santos AB, Cervantes FJ, van Lier JB (2007) Review paper on current technologies for decolourisation of textile wastewaters: perspectives for anaerobic biotechnology. Biores Technol 98:2369–2385. https://doi.org/10.1016/j.biortech.2006.11.013
29. dos Santos AJ, Sirés I, Alves APM et al (2020) Vermiculite as heterogeneous catalyst in electrochemical Fenton-based processes: application to the oxidation of Ponceau SS dye. Chemosphere 240:124838. https://doi.org/10.1016/j.chemosphere.2019.124838
30. Emami F, Tehrani-Bagha AR, Gharanjig K, Menger FM (2010) Kinetic study of the factors controlling Fenton-promoted destruction of a non-biodegradable dye. Desalination 257:124–128. https://doi.org/10.1016/j.desal.2010.02.035
31. Esteves BM, Rodrigues CSD, Boaventura RAR et al (2016) Coupling of acrylic dyeing wastewater treatment by heterogeneous Fenton oxidation in a continuous stirred tank reactor with biological degradation in a sequential batch reactor. J Environ Manag 166:193–203. https://doi.org/10.1016/j.jenvman.2015.10.008
32. Fan H-J, Huang S-T, Chung W-H et al (2009) Degradation pathways of crystal violet by Fenton and Fenton-like systems: condition optimization and intermediate separation and identification. J Hazard Mater 171:1032–1044. https://doi.org/10.1016/j.jhazmat.2009.06.117
33. Farshchi ME, Aghdasinia H, Khataee A (2018) Modeling of heterogeneous Fenton process for dye degradation in a fluidized-bed reactor: kinetics and mass transfer. J Clean Prod 182:644–653. https://doi.org/10.1016/j.jclepro.2018.01.225
34. Fenton HJH (1876) Chem News 1876(33):190
35. Fenton HJH (1984) Oxidation of tartaric acid in presence of iron. J Chem Soc 65:899–910
36. Fernandes NC, Brito LB, Costa GG et al (2018) Removal of azo dye using Fenton and Fenton-like processes: evaluation of process factors by Box-Behnken design and ecotoxicity tests. Chem Biol Interact 291:47–54. https://doi.org/10.1016/j.cbi.2018.06.003
37. Forgacs E, Cserháti T, Oros G (2004) Removal of synthetic dyes from wastewaters: a review. Environ Int 30:953–971. https://doi.org/10.1016/j.envint.2004.02.001
38. Gągol M, Przyjazny A, Boczka G (2018) Wastewater treatment by means of advanced oxidation processes based on cavitation – a review. Chem Eng J 338:599–627. https://doi.org/10.1016/j.cej.2018.01.049

39. Gallard H, De Laat J, Legube B (1999) Spectrophotometric study of the formation of iron(III)-hydroperoxy complexes in homogeneous aqueous solutions. Water Res 33:2929–2936. https://doi.org/10.1016/S0043-1354(99)00007-X
40. Groves JT (2006) High-valent iron in chemical and biological oxidations. J Inorg Biochem 100:434–447. https://doi.org/10.1016/j.jinorgbio.2006.01.012
41. Gürses A, Açikyildiz M, Güneş K (2016) Classification of dye and pigments. In: Dyes and pigments. Springer, pp 31–45
42. Haber F, Weiss J (1934) Proc. R. Soc. London 1934, A147, 332 ± 351. https://doi.org/10.1098/rspa.1934.0221
43. Hayat H, Mahmood Q, Pervez A et al (2015) Comparative decolorization of dyes in textile wastewater using biological and chemical treatment. Sep Purif Technol 154:149–153. https://doi.org/10.1016/j.seppur.2015.09.025
44. He XL, Song C, Li YY, Wang N, Xu L, Han X, Wei DS (2018) Efficient degradation of Azo dyes by a newly isolated fungus Trichoderma tomentosum under non-sterile conditions. Ecotoxicol Environ Saf 150:232–239. https://doi.org/10.1016/j.ecoenv.2017.12.043
45. Huang Y-H, Huang Y-F, Chang P-S, Chen C-Y (2008) Comparative study of oxidation of dye-Reactive Black B by different advanced oxidation processes: Fenton, electro-Fenton and photo-Fenton. J Hazard Mater 154:655–662. https://doi.org/10.1016/j.jhazmat.2007.10.077
46. Joseph J, Radhakrishnan RC, Johnson JK et al (2020) Ion-exchange mediated removal of cationic dye-stuffs from water using ammonium phosphomolybdate. Mater Chem Phys 242:122488. https://doi.org/10.1016/j.matchemphys.2019.122488
47. Khanday WA, Asif M, Hameed BH (2017) Cross-linked beads of activated oil palm ash zeolite/chitosan composite as a bio-adsorbent for the removal of methylene blue and acid blue 29 dyes. Int J Biol Macromol 95:895–902. https://doi.org/10.1016/j.ijbiomac.2016.10.075
48. Kuo WG (1992) Decolorizing dye wastewater with Fenton's reagent. Water Res 26:881–886. https://doi.org/10.1016/0043-1354(92)90192-7
49. Lin SH, Lo CC (1997) Fenton process for treatment of desizing wastewater. Water Res 31:2050–2056. https://doi.org/10.1016/S0043-1354(97)00024-9
50. Lin S-S, Gurol MD (1998) Catalytic decomposition of hydrogen peroxide on iron oxide: kinetics, mechanism, and implications. Environ Sci Technol 32:1417–1423. https://doi.org/10.1021/es970648k
51. Lu S, Liu L, Demissie H et al (2021) Design and application of metal-organic frameworks and derivatives as heterogeneous Fenton-like catalysts for organic wastewater treatment: a review. Environ Int 146:106273. https://doi.org/10.1016/j.envint.2020.106273
52. Mahtab MS, Farooqi IH, Khursheed A (2021) Sustainable approaches to the Fenton process for wastewater treatment: a review. Mater Today: Proc S2214785321031357. https://doi.org/10.1016/j.matpr.2021.04.215
53. Marrakchi F, Ahmed MJ, Khanday WA et al (2017) Mesoporous-activated carbon prepared from chitosan flakes via single-step sodium hydroxide activation for the adsorption of methylene blue. Int J Biol Macromol 98:233–239. https://doi.org/10.1016/j.ijbiomac.2017.01.119
54. Mutar HR, Jasim KK (2021) Adsorption study of disperse yellow dye on nanocellulose surface. Mater Today Proc S2214785321028443. https://doi.org/10.1016/j.matpr.2021.04.003
55. Nasuha N, Hameed BH, Okoye PU (2021) Dark-Fenton oxidative degradation of methylene blue and acid blue 29 dyes using sulfuric acid-activated slag of the steel-making process. J Environ Chem Eng 9:104831. https://doi.org/10.1016/j.jece.2020.104831
56. Neyens E, Baeyens J (2003) A review of classic Fenton's peroxidation as an advanced oxidation technique. J Hazard Mater 98:33–50. https://doi.org/10.1016/S0304-3894(02)00282-0
57. Nogueira RFP, Trovó AG, da Silva MRA et al (2007) Fundamentos e aplicações ambientais dos processos fenton e foto-fenton. Quím Nova 30:400–408. https://doi.org/10.1590/S0100-40422007000200030
58. OECD (2004) Emission scenario document on textile finishing industry Europe. OECD, Paris, pp 77
59. Ozbey Unal B, Bilici Z, Ugur N et al (2019) Adsorption and Fenton oxidation of azo dyes by magnetite nanoparticles deposited on a glass substrate. J Water Process Eng 32:100897. https://doi.org/10.1016/j.jwpe.2019.100897

60. Pignatello JJ, Oliveros E, MacKay A (2006) Advanced oxidation processes for organic contaminant destruction based on the Fenton reaction and related chemistry. Crit Rev Environ Sci Technol 36:1–84. https://doi.org/10.1080/10643380500326564
61. Pliego G, Zazo JA, Casas JA, Rodriguez JJ (2013) Case study of the application of Fenton process to highly polluted wastewater from power plant. J Hazard Mater 252–253:180–185. https://doi.org/10.1016/j.jhazmat.2013.02.042
62. Ramirez JH, Duarte FM, Martins FG et al (2009) Modelling of the synthetic dye Orange II degradation using Fenton's reagent: from batch to continuous reactor operation. Chem Eng J 148:394–404. https://doi.org/10.1016/j.cej.2008.09.012
63. Rial JB, Ferreira ML (2021) Challenges of dye removal treatments based on IONzymes: beyond heterogeneous Fenton. J Water Process Eng 41:102065. https://doi.org/10.1016/j.jwpe.2021.102065
64. Ribeiro JP, Nunes MI (2021) Recent trends and developments in Fenton processes for industrial wastewater treatment – a critical review. Environ Res 197:110957. https://doi.org/10.1016/j.envres.2021.110957
65. Rizzo L (2011) Bioassays as a tool for evaluating advanced oxidation processes in water and wastewater treatment. Water Res 45:4311–4340. https://doi.org/10.1016/j.watres.2011.05.035
66. Robinson T, Marchant R, Nigam P (2001) Remediation of dyes in textile effuent: a critical review on current treatment technologies with a proposed alternative. Biores Technol 77:247–255. https://doi.org/10.1016/s0960-8524(00)00080-8
67. ScienceDirect (2021) https://www.sciencedirect.com. Accessed 25 May 2021
68. Shanmugam BK, Easwaran SN, Mohanakrishnan AS et al (2019) Biodegradation of tannery dye effluent using Fenton's reagent and bacterial consortium: a biocalorimetric investigation. J Environ Manag 242:106–113. https://doi.org/10.1016/j.jenvman.2019.04.075
69. Sharma R, Kar PK, Dash S (2021) Efficient adsorption of some substituted styrylpyridinium dyes on silica surface from organic solvent media – analysis of adsorption-solvation correlation. Colloids Surf A 624:126847. https://doi.org/10.1016/j.colsurfa.2021.126847
70. Shin J, Bae S, Chon K (2021) Fenton oxidation of synthetic food dyes by Fe-embedded coffee biochar catalysts prepared at different pyrolysis temperatures: a mechanism study. Chem Eng J 421:129943. https://doi.org/10.1016/j.cej.2021.129943
71. Silva M, Baltrusaitis J (2021) Destruction of emerging organophosphate contaminants in wastewater using the heterogeneous iron-based photo-Fenton-like process. J Hazard Mater Lett 2:100012. https://doi.org/10.1016/j.hazl.2020.100012
72. Solís M, Solís A, Pérez HI et al (2012) Microbial decolouration of azo dyes: a review. Process Biochem 47:1723–1748. https://doi.org/10.1016/j.procbio.2012.08.014
73. Sun J-H, Shi S-H, Lee Y-F, Sun S-P (2009) Fenton oxidative decolorization of the azo dye Direct Blue 15 in aqueous solution. Chem Eng J 155:680–683. https://doi.org/10.1016/j.cej.2009.08.027
74. Sun S-P, Li C-J, Sun J-H et al (2009) Decolorization of an azo dye Orange G in aqueous solution by Fenton oxidation process: effect of system parameters and kinetic study. J Hazard Mater 161:1052–1057. https://doi.org/10.1016/j.jhazmat.2008.04.080
75. Tantak N, Chaudhari S (2006) Degradation of azo dyes by sequential Fenton's oxidation and aerobic biological treatment. J Hazard Mater 136:698–705. https://doi.org/10.1016/j.jhazmat.2005.12.049
76. Thomas N, Dionysiou DD, Pillai SC (2021) Heterogeneous Fenton catalysts: a review of recent advances. J Hazard Mater 404:124082. https://doi.org/10.1016/j.jhazmat.2020.124082
77. Vieira MM, Pereira Dornelas AS, Carlos TD et al (2021) When treatment increases the contaminant's ecotoxicity: a study of the Fenton Process in the degradation of methylene blue. Chemosphere 131117. https://doi.org/10.1016/j.chemosphere.2021.131117
78. Wakrim A, Dassaa A, Zaroual Z et al (2021) Mechanistic study of carmoisine dye degradation in aqueous solution by Fenton process. Mater Today Proc 37:3847–3853. https://doi.org/10.1016/j.matpr.2020.08.405
79. Wang C, Sun R, Huang R, Cao Y (2021) A novel strategy for enhancing heterogeneous Fenton degradation of dye wastewater using natural pyrite: kinetics and mechanism. Chemosphere 272:129883. https://doi.org/10.1016/j.chemosphere.2021.129883

80. Wang J, Shang K, Guo X, Xia X, Da L (2020) A novel and effective pretreatment to stimulate short-chain fatty acid production from waste activated sludge anaerobic fermentation by ferrous iron catalyzed peracetic acid. J Chem Technol Biotechnol 95:567–576. https://doi.org/10.1002/jctb.6232
81. Wardman P, Candeias LP (1996) Fenton chemistry: an introduction. Radiat Res 145:523. https://doi.org/10.2307/3579270
82. Xie X, Liu N, Yang F et al (2018) Comparative study of antiestrogenic activity of two dyes after Fenton oxidation and biological degradation. Ecotoxicol Environ Saf 164:416–424. https://doi.org/10.1016/j.ecoenv.2018.08.012
83. Xie Y, Chen L, Liu R (2016) Oxidation of AOX and organic compounds in pharmaceutical wastewater in RSM-optimized-Fenton system. Chemosphere 155:217–224. https://doi.org/10.1016/j.chemosphere.2016.04.057
84. Xu M, Wu C, Zhou Y (2020) Advancements in the Fenton process for wastewater treatment. In: Advanced oxidation processes - applications, trends, and prospects. https://doi.org/10.5772/intechopen.90256
85. Xu X-R, Zhao Z-Y, Li X-Y, Gu J-D (2004) Chemical oxidative degradation of methyl tert-butyl ether in aqueous solution by Fenton's reagent. Chemosphere 55:73–79. https://doi.org/10.1016/j.chemosphere.2003.11.017
86. Zimmermann BM, Peres EC, Dotto GL, Foletto EL (2020) Decolorization and degradation of methylene blue by photo-Fenton reaction under visible light using an iron-rich clay as catalyst: CCD-RSM design and LC-MS technique. REGET 24:27. https://doi.org/10.5902/2236117040800

Treatment of Textile Industrial Dyes Using Natural Sunlight-Driven Methods

Thinley Tenzin, Shivamurthy Ravindra Yashas, and Harikaranahalli Puttaiah Shivaraju

Abstract Textile industry is one of the seven major polluting industries and large amounts of different dyes and colorants are released into the environment through the effluents. Industrial dyes and colorants are very complex in structure and toxic and they need to be removed to prevent ecological damages. Due to the drawbacks in the existing conventional wastewater treatment methods and complexity of dye molecules, it is required to find an alternative method for potential treatment of dyes in the effluents. Among the wastewater treatment techniques, photocatalysis-based advanced oxidation processes have potential to remove diversified dyes and colorants in industrial wastewater. Moreover, photocatalysis applications in treatment of industrial dyes in wastewater are eco-friendly and not produce any secondary pollutants. This chapter highlights photocatalytic treatment of various industrial dyes and its advantages.

Keywords Textile industry · Ecological damages · Photocatalysis · Eco-friendly

1 Introduction

Textile industries have been present since civilization and subsequently have been growing and proportionally blooming, along with the increasing fashion trend through time [1]. The waste components from the textile industry starting with the natural compounds as the chief raw material, which include both toxic and safe natural colorants derived from plants, insects, mollusc shells, and natural-colored minerals [2]. The waste thus produced from the textile productions can be considered negligible and are easily removed or treated with dilution and natural neutralization.

T. Tenzin · S. R. Yashas · H. P. Shivaraju (✉)
Department of Environmental Sciences, JSS Academy of Higher Education and Research, Mysuru 570015, Karnataka, India
e-mail: shivarajuenvi@gmail.com

H. P. Shivaraju
Center for Water, Food and Energy, GREENS Trust, Harikaranahalli, Dombaranahalli Post, Turuvekere Taluka, Tumkur District 572215, India

S. S. Muthu and A. Khadir (eds.), *Advanced Oxidation Processes in Dye-Containing Wastewater*, Sustainable Textiles: Production, Processing, Manufacturing & Chemistry,
https://doi.org/10.1007/978-981-19-0987-0_3

But, since the introduction of the chemically engineered colorants and dyes, subsequently increasing the utilization of such cheaper and stronger dyes which replaced the natural dyes [3]. Synthetic dyes are classified into different groups based on their chemical makeup. Extensive studies on structure and toxicity of most of the dyes are well understood for various applications [4]. The contamination of synthetic dyes and colorants causes negative effects on environment and health [5]. For instance, presence of any colored inclusion in the aquatic system blocks sunlight penetration that further damages the aquatic ecosystems [6]. Chemically synthesized dyes are very complex in nature, stable, and last for an extended duration influencing the natural water system negatively [7]. The presence of industrial dyes and organic synthetic dyes in the environment leads to various health issues and it linked to many illnesses ranging from organ damage to cancer [8].

Conventional methods including chemical and biological wastewater treatment methods are used for the remediation of such textile wastes, unfortunately the synthetic organic dyes are unable to be removed by conventional treatment techniques. Even some of the synthetic dyes and industrial colorants are very lethal to the microorganisms which are involved in biological treatment methods due to their complex and toxic structure [9]. Under the chemical treatment process, more chemical additives are introduced to remove the waste in solid form, which does not work for many water-soluble dye molecules [10]. Membrane filtration is the final step in removal of the leftover dye molecules. Reverse osmosis is usually preferred for its ability to remove any material from the effluent water, but the process usually is energy-consuming and more than 75% of the water is rejected back as a concentrated solution [11]. The conventional water treatment process is energy and resource consuming and is not sustainable in the long run [12].

Sunlight-driven treatment processes such as catalyst-based photo-degradation is an alternative and potential method to treat such toxic dyes and industrial colorants. Among all renewable energy Sun is the basic source of energy for life as we know to exist and is renewable. Catalysts are a group of emerging materials, which are capable of initiating the degradation reaction under light. Photocatalysis can be a better solution for the demineralization of various textile dyes and offers a better alternative to conventional methods [13]. There are different kinds of nanocatalysts with desired properties which can be used for treatment of dyes at different levels [14]. Metal–organic framework materials are found to have a greater surface area with the presence of pockets that enhances the absorption and adsorption process in waste removal, having heavy metals and other toxic ions usually adsorbed. The presence of fibers with defined shape and structure is very good at selective filtration under modification. This leads to the introduction of nanofilters, a selective membrane that mimics the biological membrane in functionality [15]. Then comes the more sustainable and efficient nanomaterial for waste treatment nanophotocatalyst, which is from carbonaceous nanomaterials to inorganic metal oxide nanomaterials. A nanophotocatalyst is capable of demineralization of various textile industrial dyes in presence of a light source [16].

2 Classification and Toxicity of the Textile Dyes

Dyes are vaguely classified into natural and synthetic, the natural dyes are derived from the natural world such as plant, animal, insects, and even minerals [17]. A natural dye is usually very laborious with a high price tag and often the dye molecules are not stable and are easily removed or degraded under the washing and drying process [18]. Now natural colors are limited in food industries mainly and but not in textile industries. On the other hand, synthetic organic dyes are brighter and resistant to degradation and are available at a cheaper rate, making them a perfect fit for the textile industry [19]. The synthetic dyes are further subdivided into different categories based on the presence of chromophore groups in their structural framework, such as azo dyes, anthraquinone, etc. [20] or as acidic and basic as per their dissociation property in water [21].

Among all the groups mentioned, the most prominent and evasive dyes used across the globe are the aromatic azo dyes [22]. The toxicity of these dyes is more or less well known to science. Natural dyes can be considered safer compared to synthetic ones [23] with several exceptions. Certain natural dyes are retrieved from very toxic plants and ingestion of such dye is lethal [24]. The presence of such dyes bears both ecological and health deterioration effects. Many of these dyes are persistent and remain for a long duration with increased negative impacts on environment and human health [25]. The textile dyes in an aqueous medium tend to present themselves in either acidic or basic contributing to pH fluctuation in the water body and lethal to the aquatic animals [26]. Many of the dyes from the azo group are considered to be highly toxic for both environment and human, and exposure to such chemicals is found to cause skin disease to organ failure and cancer [27]. A small quantity of organic dye is enough to color the water bodies and presence of a colored compound in water can limit the penetration of sunlight for aquatic fauna. In short, untreated dumping of dyes and colorants from the textile industry certainly causes an impact on the environment and the organisms. The schematic in Fig. 1 illustrates the influence of textile dyes in the environment [28].

3 Phytoremediation

Plants and other photosynthetic organisms are the pillar of any ecosystem. They are naturally existing components capable of utilizing solar energy and converting it into chemical energy [29]. Phytoremediation is vastly studied and used in many places for remediation, such as heavy metal removal from the soil by growing plants such as cabbage. Phytoremediation can be performed with plants and also with phytoplankton [30]. The process of remediation of textile dyes using photosynthetic organisms is through different modes starting with adsorption, absorption, and enzymatic digestion, starting from the root hairs to other tissue parts in plants and through the cell itself in phytoplankton [31].

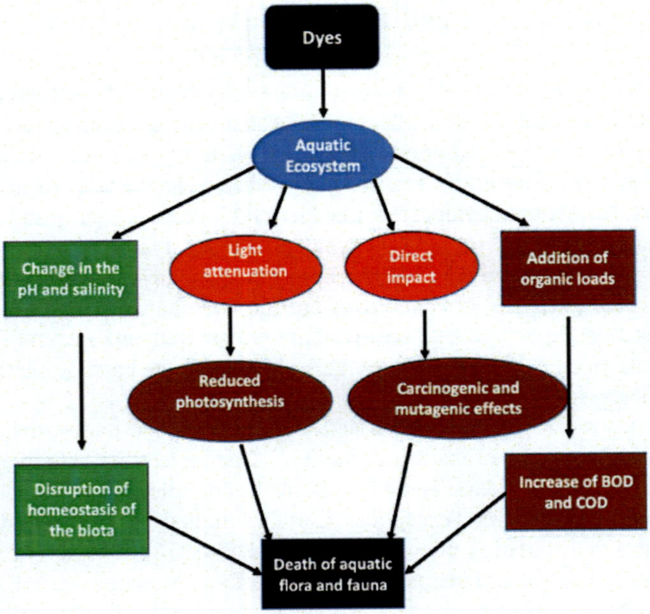

Fig. 1 Schematic representing influence of textile dye on environment

3.1 Phytoremediation with Plants

Plants are the multicellular organism with defined tissue structures. This photosynthetic mass is capable of absorption, digestion, and another process with the sun as the source of energy [32]. The plants produce different proteins and enzymes which are considered as naturally existing nanomaterials capable of catalysis [33]. Many studies have been carried throughout the decade to provide plants, which are easily available, cheaper, safer, and sustainable method for remediation of pollutants like synthetic dyes, heavy metals, etc. Usually, the setup for phytoremediation process has a different process, such as phytoextraction, phytofiltration, and phytostabilization, which is carried out for removal of inorganic metals from the waste solution in the soil [34]. Figure 2 gives a summarized view of processes involved in phytoremediation.

3.1.1 Phytotransformation

A process where the plant absorbs dye molecules from the soil with water molecules and then converts them into less toxic substances through enzymatic action [35]. *Eichhornia crassipes*, a wetland plant is shown to remove 95% of colorant Black B and Red RB through adsorption and it reduces the toxicity level of the dye molecules. Some plants used for this phytotransformation are *Lemna minor, Pistia stratiotes,*

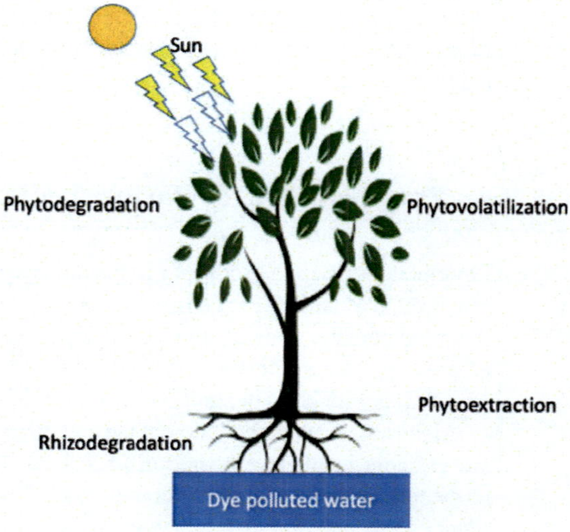

Fig. 2 Process involved in phytoremediation of dye waste effluent

Typha angustifolia, Typhonium flagelliforme, Nasturtium officinale, and Azolla filiculoides, etc. are found to be efficient in adsorbing different dye molecules from the aqueous solution and also capable of degrading the dyes with enzymes such as laccases, tyrosinases, azo-reductases, 2,6-dichlorophenol indophenol reductases, etc. which will reduce the BOD and COD providing stability to the water body [36].

3.1.2 Rhizo-degradation

A process where a plant sends a chemical signal through the root to microbes, which breaks down the organic molecules. This is a synergistic method of degrading complex organic molecules [37]. The humus layer of soil is filled with microorganisms and many form synergistic relationships with the plant root. The in vitro synergistic effect of such relation was seen in *Gaillardia pulchella* (plant) and *Pierre monteilii* (bacteria), where 100% degradation of the textile effluent was reported [38].

3.1.3 Phytovolatilization

It transforms the organic molecules into gaseous molecules and ejects out through respiration [39]. Many plant species are capable of phytoremediation without having any negative influence on the plants. *Typhonium flagelliforme* is one such plant that is capable of degradation of dye molecules from the water system even when the water has no nutritional content [40]. A study conducted for remediation of waste effluent with mono- and di-sulphonated anthraquinone was carried using *Rheum rhabarbarum* was found to be efficient under hydroponic setup. Certain species of *Aloe vera* was found to degrade toxic colors like malachite green and Congo red [41].

4 Phytoplankton for Phytoremediation

Phytoplanktons are a group of photosynthetic organisms made up of single cells or few groups of cells [42]. These are mainly composed of different algal groups. The potential benefits of using algae over larger plants are less requirement of land, ease of growing, more efficient at degradation, and rapid growth under nutrient-rich soil or solution with enough sunlight [43].

Algae are divided into macroalgae and microalgae, based on their size. Both algae are efficient in phytoremediation [42]. Lab-scale studies have proven macroalgae to be efficient in the demineralization of dyes. Macroalgae like *Streptomyces glaucescens* and *Solanum marginatum* where adsorption is studied for degradation of dye effluent and is found to remove diazo dye effectively. Many other species of macroalgae are also efficient in the removal of dye from waste effluent. Biosorption using microalgae is studied for the same purpose of removal of the dye from waste effluent. *Acutodesmus obliquus* and *Chlorella vulgaris* are species of microalgae that have shown removal efficiency of 44 mg/g and 53 mg/g of acid red dye and remazol golden yellow dyes, respectively [44]. The process of phytoremediation with algae is dependent on the solution pH. In lower pH, the organic materials get protonated and attain a positive charge, which enhances the adsorption process as the negative cell wall will attract the positively charged organic particles. The reverse happens when the pH is increased and the adsorption decreases as well [45].

5 Nanocatalyst-Based Photocatalytic Degradation of Textile Dyes

One of the most sustainable treatment methods of industrial dyes and colorants is nanoparticle-based photocatalysis. Nanoparticles are any material whose size is less than 100 nm in any one direction [46]. The material, when in nanometre, functions and presents characteristic capacity which is usually not associated with bulk or atomic element or compound [47], such nanomaterials possess enhanced optical and electrical properties [48]. Nanoparticles are comparatively stable compared to organic and inorganic catalysts. They have high surface area that contributes to increased adsorption and reaction area. Most of these materials are capable of catalytic reaction when exposed to light [49, 50]. This property can be enhanced by doping the nanomaterial with ionic metal or other nanomaterials which can be used in the various fields from sustainable energy to sustainable waste management [51]. This chapter

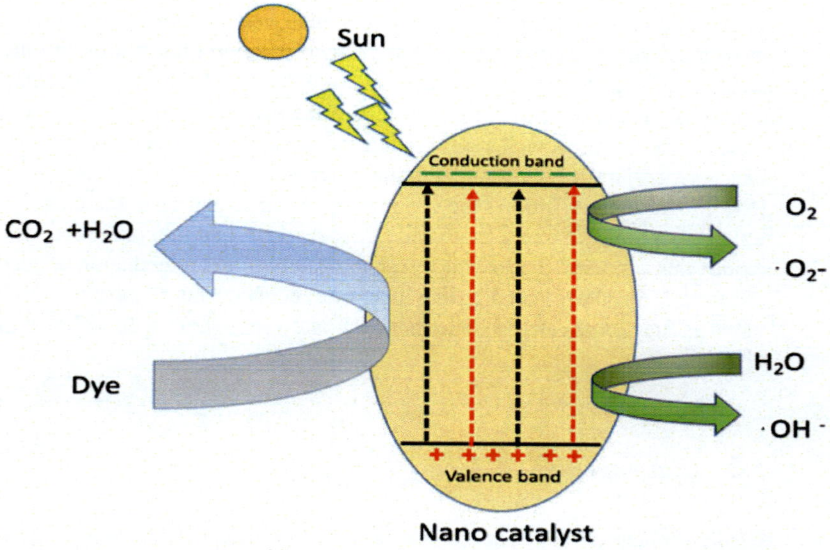

Fig. 3 Schematic representing mechanism involved in photocatalysis

will discuss materials for textile dye degradation with solar energy as the driving force rendering the process sustainable and effective as well. The nanomaterial is categorized into organic and inorganic at broad and further subdivided into a different group based on their chemical and structural makeup. Such different compositions and morphologies play an important role in their function as photocatalysts. The most well recognized and studied nanomaterial for photocatalysis are metal oxides due to their efficiency with easy synthesis and safety [52].

The principle behind photocatalysis of textile dye using nanophotocatalyst is based on the excitation of the particle under exposure to the sun. The material absorbs an amount of photon, which on excitation allows for an electron to jump from the valence band to conduction band creating an electron–hole pair, which in presence of the aqueous solution produces intrinsic hydroxide and oxygen radicals that aids in the degradation of the organic dye molecule [14]. Figure 3 summarizes the basic mechanism of photocatalysis.

5.1 Metal Oxide Nanoparticles

The metal oxide nanoparticles are one of the earliest discovered and studied materials for their catalytic property in the field of electronics and sustainable material. Metal oxide nanoparticles are easier to synthesize compared with other complex materials [53]. Such materials are synthesized using the salt of the respective metal oxide.

Titanium dioxide was the first nanomaterial discovered as a photocatalyst. Titanium dioxide is widely studied and used for remediation process, but due to their high bandgap of 3.2 eV they can effectively use only 4% of the solar spectrum, hence doping is carried out which in turn reduces the bandgap and allows photocatalysis under visible light as well [14]. Titanium dioxide is doped with organic materials like polyacrylamide, chitosan, and other carbonaceous materials for improvement [54]. Much research follows the variation in synthesis pattern to increase the visible light absorption. Tin oxide, zinc oxide, copper oxide, and iron oxide are a few metal oxides that are used for solar-driven photocatalysis for remediation of textile dye effluents [55]. Other metal oxides like zinc, copper, and manganese can also be tuned as functional catalyst under both UV and visible light with different methods of synthesis and doping [56].

5.2 *Chalcogenide*

Chalcogenides are complex cations along with elements from groups of 16 in the periodic table. Elements like oxygen, sulfur, selenium, tellurium, and polonium are branded as Chalcogen [17]. These materials are well studied and used in the field of optoelectronics. They are also excellent photocatalyst with most of them having a small bandgap within 1–2, making them a perfect photocatalyst for degradation of a pollutant under sunlight [57]. Layered metal chalcogenide is a very efficient heterogeneous catalysis and is found to degrade textile dyes efficiently. TiS_x, AgS_X are some metal chalcogenides which are used in the degradation of textile dye effluents [58].

5.3 *Metal–Organic Framework (MOF)*

A metal–organic framework is a group of nanoparticles that are also known as porous coordination polymers and are gaining interest over its highly porous and diverse structural property. Due to the presence of pores the surface area increases, which can increase the adsorption site of the material. This material is of great interest for storage of gases like hydrogen, or membrane filtration, sensors, or in drug delivery. MOF material is a photocatalyst that is mainly functional under visible range and UV light [59]. A metal–organic framework is one of the most efficient materials for photocatalytic reduction of gases and water into organic fuels and hydrogen fuel, which are the future of the energy industry. Due to such capability, it is also widely studied in degradation of dynamic dyes released from the textile industry [60]. Zr Porphyrin metal–organic framework has been used in the removal of the dyes and used as a self-cleaning material under solar irradiation. Cu- and Zn-based metal–organic frameworks are some of the most efficient and studied materials capable of improved dye degradation. Certain studies show a degradation percentage of above

90% within a very short duration [61]. The benefit of this material over metal oxide is that it is composed of both organic and inorganic materials with increased surface area and characteristic function with reduced bandgap with functionality spreading into visible to far-infrared spectrum increasing catalytic efficiency [62]. Further doping with nanoparticles can enhance its catalytic efficiency in the degradation of organic pollutants.

5.4 Carbonaceous Nanophotocatalyst

Carbonaceous nanomaterial is an organic structure derived from different organic precursor chemicals like urea, graphite, melamine, proteins, etc. and also from natural plant parts like leaf stems and barks [63]. Most of such material is either polymeric sheet or hollow tube in structure with high surface area for adsorption [64]. These materials contain great conductors of electricity unlike their bulk counterparts, e.g., graphene [65]. Usually, such materials are used along with certain metallic nanoparticles to function as an efficient catalyst but through much research. Different carbonaceous nanomaterials which are capable of photocatalysis without any dopant have also been developed [66]. These advanced materials are of great use in the sensor applications for environmental monitoring to optoelectronics and photocatalysis. Carbon nitride, graphene oxide, etc. are several engineered carbonaceous materials that are capable of demineralization of textile dyes under sunlight. Band structure engineering of graphitic carbonitride deems the material as a great material in the degradation of dye under solar irradiation [67]. Carbonaceous materials are safer for the environment and human health comparatively as it is organic based [68].

5.5 Perovskite

A perovskite is a group of material having formulae of ABO_3, where the A and B are metals [69]. The perovskite material is a very well-established material in optoelectronics application for its exceptional optical and electrical property [70]. Recent studies have revealed that perovskites are great photocatalysts and have proven very efficient in the degradation of textile dyes. Some examples are degradation of dyes with $ZnTiO_3$, $SrTiO_3$ that is doped to reduce the bandgap, which have been proven as efficient catalysts for degradation of dye under constant solar irradiation [71]. Sometimes the precursor molecule and synthesis procedure can also influence the absorption efficiency such as black $SrTiO_3$ which even is capable of photocatalysis under sunlight with high efficiency [72].

5.6 *Quantum Dots*

The discovery of quantum dots has been a milestone in the field of optoelectronics due to its small size that provided enhanced optical and electrical property [73]. Besides being used in optoelectronics, they are also a great catalyst and dopant. They are used along with effective nanoparticles to enhance absorption process to enhance degradation of textile dye waste and other organic pollutants [74]. They are used in combination with other nanoparticles like titanium dioxide to reduce the bandgap and increase their activity in the visible light allowing the material to demineralize textile dye waste using the sun as the driving energy [74].

6 Advantages of Using Material Which Uses Photon as the Driving Energy

Depending only on the sun as the major driving energy in degradation of textile dye has both negative and positive influence. The utilization of materials capable of photocatalysis using solar energy allows for a cleaner and greener alternative to the traditional remediation of textile dye. The exhaustion of solar energy won't be a concern and can provide energy for a prolonged duration. The cost and energy consumption for such a setup will be lower than the conventional method of treatment of textile dye. Conventional treatment plant used both high energy-consuming and polluting chemicals while photocatalysis requires only the catalyst and the sun to complete the remediation process. Phytoremediation is also the most efficient and cheap natural photocatalytic complex system. Throughout its lifecycle, it will keep on absorbing and degrading textile dye from wastewater. Fast reproduction and high metabolism make the phytoplankton an ideal organism for phytoremediation of textile dye-polluted water.

Nanoparticles of all kinds mentioned above are an abiotic catalyst that has been studied and developed over the years of research. Such particles are very efficient photocatalysts due to change in their physical and electrical properties as a result of small size. These materials possess a high surface area, which helps in the adsorption of the dye molecule, which is then degraded under the photocatalytic process. The utilization of nanoparticles for remediation of textile dye is efficient in both degradation and time conservation. Duration required for maximum degradation of textile dye usually takes 2–4 h allowing for efficient degradation of textile dye within half a day. In brief, the utilization of materials that can harness solar energy for remediation of the textile dye pollutant is a very sustainable technique that can provide enhanced remediation of any pollutant with no energy wastage and low capital cost.

7 Disadvantage of Using Sun as the Driving Energy

Even though the utilization of solar energy as driving energy is very sustainable, it is not always efficient. The intensity of solar energy is never constant and is dependent on environmental conditions like weather, duration of day and night, air quality, etc. The sun will be available all day on a summer day but that is not the case in the rainy and winter season. The solar energy will be abundant near to the equatorial region while it reduces as it drifts toward poles making this technology obsolete. The plants take longer duration to completely remove dye from the textile wastewater while the nanoparticles are not well studied at full scale in remediation of textile waste treatment. The nanoparticles performance usually reduces over time due to loss of the material while filtering and washing.

8 Conclusions

Nanocatalyst-based photocatalytic treatment of industrial wastewater has great advantages especially to remove organic dyes and colorants which are not able to degrade by conventional methods. Photocatalytic treatment of textile industrial dyes and colorants is environmental friendly and economic where natural sunlight can be used as an alternative driving energy without production of any secondary pollutants. Recent studies clearly revealed potential removal of various dye molecules using improved nanocatalysts. However, some of the associated problems like designing of affordable nanocatalyst and their availability for real-time applications at commercial level yet to be studied and optimization of photocatalytic process for target dyes need to be done urgently.

References

1. Kant R (2012) Textile dyeing industry an environmental hazard. Nat Sci
2. Ebrahimi R, Maleki A, Zandsalimi Y, Ghanbari R, Shahmoradi B, Rezaee R et al (2019) Photocatalytic degradation of organic dyes using WO3-doped ZnO nanoparticles fixed on a glass surface in aqueous solution. J Ind Eng Chem [Internet] 73:297–305. https://www.sciencedirect.com/science/article/pii/S1226086X19300334
3. Chan SHS, Wu TY, Juan JC, Teh CY (2011) Recent developments of metal oxide semiconductors as photocatalysts in advanced oxidation processes (AOPs) for treatment of dye waste-water. J Chem Technol Biotechnol
4. O'Neill C, Hawkes FR, Hawkes DL, Lourenço ND, Pinheiro HM, Delée W (1999) Colour in textile effluents—sources, measurement, discharge consents and simulation: a review. J Chem Technol Biotechnol
5. Pollutants of textile industry wastewater and assessment of its discharge limits by water quality standards. Turkish J Fish Aquat Sci

6. Wu Y, Xiao X, Xu C, Cao D, Du D (2013) Decolorization and detoxification of a sulfonated triphenylmethane dye aniline blue by Shewanella oneidensis MR-1 under anaerobic conditions. Appl Microbiol Biotechnol
7. Naraginti S, Yu Y-Y, Fang Z, Yong Y-C (2019) Visible light degradation of macrolide antibiotic azithromycin by novel $ZrO_2/Ag@TiO_2$ nanorod composite: transformation pathways and toxicity evaluation. Process Saf Environ Prot [Internet] 125:39–49. https://doi.org/10.1016/j.psep.2019.02.031
8. Karimi L, Zohoori S, Yazdanshenas ME (2014) Photocatalytic degradation of azo dyes in aqueous solutions under UV irradiation using nano-strontium titanate as the nanophotocatalyst. J Saudi Chem Soc [Internet] 18(5):581–588. https://doi.org/10.1016/j.jscs.2011.11.010
9. Pant D, Adholeya A (2007) Biological approaches for treatment of distillery wastewater: a review. Bioresour Technol
10. Sahu O, Chaudhari P (2013) Review on chemical treatment of industrial waste water. J Appl Sci Environ Manag
11. Van Der Bruggen B, Lejon L, Vandecasteele C (2003) Reuse, treatment, and discharge of the concentrate of pressure-driven membrane processes. Environ Sci Technol
12. Rosenfeld PE, Feng LGH (2011) The paper and pulp industry. In: Risks of hazardous wastes
13. Shivaraju HP, Yashas SR, Harini R (2020) Application of Mg-doped TiO_2 coated buoyant clay hollow-spheres for photodegradation of organic pollutants in wastewater. Mater Today Proc [Internet] 27:1369–1374. https://linkinghub.elsevier.com/retrieve/pii/S2214785320315339
14. Gunnarsson R, Helmersson U, Pilch I (2015) Synthesis of titanium-oxide nanoparticles with size and stoichiometry control. J Nanoparticle Res
15. Muhambihai P, Rama V, Subramaniam P (2020) Photocatalytic degradation of aniline blue, brilliant green and direct red 80 using NiO/CuO, CuO/ZnO and ZnO/NiO nanocomposites. Environ Nanotechnol Monit Manag [Internet] 14(September):100360. https://doi.org/10.1016/j.enmm.2020.100360
16. Devi LG, Kottam N, Murthy BN, Kumar SG (2010) Enhanced photocatalytic activity of transition metal ions Mn^{2+}, Ni^{2+} and Zn^{2+} doped polycrystalline titania for the degradation of Aniline Blue under UV/solar light. J Mol Catal A Chem [Internet] 328(1–2):44–52. https://doi.org/10.1016/j.molcata.2010.05.021
17. Wu Z, Yuan X, Zhang J, Wang H, Jiang L, Zeng G (2016) MINIREVIEW photocatalytic decontamination of wastewater containing organic dyes by metal-organic frameworks and their derivatives
18. Rossi T, Silva PMS, De Moura LF, Araújo MC, Brito JO, Freeman HS (2017) Waste from eucalyptus wood steaming as a natural dye source for textile fibers. J Clean Prod [Internet] 143:303–310. https://www.sciencedirect.com/science/article/pii/S0959652616321734
19. Gallo M, Naviglio D, Ferrara L (2014) Nasunin, an antioxidant anthocyanin from eggplant peels, as natural dye to avoid food allergies and intolerances. Eur Sci J 10(9)):1–11
20. Karimi L, Zohoori S, Yazdanshenas ME (2014) Photocatalytic degradation of azo dyes in aqueous solutions under UV irradiation using nano-strontium titanate as the nanophotocatalyst. J Saudi Chem Soc
21. Gürses A, Açıkyıldız M, Güneş K, Gürses MS (2016) Classification of dye and pigments. 31–45
22. Bera KK, Majumdar R, Chakraborty M, Bhattacharya SK (2018) Phase control synthesis of α, β and α/β Bi_2O_3 hetero-junction with enhanced and synergistic photocatalytic activity on degradation of toxic dye, Rhodamine-B under natural sunlight. J Hazard Mater [Internet] 352:182–191. https://www.sciencedirect.com/science/article/pii/S030438941830181X
23. Amini M, Arami M, Mahmoodi NM, Akbari A (2011) Dye removal from colored textile wastewater using acrylic grafted nanomembrane. Desalination
24. Abe FR, Mendonça JN, Moraes LAB, Oliveira GAR de, Gravato C, Soares AMVM et al (2017) Toxicological and behavioral responses as a tool to assess the effects of natural and synthetic dyes on zebrafish early life. Chemosphere [Internet] 178:282–290. https://www.sciencedirect.com/science/article/pii/S0045653517303831

25. El-Agez TM, Taya SA, Elrefi KS, Abdel-Latif MS (2014) Dye-sensitized solar cells using some organic dyes as photosensitizers. Opt Appl
26. Rezaei SM, Makarem S, Alexovič M, Tabani H (2021) Simultaneous separation and quantification of acidic and basic dye specimens via a dual gel electro-membrane extraction from real environmental samples. J Iran Chem Soc [Internet]. https://doi.org/10.1007/s13738-021-02167-2
27. Brown MA, De Vito SC (1993) Predicting azo dye toxicity. Crit Rev Environ Sci Technol [Internet] 23(3):249–324. https://doi.org/10.1080/10643389309388453
28. Walsh GE, Bahner LH, Horning WB (1980) Toxicity of textile mill effluents to freshwater and estuarine algae, crustaceans and fishes. Environ Pollut Ser A, Ecol Biol [Internet] 21(3):169–179. https://www.sciencedirect.com/science/article/pii/0143147180901610
29. Boardman NK (1977) Photosynthesis of sun. Ann Rev Plant Physiol 28:355–377
30. Karczewska A, Mocek A, Goliński P, Mleczek M (2015) Phytoremediation of copper- contaminated soil. In: Phytoremediation: management of environmental contaminants, vol 2, pp 143–170
31. Suresh B, Ravishankar GA (2004) Phytoremediation—a novel and promising approach for environmental clean-up. Crit Rev Biotechnol [Internet] 24(2–3):97–124. https://doi.org/10.1080/07388550490493627
32. Gomes PIA, Asaeda T (2013) Phytoremediation of heavy metals by calcifying macro-algae (Nitella pseudoflabellata): implications of redox insensitive end products. Chemosphere [Internet] 92(10):1328–1334. https://www.sciencedirect.com/science/article/pii/S0045653513007789
33. Pilon-Smits E (2005) Phytoremediation. Annu Rev Plant Biol [Internet] 56(1):15–39. https://doi.org/10.1146/annurev.arplant.56.032604.144214
34. Akansha J, Nidheesh PV, Gopinath A, Anupama KV, Suresh Kumar M (2020) Treatment of dairy industry wastewater by combined aerated electrocoagulation and phytoremediation process. Chemosphere [Internet] 253:126652. https://www.sciencedirect.com/science/article/pii/S0045653520308456
35. Gao J, Garrison AW, Hoehamer C, Mazur CS, Wolfe NL (2000) Uptake and phytotransformation of organophosphorus pesticides by axenically cultivated aquatic plants. J Agric Food Chem [Internet] 48(12):6114–6120. https://doi.org/10.1021/jf9904968
36. Tahir U, Yasmin A, Khan UH (2016) Phytoremediation: potential flora for synthetic dyestuff metabolism. J King Saud Univ Sci [Internet] 28(2):119–130. https://www.sciencedirect.com/science/article/pii/S1018364715000671
37. Wang J, Song X, Wang Y, Bai J, Li M, Dong G et al (2017) Bioenergy generation and rhizodegradation as affected by microbial community distribution in a coupled constructed wetland-microbial fuel cell system associated with three macrophytes. Sci Total Environ [Internet] 607–608:53–62. https://www.sciencedirect.com/science/article/pii/S0048969717316625
38. Tahir U, Sohail S, Khan UH (2017) Concurrent uptake and metabolism of dyestuffs through bio-assisted phytoremediation: a symbiotic approach. Environ Sci Pollut Res [Internet] 24(29):22914–22931. https://doi.org/10.1007/s11356-017-0029-8
39. Sinha RK, Valani D, Sinha S, Singh S, Herat S (2009) Solid waste management and environmental remediation bioremediation of contaminated sites: a low- cost nature, pp 1–73
40. Sinha A, Lulu S, Vino S, Osborne WJ (2019) Reactive green dye remediation by Alternanthera philoxeroides in association with plant growth promoting Klebsiella sp. VITAJ23: a pot culture study. Microbiol Res [Internet] 220(November 2018):42–52. https://doi.org/10.1016/j.micres.2018.12.004
41. Rai MS, Bhat PR, Prajna PS, Jayadev K, Rao PSV (2014) Original research article degradation of malachite green and congo red using Aloe barabadensis Mill. Extract. Int J Curr Microbiol Appl Sci 3(4):330–340
42. Pal R, Choudhury AK (2006) An introduction to phytoplanctons: diversity and ecology. Environ Pract 8:212–214. https://doi.org/10.1017/S1466046606060406%5Cn. http://search.ebscohost.com/login.aspx?direct=true&db=8gh&AN=24215819&site=ehost-live

43. Sedmak B, Eleršek T (2005) Microcystins induce morphological and physiological changes in selected representative phytoplanktons. Microb Ecol 50(2):298–305
44. Pollution A (2015) Algae and environmental sustainability. Algae Environ Sustain (Kant 2012)
45. Hinga KR, Arthur MA, Pilson MEQ, Whitaker D (1994) Carbon isotope fractionation by marine phytoplankton in culture: the effects of CO_2 concentration, pH, temperature, and species. Global Biogeochem Cycles [Internet] 8(1):91–102. https://doi.org/10.1029/93GB03393
46. George M, Ajeesha TL, Manikandan A, Anantharaman A, Jansi RS, Kumar ER et al (2021) Evaluation of Cu–$MgFe_2O_4$ spinel nanoparticles for photocatalytic and antimicrobial activates. J Phys Chem Solids [Internet] 153:110010. https://www.sciencedirect.com/science/article/pii/S0022369721000767
47. Kumar A, Khan M, Zeng X, Lo IMC (2018) Development of g-C_3N_4/TiO_2/Fe_3O_4@SiO_2 heterojunction via sol-gel route: a magnetically recyclable direct contact Z-scheme nanophotocatalyst for enhanced photocatalytic removal of ibuprofen from real sewage effluent under visible light. Chem Eng J [Internet] 353:645–656. https://www.sciencedirect.com/science/article/pii/S1385894718313937
48. Cleuziou J-P, Wernsdorfer W, Bouchiat V, Ondarçuhu T, Monthioux M (2006) Carbon nanotube superconducting quantum interference device. Nat Nanotechnol [Internet] 1(1):53–59. https://doi.org/10.1038/nnano.2006.54
49. Zhao P, Li N, Astruc D (2013) State of the art in gold nanoparticle synthesis. Coord Chem Rev [Internet] 257(3):638–665. https://www.sciencedirect.com/science/article/pii/S0010854512002305
50. Modena MM, Rühle B, Burg TP, Wuttke S (2019) Nanoparticle characterization: what to measure? Adv Mater [Internet] 31(32):1901556. https://doi.org/10.1002/adma.201901556
51. Terranova U, Viñes F, De Leeuw NH, Illas F (2020) Mechanisms of carbon dioxide reduction on strontium titanate perovskites. J Mater Chem A
52. Khan MM, Adil SF, Al-Mayouf A (2015) Metal oxides as photocatalysts. J Saudi Chem Soc
53. Gunalan S, Sivaraj R, Rajendran V (2012) Green synthesized ZnO nanoparticles against bacterial and fungal pathogens. Prog Nat Sci Mater Int
54. Teh CM, Mohamed AR (2011) Roles of titanium dioxide and ion-doped titanium dioxide on photocatalytic degradation of organic pollutants (phenolic compounds and dyes) in aqueous solutions: a review. J Alloys Compd [Internet] 509(5):1648–1660. https://www.sciencedirect.com/science/article/pii/S0925838810027118
55. Sayadi MH, Sobhani S, Shekari H (2019) Photocatalytic degradation of azithromycin using GO@Fe_3O_4/ZnO/SnO_2 nanocomposites. J Clean Prod [Internet] 232:127–136. https://doi.org/10.1016/j.jclepro.2019.05.338
56. Mishra PK, Mishra H, Ekielski A, Talegaonkar S, Vaidya B (2017) Zinc oxide nanoparticles: a promising nanomaterial for biomedical applications. Drug Discov Today
57. Arunchander A, Peera SG, Sahu AK (2017) Synthesis of flower-like molybdenum sulfide/graphene hybrid as an efficient oxygen reduction electrocatalyst for anion exchange membrane fuel cells. J Power Sources
58. Behbahani ES, Dashtian K, Ghaedi M (2020) Fe/Co-chalcogenide-stabilized Fe_3O_4 nanoparticles supported MgAl-layered double hydroxide as a new magnetically separable sorbent for the simultaneous spectrophotometric determination of anionic dyes. Microchem J [Internet] 152:104431. https://www.sciencedirect.com/science/article/pii/S0026265X19322465
59. Lv H, Zhao H, Cao T, Qian L, Wang Y, Zhao G (2015) Efficient degradation of high concentration azo-dye wastewater by heterogeneous Fenton process with iron-based metal-organic framework. J Mol Catal A Chem [Internet] 400:81–89. https://www.sciencedirect.com/science/article/pii/S1381116915000552
60. Zhu S-R, Wu M-K, Zhao W-N, Liu P-F, Yi F-Y, Li G-C et al (2017) In situ growth of metal–organic framework on BiOBr 2D material with excellent photocatalytic activity for dye degradation. Cryst Growth Des [Internet] 17(5):2309–2313. https://doi.org/10.1021/acs.cgd.6b01811
61. Zhao P, Wang J, Han X, Liu J, Zhang Y, Van der Bruggen B (2021) Zr-porphyrin metal–organic framework-based photocatalytic self-cleaning membranes for efficient dye removal. Ind Eng Chem Res [Internet] 60(4):1850–1858. https://doi.org/10.1021/acs.iecr.0c05583

62. Meng A-N, Chaihu L-X, Chen H-H, Gu Z-Y (2017) Ultrahigh adsorption and singlet-oxygen mediated degradation for efficient synergetic removal of bisphenol A by a stable zirconium-porphyrin metal-organic framework. Sci Rep [Internet] 7(1):6297. https://doi.org/10.1038/s41598-017-06194-z
63. Leary R, Westwood A (2011) Carbonaceous nanomaterials for the enhancement of TiO_2 photocatalysis. Carbon N Y [Internet] 49(3):741–772. https://www.sciencedirect.com/science/article/pii/S0008622310007207
64. Gürbüz B, Ayan S, Bozlar M, Üstündağ CB (2020) Carbonaceous nanomaterials for phototherapy: a review. Emergent Mater [Internet] 3(4):479–502. https://doi.org/10.1007/s42247-020-00118-w
65. Bai J, Zhong X, Jiang S, Huang Y, Duan X. Graphene nanomesh. Nat Nanotechnol [Internet] 5(3):190–194. https://doi.org/10.1038/nnano.2010.8
66. Tourillon G, Garnier F (1983) Effect of dopant on the physicochemical and electrical properties of organic conducting polymers. J Phys Chem [Internet] 87(13):2289–2292. https://doi.org/10.1021/j100236a010
67. Saikia M, Das T, Dihingia N, Fan X, Silva LFO, Saikia BK (2020) Formation of carbon quantum dots and graphene nanosheets from different abundant carbonaceous materials. Diam Relat Mater [Internet] 106:107813. https://www.sciencedirect.com/science/article/pii/S0925963519309021
68. Vlasova II, Kapralov AA, Michael ZP, Burkert SC, Shurin MR, Star A et al (2016) Enzymatic oxidative biodegradation of nanoparticles: mechanisms, significance and applications. Toxicol Appl Pharmacol [Internet] 299:58–69. https://www.sciencedirect.com/science/article/pii/S0041008X16300023
69. Hazen RM (1988) Perovskites. Sci Am [Internet] 258(6):74–81. http://www.jstor.org/stable/24989124
70. Long G, Sabatini R, Saidaminov MI, Lakhwani G, Rasmita A, Liu X et al (2020) Chiral-perovskite optoelectronics. Nat Rev Mater [Internet] 5(6):423–439. http://www.nature.com/articles/s41578-020-0181-5
71. Wang N, Kong D, He H (2011) Solvothermal synthesis of strontium titanate nanocrystallines from metatitanic acid and photocatalytic activities. Powder Technol [Internet] 207(1–3):470–473. https://doi.org/10.1016/j.powtec.2010.11.034
72. Zhao W, Zhao W, Zhu G, Lin T, Xu F, Huang F (2015) Black strontium titanate nanocrystals of enhanced solar absorption for photocatalysis. CrystEngComm [Internet] 17(39):7528–7534. http://xlink.rsc.org/?DOI=C5CE01263E
73. Feng H, Guo Q, Xu Y, Chen T, Zhou Y, Wang Y et al (2018) Surface nonpolarization of g-C_3N_4 by decoration with sensitized quantum dots for improved CO_2 photoreduction. ChemSusChem [Internet] 11(24):4256–4261. https://doi.org/10.1002/cssc.201802065
74. Saud PS, Pant B, Alam A-M, Ghouri ZK, Park M, Kim H-Y (2015) Carbon quantum dots anchored TiO_2 nanofibers: effective photocatalyst for waste water treatment. Ceram Int [Internet] 41(9, Part B):11953–11959. https://www.sciencedirect.com/science/article/pii/S027288421501113X

Solar Photocatalytic Treatment of Dye Removal

Li-Ngee Ho and Wan Fadhilah Khalik

Abstract Textile industries are one of the major causes of water pollution due to large amount of effluent produced. Solar photocatalytic treatment of dye could be considered as a green and sustainable method to degrade and mineralize dye to harmless products. This is in line with the Sustainable Development Goals (SDGs), i.e. goal 6: Clean water and sanitation and goal 14: Life below water. The major influencing factors on the solar photocatalytic treatment of dye wastewater include light source and intensity, functional groups of the dye molecules, pH, aeration, initial dye concentrations, photocatalyst loading, physical properties of photocatalyst (band gap, surface area, crystalline phase). Higher light intensity could enhance photocatalytic activity. Functional groups of the dye molecules are crucial in dye degradation mechanisms. Solar photocatalytic degradation of dye is more favorable at slightly acidic and alkaline condition at pH 5–9. The presence of aeration facilitates the dye removal efficiency with the formation of superoxide radicals. Excess dosage of photocatalyst hinders light penetration and reduces photodegradation efficiency. Narrow band gap leads to recombination of the electron–hole pairs. Larger specific surface area enhances the photocatalytic activity. Different crystalline phases of the semiconductor photocatalyst yield different photocatalytic efficiencies.

Keywords Solar photocatalytic · Treatment · Dye · Water pollution

L.-N. Ho (✉)
Faculty of Chemical Engineering Technology, Universiti Malaysia Perlis, 02600 Arau, Perlis, Malaysia
e-mail: lnho@unimap.edu.my

W. F. Khalik
Water Research and Environmental Sustainability Growth, Centre of Excellence (WAREG), Faculty of Civil Engineering Technology, Universiti Malaysia Perlis (UniMAP), 02600 Arau, Perlis, Malaysia

S. S. Muthu and A. Khadir (eds.), *Advanced Oxidation Processes in Dye-Containing Wastewater*, Sustainable Textiles: Production, Processing, Manufacturing & Chemistry,
https://doi.org/10.1007/978-981-19-0987-0_4

1 Introduction

Dye wastewater such as textile wastewater from industries has drawn increasing concern globally due to its intense color and large amount of chemicals that applied during the dyeing process. The textile wet processing industry consumes large amount of water and thus leads to vast discharge of wastewater [30]. Textile wastewater is usually high in chemical oxygen demand, inorganic salts, complex chemicals, total dissolved solids, turbidity, and salinity [84]. These effluents will pose severe environmental problems if they are discharged without proper treatment.

Conventionally, dye wastewater is treated by using physicochemical, biological, and membrane-based methods, such as adsorption, coagulation-flocculation, biological degradation, electrolysis, and microfiltration/ultrafiltration [17]. However, these treatment methods have a few drawbacks in terms of high chemical consumption, expensive, low efficiency, and generation of secondary waste. In order to overcome the current problems, seeking for an alternative treatment method for textile wastewater which is sustainable, simple, and highly efficient without generation of secondary sludge is of great importance. In view of this, solar photocatalytic treatment of dye could be a right solution to this problem as it is economical, sustainable, and able to mineralize the recalcitrant dye molecules to harmless products without generating secondary sludge.

2 Basic Concept of Photolysis and Solar Photocatalytic Process

2.1 Photolysis

Photolysis is a photochemical reaction initiated by light irradiation which is capable to eliminate and degrade various chemical compounds [7]. Through absorption of photons with sufficient energy in aqueous solution, radicals such as •OH and $O_2{\bullet}^-$ will be produced [32, 87]. These reactive oxygen species (ROS) will further react with the organic compounds in the solution which will lead to degradation or mineralization to H_2O and CO_2 that are harmless to the environment as illustrated in Fig. 1. Hence, photolysis is a clean technology that is simple and cost-effective without generation of secondary waste [7]. The rate of photolysis can be influenced by various factors such as the light absorption properties, reactivity of the chemicals, and the intensity of solar radiation [48].

Photolysis or photolytic degradation of dye in aqueous solution has been widely reported by researchers worldwide. A comparative study of photolysis and photocatalytic degradation of three commercial acidic dyes was reported by Márlen and Azevedo [49]. The results revealed that under photolysis of natural solar light, xanthenic dye Acid Red 51 (AR 51) could be degraded almost completely (98%) compared with azo dye Acid Yellow 23 (AY 23) and triarylmethanic dye Acid Blue

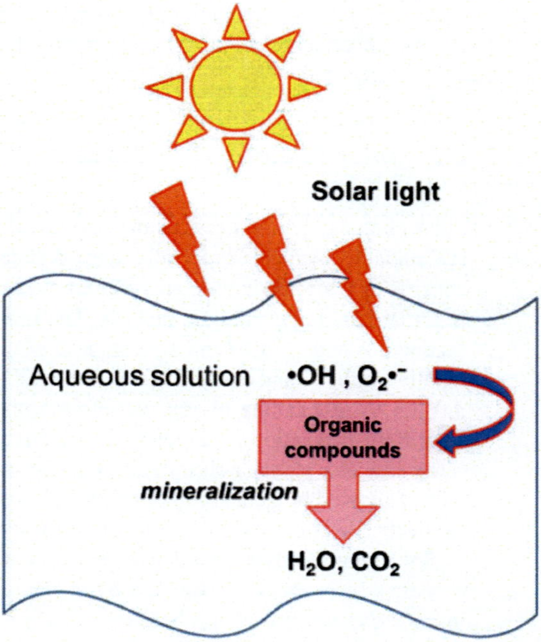

Fig. 1 Photolysis or photolytic degradation of dye

9 (AB 9) which only degraded sparingly (7%). Therefore, photolysis could be a cost-effective alternative for treatment of xanthenic dye (AR 51), while it was not favorable for the treatment of azo dye (AY 23) and triarylmethanic dye (AB 9).

In a comparative study of advanced oxidation processes by de Moraes et al. [20], three different lights such as UVA, UVC, and sunlight were used for photolysis of acid violet 17 (AV17). It was found that only UVC light slightly degraded AV17 (9.62%) while there was insignificant degradation observed under the UVA and sunlight irradiation. Besides, Lee et al. [43] investigated on the photolysis of Methylene Blue (MB) under sunlight and ultra-violet (UV) light irradiation. The findings showed that MB degraded significantly and more efficiently under photolysis of natural sunlight compared with UV light where the degradation was trivial. These findings revealed that light source also played a crucial role in photolysis. Furthermore, it was found that MB degradation could be further enhanced in photocatalytic process under both sunlight and UV light, respectively.

Pérez-Estrada et al. [62] investigated the photolytic degradation of malachite green under natural sunlight irradiation and its degradation mechanisms. However, they found that the intermediates compounds produced were more persistent and toxic than the parent compound. This indicated that the photolysis process is not powerful enough to mineralize the dye. Besides, Venkatesh et al. [83] found that the photolytic degradation of Rhodamine B under artificial sunlight irradiation was 56% while addition of SnO_2 as photocatalyst could achieve almost complete degradation efficiency at 99%. Hence, incorporating photocatalyst into photolysis is necessary to enhance the photochemical reaction by transforming it to solar photocatalytic process.

2.2 Solar Photocatalytic Process

Solar photocatalytic process is an economical way for treatment of dye. Solar light is abundantly available especially in tropical countries. Its wide range of wavelength consists of UV and visible lights that are very useful for irradiation of photocatalyst to initiate the photocatalytic process. The most commonly used photocatalysts are titanium dioxide (TiO_2) and zinc oxide (ZnO) as these semiconductor materials are readily available and cost effective. The band gap energy of TiO_2 and ZnO is about 3.0–3.3 eV depending on the particle morphology [69, 75]. As shown in Fig. 2, solar light irradiation could stimulate electron from the valance band (VB) of the semiconductor photocatalyst to the conduction band (CB) to create an electron–hole pairs, followed by generation of reactive oxygen species or radicals which are vital for further oxidation and degradation of the dye molecules.

Solar photocatalytic process for treatment of dye is a heterogeneous photocatalysis process which always carried out by application of a suspension of photocatalysts inside a dye solution. It has been intensively investigated by researchers worldwide since the early years of 1990s [19, 50, 55].

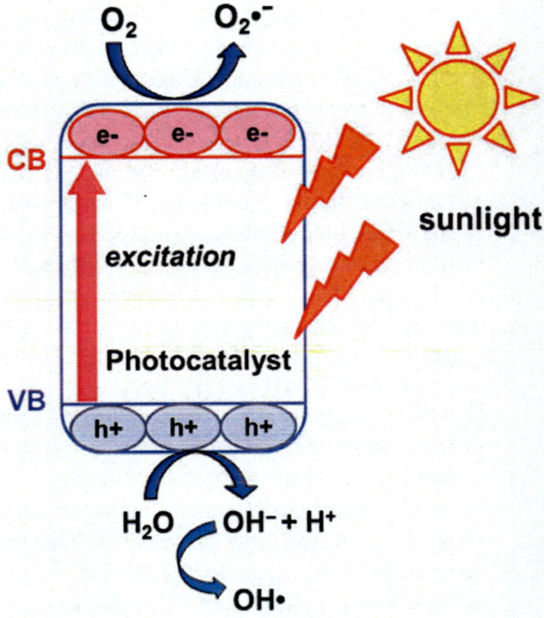

Fig. 2 Electron–hole pairs and reactive oxygen species generation in a semiconductor photocatalyst under solar light irradiation

In most of the studies of solar photocatalytic treatment of dye, researchers directly applied suspension of photocatalysts powder in the dye solutions [1, 14, 27, 58]. However, photocatalyst suspension is difficult to be recycled and it is not economical and sustainable. In order to overcome this problem, researchers have improved and modified the application of phoocatalysts through attachment or immobilization of the photocatalyst powder on a supporting material or substrate such as cement coated borosilicate glass, steel wire mesh, electro-spun membrane, and ceramic tiles, which enables the recycling of the photocatalyst more conveniently [3, 52, 92].

According to numerous studies [36, 51, 59] reported by researchers worldwide, it has been proven that solar photocatalytic degradation of dye is highly efficient which could achieve complete mineralization of dye. It could be said that solar photocatalytic treatment of dye is feasible with its high removal efficiency and simple operation. Hence, treatment of dye through solar photocatalytic process is considered as an economical and sustainable method which will benefit both the industries and environment.

3 Influencing Factors in Solar Photocatalytic Treatment of Dye Removal

Solar photocatalytic treatment of dye removal is governed by a few influencing factors that could significantly affect the photodegradation efficiency of the dye. Thus, it is of utmost importance to understand each of these influencing factors explicitly as discussed hereinafter.

3.1 Light Source and Intensity

Light is the initiator in the photocatalytic reaction to induce electron–hole pair generation in the semiconductor photocatalyst. Without light, it is impossible to further proceed to produce reactive oxygen radicals for photocatalytic degradation of dye. Thus, light source and its intensity are both important in the solar photocatalytic reaction.

Sunlight, or the solar radiation spectrum, includes bands between 100 nm and 1 mm, which encompasses ultraviolet, visible, and infrared radiation [80]. Generally, solar light or sunlight (wavelength range: 100 nm–1 mm), ultraviolet (UV) light (wavelength range: 100–380 nm), and visible light (wavelength range: 380–780 nm) [24] are the three main light sources that researchers used in photocatalytic study. Since light source is the initiator that stimulates the photocatalytic reaction in the photocatalyst, so it plays a crucial role in determining the photocatalytic activity.

Solar light that used in photocatalytic study can be divided into (i) natural sunlight and (ii) artificial solar light. Natural sunlight is available abundantly in many tropical

Table 1 Wavelength of UVA, UVB, and UVC light according to ISO standard (ISO-2138)

Name	Abbreviation	Wavelength (nm)
Ultraviolet A	UVA	315–400
Ultraviolet B	UVB	280–315
Ultraviolet C	UVC	100–280

countries. However, the intensity of the natural sunlight cannot be controlled as it depends on the weather. Thus, it may affect the overall photocatalytic activity due to the uncertainty of the weather. In order to maintain consistent sunlight intensity, some researchers utilized artificial solar light generated by Xenon lamp to initiate the photocatalytic reaction [22, 85], instead of natural sunlight. The main advantage of applying solar light as light source in photocatalytic research is that sunlight comprises of both UV and visible light that contribute to wider wavelength range and higher energy can be attained.

UV light mainly used in photocatalytic study can be mainly categorized into (i) UVA, (ii) UVB, and (iii) UVC light. Table 1 shows the wavelength of UV light according to ISO standard ISO-21348. Among these 3 types of UV lights, the most commonly used are UVA and UVC light in the photocatalytic research by employing UVA or UVC lamp as light source [40, 43].

Compared with UV light, visible light has longer wavelength (400–700 nm) and thus a lower energy. Hence, it can be considered that UV light with a shorter wavelength and higher energy is more advantageous to be used as a light source in photocatalytic degradation of dye compared with that of visible light with a longer wavelength and lower energy. Several researchers have attested that the efficiency of photocatalytic degradation of dye under UV light was higher compared with that of the visible light [10, 16, 78].

However, this phenomenon is not always true as the optical property of the photocatalyts itself is key factor which could affect the light absorbance region during light-irradiation in the photocatalytic process. When the light absorbance region of the photocatalyst is near to the UV region, more UV light will be absorbed, and vice versa in the case of visible light [45]. Thus, it can be said that the photocatalytic activity is also determined by the photoresponse of the photocatalyst which is governed by the band gap of the semiconductor photocatalyst.

Besides comparison between UV and visible light, comparative studies on the photodegradation of dyes under the influence of UV and solar light (natural or artificial) have also been widely explored. Most of the results revealed that enhanced photocatalytic degradation efficiency of dye could be observed under solar light irradiation compared with UV light alone [45, 63]. This is due to the electromagnetic spectrum of solar light includes both UV and visible light region. This wider range of spectrum possesses greater energy to induce excitation of the electrons from valence band to conduction band in the semiconductor photocatalyst for stimulation of the photocatalytic reaction.

On the other hand, light intensity is equally important in determining the photocatalytic activity. Higher light intensity means higher power or light energy which will lead to enhanced photocatalytic performance. Larbi et al. [42] reported that the increasing in UV power from 8 to 16 W increased the MB removal from 34 to 64% with the sample doped with 6% of Ce. This result indicated that UV light intensity is important in photocatalytic degradation of dye. Enhancing the power of light radiation provides more energy to excite the electron of the semiconductor photocatalyst and creates more electron–hole pairs, thus improves the photocatalytic process [31].

Hence, it could be considered that both light source and intensity play critical roles in the photocatalytic degradation of dye. As solar light includes both spectrum of UV and visible light range (wavelength: 100 nm–1 mm), it is advantageous for various types of photocatalysts with different band gap energy and dye solutions of different wavelength (in the UV–Vis spectrum). Meanwhile, higher light intensity could enhance photocatalytic activity by larger power of light irradiation for electron–hole pair generation in the photocatalyst.

3.2 *Molecular Structure and Functional Group of Dye*

Degradation of dye in photocatalytic system depends on the molecular structure and functional group of dye itself. These include number of azo bonds of dye, sulfonic groups or sulfonic substituents, methyl groups, nitrite groups, alkyl side chains, chloro groups, carboxylic groups, and number of hydroxyl groups.

3.2.1 Monoazo and Diazo Dye

Azo dyes are extensively used as coloring agent in most of industries especially textile, paint, and leather. They are one of the organic compounds that industrially synthesized which represent major group (60–70%) of more than 10,000 dyes that currently manufactured [11, 13, 93]. Due to their chromophores (–N=N–) which also known as azo bond, the wastewaters are often recalcitrant and carcinogenic which then became hazard to the human as well as aquatic life. The number of chromophores for each of azo dyes distinguished their name as well as degradation rate. For example, azo dye that contains one –N=N– bond namely monoazo, meanwhile azo dye which had two –N=N– bond namely as diazo. The difference between number of azo bonds is shown in (Fig. 3a–c). In addition, other than napthalenes, benzene rings, or aromatic heterocycles, the azo group may also be bonded to enolizable aliphatic groups which all of them essential to give the color of the dye with their shades of different intensities [8].

Studies showed that dye attached to many azo bonds were easily to be degraded compared to less azo bonds [60, 67]. During photocatalytic process, hydroxyl radicals may firstly attack the azo bond or site near the chromophore (for example, C–N=N– bond). The cleavage of azo bond and C–N= then may result in the decolorization of the dye.

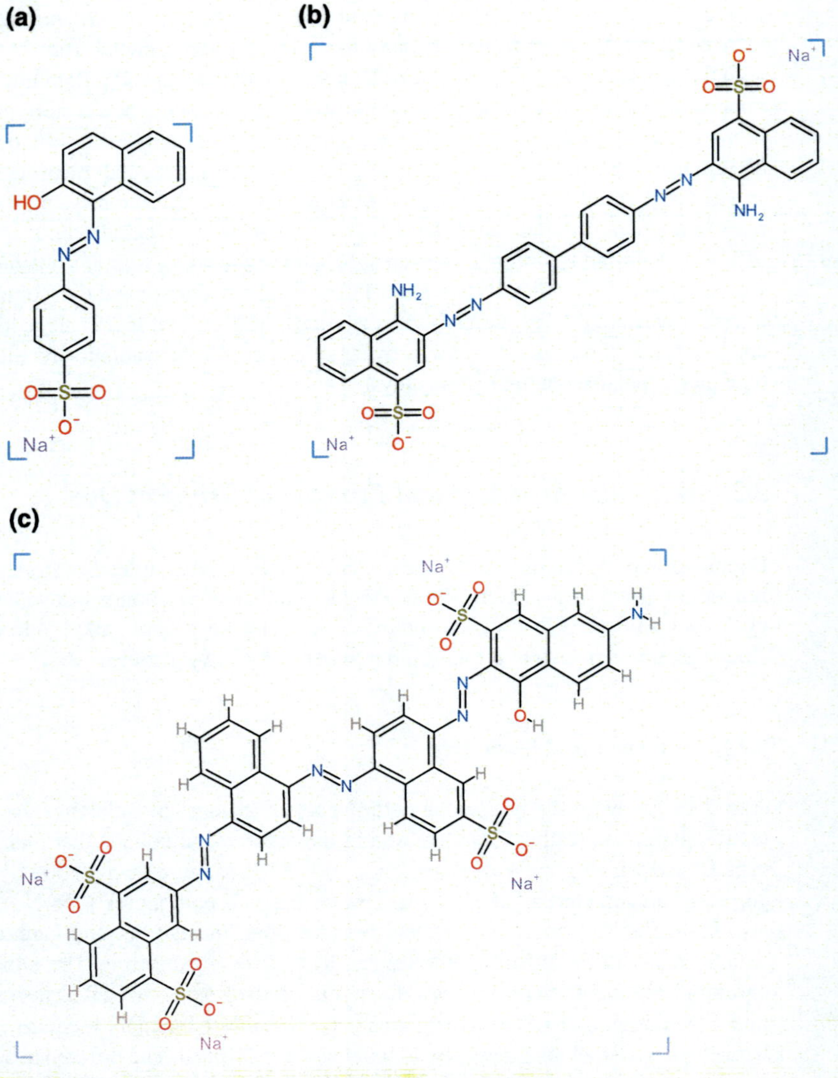

Fig. 3 Molecular structure **a** acid orange 7; **b** Congo red; **c** direct blue 71

Compared the decolorization or degradation rate between Reactive Black 5 (diazo dye) and Reactive Orange 4 (monoazo dye), it was observed that Reactive Black 5 completely decolorized after 7 min, meanwhile for Reactive Orange 4, the decolorization only achieved 92% in the same duration by using ZnO as photocatalyst Kansal et al. [33]. Ljubas et al. [47] reported in their study regarding degradation of azo dyes by sol-gel TiO_2 films found that rate constant for diazo dye (Congo Red) was much higher compared to monoazo dye (Methyl Orange). A study between various type of dyes including Congo Red and monoazo dye (Orange G) using TiO_2-CNT as photocatalyst showed rapid degradation rate of Congo Red rather than Orange G [26].

3.2.2 Sulfonate Group

Other than number of azo bonds, sulfonate group that substitutes to the molecular structure of dye also may responsible to the rapid degradation of dye in photocatalytic. Once the azo bond of dye is destructed, it then leads to the destruction of sulfonate substituent of the dye. It believed that sulfonic group may act as another electron acceptor in photocatalytic leading to the higher degradation rate of dye. One of the methods or techniques to determine the destruction of sulfonate group in dye is through ion-chromatography analysis. Once the sulfonate group is break down, the concentration of sulfate ion in the dye may increase. This phenomenon was reported by Khalik et al. [38] in a comparative study between monoazo dye Orange G and New Coccine using zinc oxide as catalyst. The breakdown of bonding between sulfonate and benzene ring led to the increased in sulfate concentration. In addition, New Coccine with three sulfonate groups showed higher degradation in photocatalytic compared to Orange G which only attached to two sulfonate groups. Several studies which investigated dyes with sulfonate groups also found that removal of these group happened at early stage in the photocatalytic degradation process [29, 73, 86].

However, at some point, study of the influence of sulfonate groups is very difficult due to its operation in different fields. For example, sulfonate groups may decrease the electron density in the aromatic rings and β nitrogen atom of the azo bond through –I and +M effects.

3.2.3 Other Functional Groups (–OH, $-NO_2$, $-CH_3$ and –Cl)

Other than sulfonate groups, destruction of hydroxyl groups (–OH) and nitrite groups ($-NO_2$) also may lead to the rapid degradation rate of dye. These –OH groups have electronic properties which are –I and +M effects. The reactivity of the dyes due to effect of –OH groups can be determined through field effect (–I). The resonance effect and degradation rate of dye can be increased depending on the number of hydroxyl groups in dye molecule. The para position of Chromotrope 2B contains a nitrite group and delocalization of the *p* electrons of phenyl ring and unpaired electrons of the heteroatom may occur once this substituent interacts with the ring. Due to the enriched electrons at the phenyl ring, an electrophilic entity favors to be attacked by nitrite group [39].

Meanwhile, reactivity of the dye may slightly decrease due to the existence of methyl groups ($-CH_3$) and chloro groups (–Cl) that attached to their structure. This is due to the competition of hydroxyl radicals between parent of dye molecule and the substitution ($-CH_3$, –Cl). For instance, hydroxyl radicals that generate during photocatalytic process may react with chloro groups to form chloride anions in the solution. However, due to short life of hydroxyl radicals, these radicals tend to attack the dye chromophore compared to the chloro groups.

Overall, it could be concluded that functional groups of the dye molecules play pivotal role in influencing the dye degradation mechanisms in the photocatalytic reaction which will indirectly affect the dye degradation efficiency.

3.3 pH

pH of azo dyes solution is a crucial parameter in the photocatalytic reactions since it ascertains the surface charge properties of photocatalyst, aggregates' size formation as well the adsorption behavior of photocatalyst [64]. The ionization state or also known as point of zero charge (PZC) of surface for each type of photocatalyst may be differ under alkaline or acidic medium, respectively. The point of zero charge can be described as the limiting pH when the net charge of catalyst surface is zero [41, 89]. For example, point of zero charge for TiO_2 (Degussa P25), ZnO, and Fe_2O_3 is widely reported at pH ~6.25, pH 9, and pH 6.6, respectively [54, 68, 90]. When pH of azo dye solutions less than pH at point of zero charge, the surface of photocatalyst become positively charged, meanwhile, the surface become negatively charged when pH of azo dye solutions higher than pH at point of zero charge. Due to different molecular structure, the degradation rate of azo dye for effect of pH may be different.

For instance, using Reactive Black 5 and ZnO as model of azo dye and photocatalyst, respectively, below PZC, due to dye molecules attached to four sulphonate anions, surface concentration of the dye is relatively high, meanwhile OH^- and •OH has low concentration. On the other hand, above pH 9, catalyst surface is negatively charged by means of metal-bound OH^-, resulted in decreased the surface concentration of dye and increased the amount of •OH [2]. The enhancement of photocatalytic degradation rate of dyes in higher pH due to the higher concentration of OH^- to react with holes, which then may form •OH [34]. In addition, Tang et al. [77] found that Methylene Blue dye showed highest degradation in alkaline solution and lowest in the acidic solution with $CaIn_2O_4$ as photocatalyst. However, when the dye solution pH is too alkaline, it may reduce the dye removal efficiency as reported by some of the researchers [35, 56]. It was found that in highly alkaline condition (pH 10 and above), there is electrostatic repulsion between the OH^- and negatively charged surface of ZnO which will hinder the oxidation reaction and increase the chances of the recombination of electron–hole pairs [25, 44].

Most studies reported that the degradation of dye less effective in acidic solution [37, 74]. This might be occurred due to the dissolution of photocatalyst in strongly acidic solution, where photocatalyst may react with acids to produce the corresponding salt, which then hindered the degradation rate of dye. Furthermore, photocatalyst also may dissolve, disappear, or tend to agglomerate in dye solution which has too low pH (for example, pH 1), resulted in reduced the availability of surface area for the dye and photon to be absorbed [23].

Accordingly, solar photocatalytic degradation of dye is more favorable at slightly acidic and alkaline condition at the range of pH 5–9. As dissolution of photocatalyst may occur when the dye solution is too acidic (below pH 3), while too alkaline condition (above pH 10) may lead to electrostatic repulsion between the photocatalyst and the dye molecules. Thus, adjusting the pH of the dye solution to a suitable pH range is important in solar photocatalytic degradation of dye.

3.4 Aeration

The critical operational parameters which may affect the photocatalytic efficiency of organic contaminants especially dyes are aeration [5]. Aeration means air, pure oxygen or nitrogen is purged into the organic compounds before or during the photocatalysis. The interaction between photocatalyst and light irradiation may cause the electrons to be excited from valence band to the conduction band and then leave holes in the valence band. The reaction between holes and either with water molecules (H_2O) or hydroxide ion (OH_-) may form •OH, meanwhile, electrons will react with oxygen molecules (O_2) to form superoxide radicals (•O_2) as shown in following reaction [57]:

$$e^- + O_2 \rightarrow \bullet O_2^- \quad (1)$$

$$\bullet O_2^- + \text{dye} \rightarrow \text{dye-OO}\bullet \quad (2)$$

$$\bullet O_2^- + H^+ \rightarrow \bullet HO_2 \quad (3)$$

$$\bullet O_2^- + \bullet HO_2 + H^+ \rightarrow H_2O_2 + O_2 \quad (4)$$

Other than hydroxyl radicals, these superoxide radicals may enhance the degradation rate of dye. By adding aeration into the photocatalytic process, it will boost up the degradation rate of dye. The recombination of electron–hole pair may be reduced when oxygen is purged into the system as it performs as an effective conduction band "electron-trap" [23]. According to [5], mixing in the photocatalytic process also can be improved with the presence of aeration. The oxygen molecule acts as an electron acceptor which will minimize the chance for electron and hole to recombine in photocatalytic process [46, 64].

In a study by Bizani et al. [9] regarding decolorization and degradation of dye solutions and wastewater in the presence of titanium dioxide, they found that aeration improved the decolorization rate of Cibacron Red FNR and Cibacron Yellow FN2R.

In view of the role of oxygen in photocatalysis, it could be considered that application of aeration in solar photocatalytic removal of dye will facilitate the dye removal efficiency with the formation of superoxide radicals in the photocataytic reaction.

3.5 Initial Dye Concentration

One of the important factors in degradation of azo dyes is initial concentration. The concentration level of dye wastewater of the textile industries normally is very high which may affect the environment and aquatic lives without proper treatment. The decolorization of dye depends on its concentration where the reduction rate decreases as its concentration increases. In other words, dye with low concentration only requires less time to degrade compared to dye with high concentration. The increase in initial concentration of dye takes more time to degrade results in decrease of the path length of light penetration into the solution. The degradation rate relates to the formation of •OH on the catalyst surface as well as probability of reaction between •OH and dye molecules [76]. The chance of reaction between dye molecules and oxidizing species will be increased as the initial dye concentration increases, which then lead to enhancement of degradation rate of the dye. On the other hand, the degradation rate of dye decreased as increased in the dye concentration is due to the restricted production of •OH on the surface of photocatalyst since dye ions had fully covered the active sites. For example, initial dye concentration of 20 and 100 mg/L with constant amount of catalyst, irradiation time and light intensities, may form similar amount of •OH. Since the formation of •OH and $\bullet O_2^-$ remains constant in both initial dye concentration of 20 and 100 mg/L, the probability of dye molecule reacting with •OH decrease in high concentration.

Similar results were reported earlier for the degradation of dye [15, 71, 91]. Brites et al. [12] reported that higher concentration of textile dye required longer time to be degraded by using ZnO as photocatalyst. In addition, Dubey et al. [21] evaluated the degradation of methyl orange using TiO_2/NaY and found that decreased in dye degradation rate at higher concentration was due to prevention of light intensity into the solution which was blocked by adsorbed dye. Thennarasu et al. [79] investigated the photocatalytic degradation of Orange G dye under solar light using ZnO as photocatalyst stated that reduction in degradation rate at higher dye concentration was due to the increased in dye concentration as well as number of dye molecules adsorbed on the ZnO surface, but amount of •OH that formed on the surface of ZnO was constant.

In brief, controlling the initial dye concentration in solar photocatalytic process is important to achieve optimum dye degradation efficiency. When the initial dye concentration is too high, dye degradation rate decreases due to the absorption of light by dye molecules which blocks the light transmission reaching the photocatayst. Thus, it is more feasible to dilute the industrial textile wastewater with high concentration of dye to lower concentration in order to decolourize and mineralize the dye molecules efficiently during the solar photocatalytic process.

3.6 Photocatalyst Loading (Dosage)

Photocatalyst loading or dosage is one of the key factors in determining the efficiency of the solar photocatalytic degradation of dye. Photocatalyst provides the active sites for photocatalytic reaction to take place. Higher photocatalyst loading will increase the number of active sites for adsorption of dye molecules, which is then followed by the photocatalytic reaction upon solar light irradiation. However, photocatalyst loading closely correlates with both initial dye concentration and volume of the dye solution. When the initial dye concentration remains constant, increasing the photocatalyst loading will initially accelerate the photocatalytic efficiency until it achieves saturation of the dye molecules. Subsequently, further increasing the photocatalyts loading will only lead to excessive active sites for adsorption of dye molecules which will not further enhance the photocatalytic efficiency [27, 70].

Meanwhile, by increasing the volume of the dye solution with constant dye concentrations, the increase of photocatalyst loading will provide more active sites for the larger number of dye molecules due to increased volume of the dye solution. Thus, it will elevate the photocatalytic process. If the volume of the dye solution remains constant, additional photocatalyst loading will increase the photocatalytic efficiency up to a certain level, where all of the dye molecules have already been consumed. Eventually, further increasing the photocatalyst loading will not have significant effect on boosting the photocatalytic efficiency due to limitation of the number of dye molecules exist in the dye solution.

In addition, over dose of photocatalyts may deteriorate the photodegradation efficiency as reported by some of the researchers [6, 28]. It may be ascribed to the increase of light reflectance by the catalyst and reduction in light penetration [28, 35]. Furthermore, high dosage of photocatalyst will increase the turbidity of the dye solution which hinders the light irradiation for photocatalytic reaction, thus degradation efficiency depleted [66, 79].

Hence, optimum photocatalyst loading (dosage) is very important to ensure the efficiency of the photocatalytic activity. Excess dosage of photocatalyst increases light reflectance by the catalyst and hinders light penetration which leads to depletion in photodegradation efficiency. From the economical viewpoint, it is rational to evaluate and determine the optimum photocatalyst loading or dosage in order to avoid excessive photocatalyst dosage and wastage.

3.7 Physical Properties of Photocatalysts

The physical properties of photocatalysts are one of the crucial factors in solar photocatalytic degradation of dye. The main physical properties of semiconductor photocatalyst that have profound influence on the solar photocatalytic activity are: (i) band gap energy; (ii) specific surface area; (iii) crystalline phase.

3.7.1 Band Gap Energy

The band gap energy of semiconductor photocatalyst indicates the amount of energy required to be absorbed from the light source to generate the electron-hole pair to initiate the photocatalytic reaction. Thus, band gap energy plays the most important role in determining the efficiency of the photocatalytic process. Titanium dioxide (TiO_2) and zinc oxide (ZnO) are the two most widely used semiconductor photocatalysts which possess band gap energy within 3.0–3.2 eV and 3.1–3.3 eV, respectively [69, 75].

A study by Al-Mamun et al. [4] compared the photocatalytic performance of the pure synthesized TiO_2 (3.17 eV) with the nanocomposites of TiO_2 doped with graphene oxide (3.10 eV) and nanocomposites TiO_2 doped with graphene oxide and Ag (2.96 eV). The results revealed that removal efficiency of methyl orange dye under solar irradiation decreased following the increase of the band gap energy. This finding indicated that smaller band gap energy could accelerate the photocatalytic reaction by reducing the energy required to stimulate the electron transfer from valance band to the conduction band in the semiconductor photocatalyst.

However, this phenomenon is not always true in all band gap energy range. In a solar photocatalytic degradation of New Coccine dye by using four types of different photocatalysts, i.e. TiO_2, ZnO, CuO and Fe_2O_3 [36], it was found that both TiO_2 and ZnO with similar band gap energy (3.2 eV) achieved nearly complete dye removal compared with relatively low removal efficiency attained by CuO (1.4 eV) and Fe_2O_3 (2.2 eV) which possessed relatively narrow band gap. It was reported that photocatalysts with narrow band gap such as CuO and Fe_2O_3 will permit rapid recombination of hole and electron, which will retard the photocatalytic activity for decolorization and degradation of dye [53]. Thus, it could be concluded that band gap energy of the semiconductor photocatalyst plays a pivotal role in solar photocatalytic degradation of dye.

3.7.2 Specific Surface Area

Specific surface area has always become a matter of concern in photocatalysis. High specific surface area provides larger number of active sites for dye molecule attachment to undergo photocatalytic reaction. Hence, there are considerable efforts in synthesis of porous materials to increase the porosity and enlarge the surface area. In a typical study by Dash et al. [18], a range of mesoporous titanium dioxide (TiO_2) nanostructure were fabricated with different surface area and pore volume. It was found that the TiO_2 with the highest surface area (157.35 m^2/g) and pore volume (0.31 cm^3/g) yielded the highest photocatalytic degradation of Rhodamine 6G dye (98%) compared to that of the TiO_2 with lowest surface area (3.52 m^2/g) and pore volume (0.09 cm^3/g) which has only 15% of photocatalytic degradation. Meanwhile, a comparative study on the photocatalytic dye removal between mesoporous titania and non-porous titania showed that 68% dye removal was achieved by mesoporous titania compared with the non-porous titania which only showed 57% removal. It may be ascribed to the porous nature of the mesoporous titania surface that could provide sufficient surface area for the adsorption of dyes and faster migration or diffusion of parent as well as intermediate products and thus enhancing the overall removal efficiency [65]. Therefore, it could be stated that specific surface area of the photocatalyst plays a key role in determining the photocatalytic efficiency.

3.7.3 Crystalline Phase

Besides band gap energy, crystalline phase of the semiconductor photocatalyst is also crucial in photocatalytic reaction. For instant, a study by Singh et al. [72] reported that the TiO_2 with mixture of 48% rutile and 52% anatase phase showed higher solar photocatalytic degradation of Rhodamine B and methyl orange dyes, respectively, compared with that of the TiO_2 with 100% of anatase phase. Besides, a comparative study of three different phases of TiO_2 (brookite, anatase and rutile) showed that photocatalytic decolourization of Rhodamine B was almost completely achieved (97%) by using brookite and anatase as photocatalyst, respectively. However, only about 48% decolourization could be attained by rutile [81]. These findings revealed that the crystalline phase of semiconductor photocatalyst has significant influence on the photocatalytic degradation of dye.

Overall, it could be concluded that physical properties of the photocatalyst such as band gap energy, specific surface area and crystalline phase, significantly affect the performance of the solar photocatalytic treatment of dye. When the band gap is too narrow, it will lead to recombination of the electron-hole pairs. Meanwhile, larger specific surface area with higher amount of active sites will enhance the photocatalytic activity. Last but not least, different crystalline phases of the semiconductor photocatalyst yield different efficiencies in solar photocatalytic degradation of dye.

3.8 Applications—Solar Photocatalytic Treatment of Real Textile Dye Wastewater

In the research of solar photocatalytic treatment of dye, synthetic wastewater has always been applied in most of the studies. As the synthetic dye wastewater is more convenient to prepare and the composition of this synthetic dye solution is not as complicated as that of the real textile dye wastewater. However, application of real textile dye wastewater in solar photocatalytic treatment is vital and important for evaluation of the feasibility of this treatment method for application in the industries.

Although the application of real textile wastewater in solar photocatalytic process is not as direct and simple as that of the synthetic dye wastewater, there are still quite a number of studies on the solar photocatalytic treatment of local textile dye wastewater. For example, Khalik et al. [35] from Malaysia investigated on the solar photocatalytic degradation of a local *Batik* textile wastewater by applying ZnO as photocatalyst. The findings showed that about 92% of decolourization efficiency and 91% of chemical oxygen demand removal could be achieved. Besides, Tijani et al. [82] from Nigeria compared the photocatalytic mineralization of local dye wastewater under artificial visible light and natural sunlight irradiation using $Ag_2O/B_2O_3/TiO_2$ nanocomposite as photocatalyst. The results revealed that natural sunlight showed significant higher removal efficiency compared with artificial visible light. Under natural sunlight, chemical oxygen demand removal of 96.2% and total organic carbon removal of 86.1% could be attained. In a study by Zahoor et al. [88], CeO_2–TiO_2 nanocomposites were used as the photocatalysts to treat a dye wastewater collected from local textile industry in the Faisalabad region of Pakistan. Significant degradation of dye could be observed within 60 $_{min}$ of reaction time. In addition, Parvin et al. [61] reported on the application of synthesized Fe_2O_3 nanoparticles under sunlight for degradation of dissolved organic matter from a local textile wastewater obtained from Savar, Bangladesh. The real textile wastewater was solar irradiated without any dilution in the presence of photocatalyst Fe_2O_3 nanoparticles. The results showed that after 40 h of photocatalytic irradiation, TOC, COD and biological oxygen demand were reduced by 42.2, 44.68 and 37.89%, respectively.

Overall, from the results reported by researchers from different countries in the world, it could be concluded that the solar photocatalytic degradation of real textile dye wastewater is feasible. The main factor to be concerned when dealing with the photocatalytic treatment of real textile wastewater is the pH condition and the initial concentrations of the raw textile wastewater. Thus, adjustment of the pH of the raw textile wastewater to a suitable pH could facilitate and enhance the dye removal efficiency. Besides, the raw textile wastewater should always be diluted before undergoing solar photocatalytic irradiation in order to boost up the removal efficiency. As the highly concentrated textile wastewater reduces the penetration of the solar light into the solution which leads to lower photocatalytic reactivity for degradation of dye.

4 Conclusions

In summary, it could be concluded that solar photocatalytic degradation of dye is a green and sustainable treatment method that not only able to decolourize the dye wastewater, but also mineralize the recalcitrant dye molecules to harmless products such as carbon dioxide and water. In order to achieve the optimum condition for solar photcatalytic degradation of dye, one must consider the main influencing factors in solar photocatalytic process such as light source and intensity, molecular structure of dye, pH, aeration, initial dye concentrations, physical properties of the photocatalyst, and photocatalyst dosage. It has been proven that real textile wastewater could be treated efficiently through solar photocatalytic process. Therefore, solar photocatalytic treatment for dye removal is considered as a sustainable, simple, feasible, and economical method for application in the textile industries.

References

1. Abel MJ, Pramothkumar A, Archana V, Senthilkumar N, Jothivenkatachalam K, Prince JJ (2020) Facile synthesis of solar light active spinel nickel manganite ($NiMn_2O_4$) by co-precipitation route for photocatalytic application. Res Chem Intermed 46(7):3509–3525. https://doi.org/10.1007/s11164-020-04159-y
2. Akyol A, Bayramoğlu M (2005) Photocatalytic degradation of remazol red F3B using ZnO catalyst. J Hazard Mater 124:241–246. https://doi.org/10.1016/j.jhazmat.2005.05.006
3. Al-Mamun MR, Kader S, Islam MS (2021a) Solar-TiO_2 immobilized photocatalytic reactors performance assessment in the degradation of methyl orange dye in aqueous solution. Environ Nanotechnol Monit Manag 16:100514. https://doi.org/10.1016/j.enmm.2021.100514
4. Al-Mamun MR, Karim MN, Nitun NA, Kader S, Islam MS, Hossain Khan MZ (2021b) Photocatalytic performance assessment of GO and Ag co-synthesized TiO_2 nanocomposite for the removal of methyl orange dye under solar irradiation. Environ Technol Innov 22:101537. https://doi.org/10.1016/j.eti.2021.101537
5. Araña J, Doña-Rodríguez JM, Portillo-Carrizo D, Fernández-Rodríguez C, Pérez-Peña J, Díaz OG, Navio JA, Macías M (2010) Photocatalytic degradation of phenolic compounds with new TiO_2 catalysts. Appl Catal B 100(1):346–354. https://doi.org/10.1016/j.apcatb.2010.08.011
6. Barzgari Z, Ghazizadeh A, Askari SZ (2016) Preparation of Mn-doped ZnO nanostructured for photocatalytic degradation of Orange G under solar light. Res Chem Intermed 42(5):4303–4315. https://doi.org/10.1007/s11164-015-2276-y
7. Beiknejad D, Chaichi MJ (2014) Estimation of photolysis half-lives of dyes in a continuous-flow system with the aid of quantitative structure-property relationship. Front Environ Sci Eng 8:683–692. https://doi.org/10.1007/s11783-014-0680-y
8. Benkhaya S, M'rabet S, El Harfi A (2020) Classifications, properties, recent synthesis and applications of azo dyes. Heliyon 6(1):e03271. https://doi.org/10.1016/j.heliyon.2020.e03271
9. Bizani E, Fytianos K, Poulios I, Tsiridis V (2006) Photocatalytic decolorization and degradation of dye solutions and wastewaters in the presence of titanium dioxide. J Hazard Mater 136:85–94. https://doi.org/10.1016/j.jhazmat.2005.11.017
10. Bojinova A, Kaneva N, Papazova K, Eliyas A, Stoyanova-Eliyas E, Dimitrov D (2017) Green synthesis of UV and visible light active TiO_2/WO_3 powders and films for malachite green and ethylene photodegradation. React Kinet Mech Catal 120:821–832. https://doi.org/10.1007/s11144-016-1128-0
11. Brás R, Gomes A, Ferra MIA, Pinheiro HM, Gonçalves IC (2005) Monoazo and diazo dye decolourisation studies in a methanogenic UASB reactor. J Biotechnol 115:57–66. https://doi.org/10.1016/j.jbiotec.2004.08.001
12. Brites FF, Santana VS, Fernandes-Machado NRC (2011) Effect of support on the photocatalytic degradation oftextile effluents using Nb_2O_5 and ZnO: Photocatalytic degradation of textile dye. Top Catal 54:264–269. https://doi.org/10.1007/s11244-011-9657-2
13. Carliell CM, Barclay SJ, Buckley CA (1996) Treatment of exhausted reactive dyebath effluent using anaerobic digestion: laboratory and full-scale trials. Water SA 22(3):233–275. https://hdl.handle.net/10520/AJA03784738_1419
14. Chaudhari SM, Gawal PM, Sane PK, Sontakke SM, Nemade PR (2018) Solar light-assisted photocatalytic degradation of methylene blue with Mo/TiO_2: a comparison with Cr- and Ni-doped TiO_2. Res Chem Intermed 44(5):3115–3134. https://doi.org/10.1007/s11164-018-3296-1

15. Chen CY (2009) Photocatalytic degradation of azo dye reactive Orange 16 by TiO_2. Water Air Soil Pollut 202:335–342. https://doi.org/10.1007/s11270-009-9980-4
16. Chen P, Zhu L, Chang Z, Gao H, Chen D, Qiu M (2020) Spherical Cu_2O assembled by small nanoparticles and its high efficiency in photodegradations of methylene blue under different light sources. Chem Res Chin Univ 36(6):1108–1115. https://doi.org/10.1007/s40242-020-0309-6
17. Dasgupta J, Sikder J, Chakraborty S, Curcio S, Drioli E (2015) Remediation of textile effluents by membrane based treatment techniques: a state of the art review. J Environ Manag 147:55–72. https://doi.org/10.1016/j.jenvman.2014.08.008
18. Dash L, Biswas R, Ghosh R, Kaur V, Banerjee B, Sen T, Patil RA, Ma YR, Haldar KK (2020) Fabrication of mesoporous titanium dioxide using azadirachta indica leaves extract towards visible-light-driven photocatalytic dye degradation. J Photochem Photobiol A 400:112682. https://doi.org/10.1016/j.jphotochem.2020.112682
19. Davis RJ, Gainer JL, O'Neal G, Wenwu I (1994) Photocatalytic decolorization of wastewater dyes. Water Environ Res 66:50–53. https://doi.org/10.2175/WER.66.1.8
20. de Moraes NFS, Santana RMR, Gomes RKM, Santos Júnior SG, de Lucena ALA, Zaidan LEMC, Napoleão DC (2021) Performance verifcation of different advanced oxidation processes, in the degradation of the dye acid violet 17: reaction kinetics, toxicity and degradation prediction by artificial neural networks. Chem Pap 75:539–552. https://doi.org/10.1007/s11696-020-01325-9
21. Dubey N, Rayalu SS, Labhsetwar NK, Naidu RR, Chatti RV, Devotta S (2006) Photocatalytic properties of zeolite-based materials for the photoreduction of methyl orange. Appl Catal A 303:152–157. https://doi.org/10.1016/j.apcata.2006.01.043
22. Feng S, Li F (2021) Photocatalytic dyes degradation on suspended and cement paste immobilized $TiO_2/g\text{-}C_3N_4$ under simulated solar light. J Environ Chem Eng 9(4):105488. https://doi.org/10.1016/j.jece.2021.105488
23. Fox MA, Dulay MT (1993) Heterogeneous photocatalysis. Chem Rev 93(1):341–357. https://doi.org/10.1021/cr00017a016
24. Fu SC, Zhong XL, Zhang Y, Lai TW, Chan KC, Lee KY, Chao CYH (2020) Bio-inspired cooling technologies and the applications in buildings. Energy Build 225:110313. https://doi.org/10.1016/j.enbuild.2020.110313
25. Habibi MH, Askari E (2011) Photocatalytic degradation of an azo textile dye with manganese-doped ZnO nanoparticles coated on glass. Iran J Catal 1:41–44. http://ijc.iaush.ac.ir/article_551237.html
26. Hemalatha K, Ette PM, Madras G, Ramesha K (2015) Visible light assisted photocatalytic degradation of organic dyes on TiO_2-CNT nanocomposites. J Sol-Gel Sci Technol 73:72–82. https://doi.org/10.1007/s10971-014-3496-0
27. Ho LN, Ong SA, Osman H, Chong FM (2012a) Enhanced photocatalytic activity of fish scale loaded TiO_2 composites under solar light irradiation. J Environ Sci 24(6):1142–1148. https://doi.org/10.1016/S1001-0742(11)60872-3
28. Ho LN, Ong SA, See YL (2012b) Photocatalytic degradation of reactive black 5 by fish scale-loaded TiO_2 composites. Water Air Soil Pollut 223(7):4437–4442. https://doi.org/10.1007/s11270-012-1207-4
29. Houas A, Lachheb H, Ksibi M, Elaloui E, Guillard C, Herrmann JM (2001) Photocatalytic degradation pathway of methylene blue in water. Appl Catal B 31(2):145–157. https://doi.org/10.1016/s0926-3373(00)00276-9
30. Hussain T, Wahab A (2018) A critical review of the current water conservation practices in textile wet processing. J Clean Prod 198:806–819. https://doi.org/10.1016/j.jclepro.2018.07.051
31. Hussin F, Lintang HO, Lee SL, Yuliati L (2017) Photocatalytic synthesis of reduced graphene oxide-zinc oxide: effects of light intensity and exposure time. J Photochem Photobiol A 340:128–135. https://doi.org/10.1016/j.jphotochem.2017.03.016
32. Joseph CG, Taufiq-Yap YH, Krishnan V (2017) ultrasonic assisted photolytic degradation of reactive black 5 (RB5) simulated wastewater. ASEAN J Chem Eng 17(2):37–50. https://doi.org/10.22146/ajche.49554

33. Kansal SK, Kaur N, Singh S (2009) Photocatalytic Degradation of Two Commercial Reactive Dyes in Aqueous Phase Using Nanophotocatalysts. Nanoscale Res Lett 4:709. https://doi.org/10.1007/s11671-009-9300-3
34. Kazeminezhad I, Sadollahkhani A (2016) Influence of pH on the photocatalytic activity of ZnO nanoparticles. J Mater Sci Mater Electron 27:4206–4215. https://doi.org/10.1007/s10854-016-4284-0
35. Khalik WF, Ho LN, Ong SA, Wong YS, Yusoff NA, Ridwan F (2015) Decolorization and mineralization of Batik wastewater through solar photocatalytic process. Sains Malaysiana 44(4):607–6121. https://doi.org/10.17576/jsm-2015-4404-16
36. Khalik WF, Ho LN, Ong SA, Wong YS, Yusoff NA, Ridwan F (2016) Solar photocatalytic mineralization of dye new coccine in aqueous phase using different photocatalysts. Water Air Soil Pollut 227:118. https://doi.org/10.1007/s11270-016-2822-2
37. Khalik WF, Ho LN, Ong SA, Voon CH, Wong YS, Yusoff NA, Lee SL, Yusuf SY (2017) Optimization of degradation of reactive black 5 (RB5) and electricity generation in solar photocatalytic fuel cell system. Chemosphere 184:112–119. https://doi.org/10.1016/j.chemosphere.2017.05.160
38. Khalik WF, Ong SA, Ho LN, Wong YS, Yusoff NA, Ridwan F (2018) Comparison on solar photocatalytic degradation of orange g and new coccine using zinc oxide as catalyst. Environ Eng Manag J 17(2):301–308. www.eemj.icpm.tuiasi.ro/pdfs/vol17/full/no2/15_527_Khalik_13.pdf
39. Khataee AR, Kasiri MB (2010) Photocatalytic degradation of organic dyes in the presence of nanostructured titanium dioxide: Influence of the chemical structure of dyes. J Mol Catal A Chem 328:8–26. https://doi.org/10.1016/j.molcata.2010.05.023
40. Kocakuşakoğlu A, Dağlar M, Konyar M, Yatmaz CH, Öztürk K (2015) Photocatalytic activity of reticulated ZnO porous ceramics in degradation of azo dye molecules. J Eur Ceram Soc 35(10):2845–2853. https://doi.org/10.1016/j.jeurceramsoc.2015.03.042
41. Koe WS, Lee JW, Chong WC, Pang YL, Sim LC (2020) An overview of photocatalytic degradation: photocatalysts, mechanisms, and development of photocatalytic membrane. Environ Sci Pollut Res 27:2522–2565. https://doi.org/10.1007/s11356-019-07193-5
42. Larbi T, Amara MA, Ouni B, Amlouk M (2017) Enhanced photocatalytic degradation of methylene blue dye under UV-sunlight irradiation by cesium doped chromium oxide thin films. Mater Res Bull 95:152–162. https://doi.org/10.1016/j.materresbull.2017.07.024
43. Lee SL, Ho LN, Ong SA, Lee GM, Wong YS, Voon CH, Khalik WF, Yusoff NA, Nordin N (2016) Comparative study of photocatalytic fuel cell for degradation of methylene blue under sunlight and ultra-violet light irradiation. Water Air Soil Pollut 227(12):445. https://doi.org/10.1007/s11270-016-3148-9
44. Lee SL, Ho LN, Ong SA, Wong YS, Voon CH, Khalik WF, Yusoff NA, Nordin N (2017) A highly efficient immobilized ZnO/Zn photoanode for degradation of azo dye Reactive Green 19 in a photocatalytic fuel cell. Chemosphere 166:118–125. https://doi.org/10.1016/j.chemosphere.2016.09.082
45. Lee SL, Ho LN, Ong SA, Wong YS, Voon CH, Khalik WF, Yusoff NA, Nordin N (2018) Exploring the relationship between molecular structure of dyes and light sources for photodegradation and electricity generation in photocatalytic fuel cell. Chemosphere 209:935–943. https://doi.org/10.1016/j.chemosphere.2018.06.157
46. Litter MI (1999) Heterogeneous photocatalysis transition metal ions in photocatalytic systems. Appl Catal B 23:89–114. https://doi.org/10.1016/S0926-3373(99)00069-7
47. Ljubas D, Ćurković L, Marinović V, Bačić I, Tavčar B (2015) Photocatalytic degradation of azo dyes by sol-gel TiO_2 films: effects of polyethylene glycol addition, reaction temperatures and irradiation wavelengths. React Kinet Mech Catal 116(2):563–576. https://doi.org/10.1007/s11144-015-0917-1
48. Lyman WJ, Reehl WF, Rosenblatt DH (eds) (1982) Handbook of chemical property estimation methods, environmental behaviour of organic compounds. McGraw-Hilll, New York, USA. https://doi.org/10.1002/jps.2600720132

49. Márlen GD, Azevedo EB (2009) Photocatalytic decolorization of commercial acid dyes using solar irradiation. Water Air Soil Pollut 204:79. https://doi.org/10.1007/s11270-009-0028-6
50. Matthews RW (1991) Photooxidative degradation of coloured organics in water using supported catalysts TiO_2 on sand. Water Res 25(10):1169–1176. https://doi.org/10.1016/0043-1354(91)90054-T
51. Min OM, Ho LN, Ong SA, Wong YS (2015) Comparison between the photocatalytic degradation of single and binary azo dyes in TiO_2 suspensions under solar light irradiation. J Water Reuse Desalin 5(4):579–591. https://doi.org/10.2166/wrd.2015.022
52. Mohsin M, Bhatti IA, Ashar A, Mahmood A, Maryam HQ, Iqbal M (2020) Fe/ZnO@ceramic fabrication for the enhanced photocatalytic performance under solar light irradiation for dye degradation. J Market Res 9(3):4218–4229. https://doi.org/10.1016/j.jmrt.2020.02.048
53. Muruganandham M, Swaminathan M (2004) Solar photocatalytic degradation of a reactive azo dye in TiO_2-suspension. Sol Energy Mater Sol Cells 81:439–457. https://doi.org/10.1016/j.solmat.2003.11.022
54. Mustafa S, Tasleem S, Naeem A (2004) Surface charge properties of Fe_2O_3 in aqueous and alcoholic mixed solvents. J Colloid Interface Sci 275:523–529. https://doi.org/10.1016/j.jcis.2004.02.089
55. Nasr C, Vinodgopal K, Fisher L, Hotchandani S, Chattopadhyay AK, Kamat PV (1996) Environmental photochemistry on semiconductor surfaces. Visible light induced degradation of a textile diazo dye, naphthol blue black, on TiO_2 nanoparticles. J Phys Chem 100(20):8436–8442. https://doi.org/10.1021/jp953556v
56. Nawaz A, Khan A, Ali N, Ali N, Bilal M (2020) Fabrication and characterization of new ternary ferrites-chitosan nanocomposite for solar-light driven photocatalytic degradation of a model textile dye. Environ Technol Innov 20:101079. https://doi.org/10.1016/j.eti.2020.101079
57. Nishio J, Tokumura M, Znad HT, Kawase Y (2006) Photocatalytic decolorization of azo-dye with zinc oxide powder in an external UV light irradiation slurry photoreactor. J Hazard Mater 138(1):106–115. https://doi.org/10.1016/j.jhazmat.2006.05.039
58. Ong SA, Ho LN, Wong YS, Min OM, Lai LS, Khiew SK, Murali V (2012a) Photocatalytic mineralization of azo dye acid orange 7 under solar light irradiation. Desalin Water Treat 48(1–3):245–251. https://doi.org/10.1080/19443994.2012.698820
59. Ong SA, Min OM, Ho LN, Wong YS (2012b) Comparative study on photocatalytic degradation of mono azo dye acid orange 7 and methyl orange under solar light irradiation. Water Air Soil Pollut 223(8):5483–5493. https://doi.org/10.1007/s11270-012-1295-1
60. Ong SA, Min OM, Ho LN, Wong YS (2013) Solar photocatalytic degradation of mono azo methyl orange and diazo reactive green 19 in single and binary dye solutions: adsorbability vs photodegradation rate. Environ Sci Pollut Res 20:3405–3413. https://doi.org/10.1007/s11356-012-1286-1
61. Parvin F, Nayna OK, Tareq SM, Rikta SY, Kamal AK (2018) Facile synthesis of iron oxide nanoparticle and synergistic effect of iron nanoparticle in the presence of sunlight for the degradation of DOM from textile wastewater. Appl Water Sci 8:73. https://doi.org/10.1007/s13201-018-0719-5
62. Pérez-Estrada LA, Agüera A, Hernando MD, Malato S, Fernández-Alba AR (2008) Photodegradation of malachite green under natural sunlight irradiation: kinetic and toxicity of the transformation products. Chemosphere 70(11):2068–2075. https://doi.org/10.1016/j.chemosphere.2007.09.008
63. Pirsaheb M, Hossaini H, Azizi N, Khosravi T (2020) Synthesized Cr/TiO_2 immobilized on pumice powder for photochemical degradation of acid orange-7 dye under UV/visible light: influential operating factors, optimization, and modeling. J Environ Health Sci Eng 18(2):1329–1341. https://doi.org/10.1007/s40201-020-00550-4
64. Qamar M, Muneer M (2009) A comparative photocatalytic activity of titanium dioxide and zinc oxide by investigating the degradation of vanillin. Desalination 249:535–540. https://doi.org/10.1016/j.desal.2009.01.022
65. Qamar M, Merzougui B, Anjum D, Hakeem AS, Yamani ZH, Bahnemann D (2014) Synthesis and photocatalytic activity of mesoporous nanocrystalline Fe-doped titanium dioxide. Catal Today 230:158–165. https://doi.org/10.1016/j.cattod.2013.10.040

66. Rauf MA, Ashraf SS (2009) Fundamental principles and application of heterogeneous photocatalytic degradation of dyes in solution. Chem Eng J 151(1–3):10–18. https://doi.org/10.1016/j.cej.2009.02.026
67. Sajid MM, Shad NA, Javed Y, Khan SB, Zhang Z, Amin N, Zhai H (2020) Preparation and characterization of Vanadium pentoxide (V_2O_5) for photocatalytic degradation of monoazo and diazo dyes. Surf Interfaces 19:100502. https://doi.org/10.1016/jsurfin.2020.100502
68. Sakthivel S, Neppolian B, Shankar MV, Arabindoo B, Palanichamy M, Murugesan V (2003) Solar photocatalytic degradation of azo dye: comparison of photocatalytic efficiency of ZnO and TiO_2. Sol Energy Mater Sol Cells 77:65–82. https://doi.org/10.1016/S0927-0248(02)00255-6
69. Sclafani A, Herrmann JM (1996) Comparison of the photoelectronic and photocatalytic activities of various anatase and rutile forms of titania in pure liquid organic phases and in aqueous solutions. J Phys Chem 100(32):13655–13661. https://doi.org/10.1021/jp9533584
70. Selvaraj A, Sivakumar S, Ramasamy AK, Balasubramanian V (2013) Photocatalytic degradation of triazine dyes over N-doped TiO_2 in solar radiation. Res Chem Intermed 39(6):2287–2302. https://doi.org/10.1007/s11164-012-0756-x
71. Shaban M, Abukhadra MR, Ibrahim SS, Shahien MG (2017) Photocatalytic degradation and photo-Fenton oxidation of Congo red dye pollutants in water using natural chromite – response surface optimization. Appl Water Sci 7:4743–4756. https://doi.org/10.1007/s13201-017-0637-y
72. Singh J, Sahu K, Satpati B, Mohapatra S (2019) Facile synthesis, structural, optical and photocatalytic properties of anatase/rutile mixed phase TiO_2 ball-like sub-micron structures. Optik 188:270–276. https://doi.org/10.1016/j.ijleo.2019.05.053
73. So CM, Cheng MY, Yu JC, Wong PK (2002) Degradation of azo dye Procion Red MX-5B by photocatalytic oxidation. Chemosphere 46:905–912. https://doi.org/10.1016/s0045-6535(01)00153-9
74. Sobana N, Swaminathan M (2007) The effect of operational parameters on the photocatalytic degradation of acid red 18 by ZnO. Sep Purif Technol 56:101–107. https://doi.org/10.1016/j.seppur.2007.01.032
75. Srikant V, Clarke DR (1998) On the optical band gap of zinc oxide. J Appl Phys 83:5447–5451. https://doi.org/10.1063/1.367375
76. Subash B, Krishnakumar B, Swaminathan M, Shanthi M (2013) Solar-light-assisted photocatalytic degradation of NBB dye on Zr-codoped Ag-ZnO catalyst. Res Chem Intermed 39:3181–3197. https://doi.org/10.1007/s11164-012-0831-3
77. Tang J, Zou Z, Ye J (2005) Kinetics of MB degradation and effect of pH on the photocatalytic activity of MIn_2O_4 (M=Ca, Sr, Ba) under visible light irradiation. Res Chem Intermed 31:513–519. https://doi.org/10.1163/1568567053956699
78. Taymaz BH, Eskizeybek V, Kamış H (2021) A novel polyaniline/NiO nanocomposite as a UV and visible-light photocatalyst for complete degradation of the model dyes and the real textile wastewater. Environ Sci Pollut Res 28(6):6700–6718. https://doi.org/10.1007/s11356-020-10956-0
79. Thennarasu G, Kavithaa S, Sivasamy A (2012) Photocatalytic degradation of Orange G dye under solar light using nanocrystalline semiconductor metal oxide. Environ Sci Pollut Res 19(7):2755–2765. https://doi.org/10.1007/s11356-012-0775-6
80. Thomas GE, Stamnes K (1999) Radiative transfer in the atmosphere and ocean. Cambridge University Press, Cambridge, UK. https://doi.org/10.1017/CBO9780511613470
81. Thuong HTT, Kim CTT, Quang LN, Kosslick H (2019) Highly active brookite TiO_2-assisted photocatalytic degradation of dyes under the simulated solar–UVA radiation. Progr Natl Sci Mater Int 29(6):641–647. https://doi.org/10.1016/j.pnsc.2019.10.001
82. Tijani JO, Momoh UO, Salau RB, Bankole MT, Abdulkareem AS, Roos WD (2019) Synthesis and characterization of $Ag_2O/B_2O_3/TiO_2$ ternary nanocomposites for photocatalytic mineralization of local dyeing wastewater under artificial and natural sunlight irradiation. Environ Sci Pollut Res 26(19):19942–19967. https://doi.org/10.1007/s11356-019-05124-y

83. Venkatesh D, Pavalamalar S, Anbalagan K (2019) Selective photodegradation on dual dye system by recoverable nano SnO_2 photocatalyst. J Inorg Organomet Polym Mater 29:939–953. https://doi.org/10.1007/s10904-018-01069-w
84. Verma AK, Dash RR, Bhunia P (2012) A review on chemical coagulation/flocculation technologies for removal of colour from textile wastewaters. J Environ Manag 93:154–168. https://doi.org/10.1016/j.jenvman.2011.09.012
85. Wei P, Yin S, Zhou T, Peng C, Xu X, Lu J, Liu M, Jia J, Zhang K (2021) Rational design of Z-scheme $ZnFe_2O_4/Ag@Ag_2CO_3$ hybrid with enhanced photocatalytic activity, stability and recovery performance for tetracycline degradation. Sep Purif Technol 266:118544. https://doi.org/10.1016/j.seppur.2021.118544
86. Wu CH (2004) Comparison of azo dye degradation efficiency using UV/single semiconductor and UV/couples semiconductor systems. Chemosphere 57:601–608. https://doi.org/10.1016/j.chemosphere.2004.07.008
87. Yong L, Zhanqi G, Yuefei J, Xiaobin H, Cheng S, Shaogui Y, Lianhong W, Qingeng W, Die F (2015) Photodegradation of malachite green under simulated and natural irradiation: kinetics, products, and pathways. J Hazard Mater 285:127–136. https://doi.org/10.1016/j.jhazmat.2014.11.041
88. Zahoor M, Arshad A, Khan Y, Iqbal M, Bajwa SZ, Soomro RA, Ahmad I, Butt FK, Iqbal MZ, Wu A, Khan WS (2018) Enhanced photocatalytic performance of CeO_2–TiO_2 nanocomposite for degradation of crystal violet dye and industrial waste effluent. Appl Nanosci 8(5):1091–1099. https://doi.org/10.1007/s13204-018-0730-z
89. Zawawi A, Ramli R, Yub Harun N (2017) Photodegradation of 1-Butyl-3-methylimidazolium Chloride [Bmim]Cl via synergistic effect of adsorption-photodegradation of Fe-TiO_2/AC. Technologies 5(4):82. https://doi.org/10.3390/technologies5040082
90. Zhao J, Chen C, Ma W (2005) Photocatalytic degradation of organic pollutants under visible light irradiation. Top Catal 35:269–278. https://doi.org/10.1007/s11244-005-3834-0
91. Zhang J, Zou K, Yang W, Wang Y (2019) Synthesis, photoluminescene properties and photocatalytic activity of a novel Y_2O_3/Co_3O_4 nano-composite catalyst in the degradation of methyl red dye. Russ J Phys Chem A 93(13):2824–2833. https://doi.org/10.1134/s0036024419130399
92. Zhang X, Fu K, Su Z (2021) Fabrication of 3D MoS_2-TiO_2@PAN electro-spun membrane for efficient and recyclable photocatalytic degradation of organic dyes. Mater Sci Eng B 269:115179. https://doi.org/10.1016/j.mseb.2021.115179
93. Zollinger H (1987) Color chemistry: syntheses, properties and applications of organic dyes and pigment. VCH, Weinheim, Germany. https://doi.org/10.1002/nadc.19870351215

Photo (Catalytic) Oxidation Processes for the Removal of Dye: Focusing on TiO_2 Performance

Jayato Nayak

Abstract Production, exploitation and discharge of a wide variety of organic dyes by different industrial houses have thrown a critical challenge to maintain the effluent quality. Industrial reluctance in third world countries made significant deterioration of ground and fresh water quality and overall environment. Though different technological approaches were forwarded, but photon induced catalytic decay of organic dye particles by titanium dioxide (TiO_2) has opened up newer possibilities to the global scientific communities due to its diversified utilization. In-fact, TiO_2 is a preferred photocatalytic-oxidative agent due to the salient features of its excellent photocatalysis activity such as its narrow band gap with high thermo-chemical stability. Moreover, its water insolubility, environmentally non-toxic and non-reactive nature, less energy intake during reaction with room temperature operating conditions, made it a highly preferential amongst other photocatalysts. Such heterogeneous catalytic activity is activated by the ultraviolet radiation on the TiO_2 particles while contact with pollutants present in waste effluent stream. After several researches, integrated photo(catalytic) oxidation is being slowly implemented for degradation of dyes in waste water in some of the developed countries. A number of diversified methods could be adopted to modify the structure of TiO_2 integrated nano-composite photocatalyst by varying the dopant, particle size and irradiation, which can improve the photo-oxidative performance. The chapter contains the comprehensive and fundamental aspects with thorough scrutiny of recent researches regarding the photo-oxidation of organic dye compounds by titanium-di-oxide nanoparticles which paves the pathway towards the use of such photo-oxidative catalysts more anticipated and conducive in imminent R&D and commercial applications.

Keywords Photo-oxidation · Nano-catalyst · TiO_2 · Low band gap · Dye abatement

J. Nayak (✉)
Department of Chemical Engineering, School of Bio and Chemical Engineering (SBCE), Kalasalingam Academy of Research and Education (Deemed to be University), Krishnankoil, Srivilliputhur, Tamil Nadu 626126, India

S. S. Muthu and A. Khadir (eds.), *Advanced Oxidation Processes in Dye-Containing Wastewater*, Sustainable Textiles: Production, Processing, Manufacturing & Chemistry,
https://doi.org/10.1007/978-981-19-0987-0_5

1 Introduction

During the present era of booming industrialization of chemical and allied processing sectors like textile, pharmaceutical, automobile, textile, paint, and leather industries are reckoned as the indicators of economic development of a country. However, the use of organic dyes as colouring agent for different purposes of applications is bringing severe threat when released as waste water. Though the developed countries are critically pondering over the issue of dye mitigation from waste effluents, but the developing countries need to put more efforts regarding such environmentally vulnerable problems. Separation and treatment of dyeing chemicals received considerable attention because of intensification of colour and toxic characteristics in aqueous solutions. These are treated as critically emerging pollutants because even a minute amount of about <1 mg/L for some dyes are even scaringly harmful [1]. In the present times, about 15,000 types of synthetic dyes are commercialized with a global production rate of more than 800,000 ton per year. A huge quantity of dyes of about 200,000 tonnes per year get mixed with the in the open environment as solid wastes or liquid effluents. In fact, about 20% of textile industry waste effluent volume gets generated from the dyeing and finishing process [2, 3]. Such wastewaters comprise of highly poisonous, cancer causing and mutagenic highly reactive organic and azo dyes [4, 5]. Contaminating dyeing chemicals due to the direct discharge results in detrimental effects to the aquatic ecosystem and the biodiversity with the indirect effects on human civilization [6, 7].

Though a wide variety of treatment methodologies were proposed involving physical, chemical and biological routes but none of those could overcome the inherent drawbacks. As an example, sedimentation cannot be efficient as a standalone process, membrane separation suffers from frequent pore blocking and high cost. Though adsorption and coagulation-precipitation methods are effective, but the contaminated sludge disposal turns out to be a huge environmental concern [8–10]. Moreover, advanced oxidative technologies using chlorine or ozone can degrade some the dyes but the cost of process is usually high enough to be implemented easily [11]. Amongst all, for the mitigation of such harmful dyes, photodegradation by TiO_2 is gaining overwhelming response from the researchers in the current decades [12, 13]. Such nano-catalytic semiconductor materials exhibit excellent ability towards photocatalytic remediation under solar or UV irradiation [14, 15]. In the galore of different diversified transition metal oxides, titanium-di-oxide due to its narrow energy band gap with a sufficiently positive valence band initiates the oxidative degradation, even by exposure under solar irradiation [16]. Exhaustive experimental investigations for the degradation and mineralization of specific organic dyes indicated that that TiO_2 used to be the most efficient photo active catalyst in the abatement of azo and phenol group containing dyes from waste effluents [17].

Due to the lower cost and easier lab-based preparation methodologies with respect to other materials for discarded discharge treatment [18], TiO_2 is always preferred due to its alteration in oxidation level even at low light intensities which is responsible for quick degradation of pollutants with elevated efficiency [19]. Thus, it could be referred as an emerging advanced oxidising agent for complete mineralisation, while consuming reduced energy and stopping the generation of unwanted intermediate products [20]. In case of cadmium sulphate and zinc oxide photocatalysis, some further formation of toxic intermediates used to occur while being used for waste water treatments. TiO_2 nanomaterials are having efficient photoactive antibacterial properties against *E. Coli*, where the microbial cell is killed without damaging the parent tissues. The structural modifications and surface morphologies could be easily tailor made for TiO_2 including the formation of nanotubes, nanosheets spheres and crystalline structures. Structural alterations offer huge numbers of active sites, extremely efficient mass transfer rates and effective degradability to the target pollutants [21–23]. The most common commercial form of TiO_2 is Degussa P25, containing 25% rutile and 75% anatase, which inhibits the electron–hole recombination phenomena [24]. Such photoactive reactions are dependent on the pH of the medium [25] and by the utilization of co-catalytic systems [26]. Improvement of easy overcome of band gap energy of TiO_2 could be done by the incorporation of doping substances like platinum, iron, sulphur, gold and copper [27]. After doping with TiO2, under light irradiations, new energy levels are created which enables faster, highly improved, effective and efficient degradation of pollutants [28].

The present chapter summarizes the mechanism, development and recent photocatalytic mitigation studies employing TiO_2 nanoparticles for the non-toxic transformation of pigments under light irradiation.

2 Photo-reactive Mechanism of TiO_2 for Degradation of Organics in Wastewater

The general light induced reaction stages for TiO_2 nano-catalysts for degradation of organics in wastewater are as following [28], [13]: (i) mass transfer of contaminating agents from bulk aqueous volume to the catalyst surface; (ii) chemisorption of contaminants on the photo exited superficial area of TiO_2; (iii) oxidative mineralization over photocatalyst surface; (iv) desorptive process of the decayed contaminants in form of non-toxic materials from nano-catalyst surface; and finally the (v) interfacial mass transfer of the mineralized products to the bulk liquid. As usual, the slowest stage used to be the considered as the rate determining one, where the mass transfer stages are much quicker than the stages involving chemical reactions. Such photochemical degradation is dependent on the generation and reactivity of extremely active OH^- ions having oxidative potential 2.80 V and relative oxidative strength of 2.06, which is mainly responsible for the effective destruction of toxic contaminating organic materials present in industrial effluents, onto the surface of TiO_2 [29]. As a matter of fact, photo-induced holes get generated after exposure of TiO_2 nanocatalysts under active photons. These holes participate in the oxidative activity with water and produce negative hydroxyl radicals which is extremely active in oxidative destruction of organic materials. Furthermore, availability of oxygen averts the recoupling of electrons and holes, where under dearth of oxygen, the efficacy of the photoactive degradation reduces drastically. Photocatalytic mechanism for organic dye degradation by TiO_2 with governing reactions are shown as following [13, 30]:

$$TiO_2 + h\upsilon \rightarrow TiO_2(e^-_{CB} + h^+_{vB})$$
$$TiO_2(h^+_{vB}) + H_2O \rightarrow TiO_2 + H^+ + OH^*$$
$$TiO_2(h^+_{vB}) + OH^- \rightarrow TiO_2 + OH^*$$
$$TiO_2(e^-_{CB}) + O_2 \rightarrow TiO_2 + O_2^{*-}$$
$$O_2^{*-} + H^+ \rightarrow HO_2^*$$
$$HO_2^* + HO_2* \rightarrow H_2O_2 + O_2$$
$$TiO_2(e^-_{CB}) + H_2O_2 \rightarrow OH^* + OH^-$$
$$H_2O_2 + O_2^{*-} \rightarrow OH^* + OH^- + O_2$$
$$\text{Dye} + OH^* \rightarrow \text{Degraded to non-toxic chemicals}$$
$$\text{Dye} + TiO_2(h^+_{vB}) \rightarrow \text{oxidised non-toxic chemicals}$$

Ultraviolet-A light of wavelength range of 310–400 nm induces excitation of electron of valance band to get transferred to the conduction band. With an approximate λ_{max} value of 400 nm, the energy difference between the bands of TiO_2 is about 3.1 eV [31]. This band gap reduces by the use of doping agents enabling easy movement of electrons from valance band to conduction band with fast generation of huge number of mobile holes and electrons. Diagrammatic representation of electron transfer within the energy bands for pure and doped TiO_2 has been shown in Fig. 1. Holes and electrons can recouple and sifted by proper scavenging agents like hydroxyl ions, water molecules or other oxidizing molecules. Thereafter, by the formation of negatively charged hydroxyl, hydroperoxyl and superoxide radicals on the photocatalyst surface take active part in degradation of organic dye species of wastewater [32, 33].

3 Doping with TiO_2 to Enhance Reactivity Towards Dye Contaminated Liquid Discharge

In the recent times, a lot of researches were carried out to intensify the photo-reactivity of TiO_2 through the incorporation of dopant materials with TiO_2 for the degradation of dye contaminated clothing industrial discharges. Researches were reported on the incorporation of metal dopants (like, iron (Fe), manganese (Mn), copper (Cu), aluminum (Al), chromium (Cr), silver (Ag), gold (Au), platinum (Pt), Titanium (Ti), palladium(Pd),dysprosium (Dy),yttrium (Y),zinc (Zn), bismuth (Bi), molybdenum(Mo), cobalt (Co), nickel (Ni), europium (Eu), cerium (Ce) etc.) [34–40], non-metal dopants (like, nitrogen (N), carbon (C), sulphur (S),phosphorus (P),fluorine(F), iodine (I),boron(B) etc.). Combinations of metal–metal (like, Zn-Cu, Fe–Ni, Y-Dy,

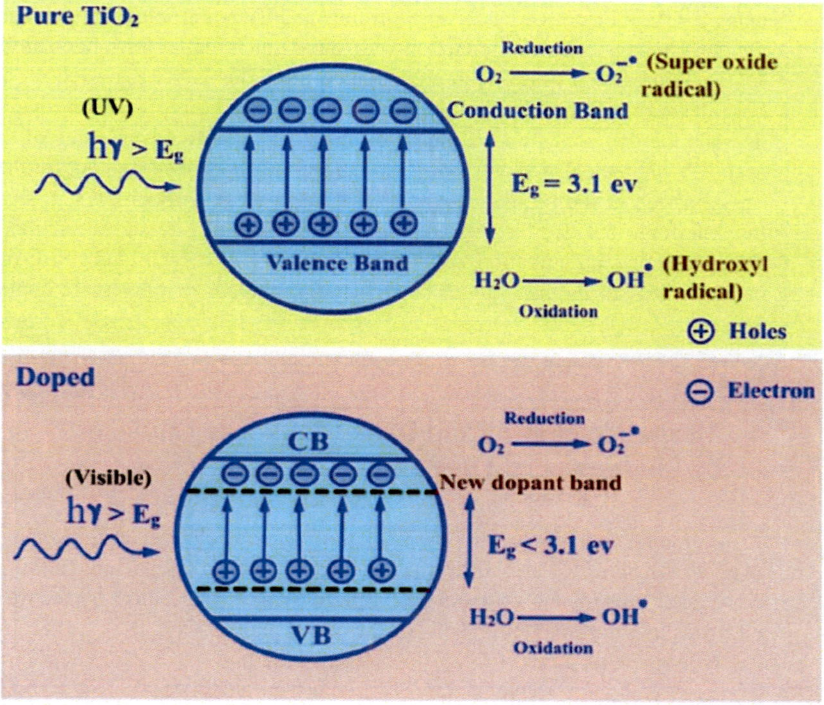

Fig. 1 Energy band gap representation for pristine and doped TiO_2. Reproduced with copyright from "Kannappan Panchamoorthy Gopinath, Nagarajan Vikas Madhav, Abhishek Krishnan, Rajagopal Malolan, Goutham Rangarajan, Present applications of titanium dioxide for the photocatalytic removal of pollutants from water: A review, Journal of Environmental Management 270 (2020) 110,906"

Cr-Co, Co–Ni, Ag-Mo, Zn- Eu, etc.), metal/non-metal (Pt–N, Mn-P, Y-N, Mo-C, Cu–N, etc.) and nonmetal/nonmetal (N-S, B-N, C-F, N-I, etc.) were also integrated with TiO_2 [30, 41–44]. Use of different types of dopants integrated with TiO_2 under Ultraviolet irradiation could be referred as the most efficient route to degrade the contaminating chemicals by reducing the energy of band gaps while increasing the photoactive catalytic surface [45]. During such photo-degradative activity, TiO_2 crystal particles are employed in suspended forms in liquid or fine films. In those forms, the metallic dopants could be effectively incorporated at ease inside the TiO_2 lattice because of their nearly equal radius of ions [46].

In-fact, such efforts developed with photocatalytic degradation of contaminants present in blended/sewage/clothing industrial liquid discharges employing doped TiO_2 nanomaterials have been reported to be extremely efficient from the perspectives of mitigation of colour and degradation of contaminating organics [47]. Supporting reports have been found on mitigation of COD, colour and black sulphur pigments from textile industrial discharge containing organics and compounds of sodium, using a UV-photoactive batch reaction unit and TiO_2 particles. Effectiveness and the performance of colour reduction was found much higher for a fine-film layered TiO_2 photoreaction vessel than an uncoated one. This is the main reason of grabbing devotion of scientists to use the nano-catalyst layered reaction chambers to degrade the dye contaminated liquid discharges [48]. Moreover, some of the pilot scale UV-photocatalyzed reactive degradation studies were carried out for the scale-up validation of the bench-scale results. Though these works represent significant breakthrough, but, for the degradation of dye containing wastewaters, modelling-simulation and optimization studies, quick partition, recycle and reuse of nano-catalysts post to the use should be rigorously studied for pilot scale applications [49, 50].

4 Mitigation of Dye from Industrial Waste Effluents by Photoreactive TiO_2 Nano-catalyst

In the indirect photodegradation by TiO_2, the dye molecules get excited by the energy of photons leading towards the formation of triplet excited state, which follows further conversion towards the generation of a partially oxidized radical by the transfer of electrons into the conduction band of the photocatalyst [51].

$$\text{Dye} + \text{hv} \rightarrow \text{Dye} * \text{(intermediate)}$$

$$\text{Dye} + TiO_2 \rightarrow \text{Dye}^+ + TiO_2^-$$

In the direct photoreactive mitigation, dye molecules react with the strongly oxidative OH^- radicals along with the generated electrons and holes that is responsible for the redox reaction of the dye chemicals [29]. But, the indirect degradation used to be quicker and dominating than the other one as observed in case of bioaccumulation and destruction of bodies of living species, deterioration of ecology and environment. Cancer causing azo pigment removal has opened up newer challenges due to its recalcitrant properties. Thus, some potential catalysts are required to get developed and implemented that is effective, less costly and possess the ability to degrade bulk volume of dyeing chemicals at a quicker rate [52, 53]. This is where the applications of TiO_2 nanocatalysts come to the forefront showing remarkable catalytic degradation potential, extraordinary surface area, cost effectiveness and null toxicity in aqueous solutions. Immobilization of such supported photoactive catalysts was proved to be efficient in degradation of commercial C.I Acid Orange 10, 12 and 8, azo dyes. More than 95% colour and total organic matters were mitigated within 6 h by the breaking of azo bonds and formation of ammonium radicals [54]. TiO_2 amalgamated with zeolite ZIF-8 showed about 15% more efficiency than the pristine while photo-degradation of the dye rhodamine blue because of the bigger pore structures of the zeolite framework [55]. Methylene blue could be fully degraded at ease within a quicker time period employing TiO_2 nanocatalysts. In-fact, at the superfine structural levels, effectiveness of the decaying transformation of pigments got improved linearly with the mass of TiO_2 and the use of acid dosing [56]. Through solvothermal synthesis of TiO_2 crystals with egg shell support in 1:1 ratio, the surface area features and amount was improvised by dispersion and synergic effects, which exhibited remarkable degradation efficiency (>90%) to methylene blue and rhodamine R6 dyes [57]. Hydrothermal synthesis of TiO_2 nanotubes in NaOH solution resulted high mitigation of commercial pigment Orange II at elevated temperatures [58]. It was found that 500 and 1500 ppm TiO_2 photocatalyst concentrations were optimum for the 99% decay of methyl orange and Congo red dyes with 100% colour elimination under solar irradiation [59]. Doping of TiO_2 in the metal framework of Pt, Ni and Pt-Ni produced huge surface area which enriched stability and photocatalytic potential with recycle characteristics [60]. Use of transition metal dopants like Mn, Co, Zr, Cr, Fe, Ni and Cu with TiO_2 exhibited enlarged active surface area, decrease of energy band gap and fast degradation to methylene blue, methylene orange and malachite green under visible conditions. Copper doping showed the best outcome because of its comparatively lesser energy necessities and reduced charge carrier recombination potential for the transfer of electrons [61–63]. For the Reactive Red dye 198, use of dopants like iron, sulphur and nitrogen generate superoxide radicals boosted the reactivity of developed catalyst due to the generation of N2p and S2p electronic configurations which decreases the bandgap energy [64]. Graphene supported TiO_2 enabled the generation of superoxide radicals which mitigated Congo red and methylene blue [65]. Employing the carbon doped TiO_2 nanorods showed more than 85% degradation performance against methylene blue and rhodamine B at temperature higher than 400 °C [66]. Use of recycled Yttrium dopant in TiO_2 nanosheets reduced hexavalent chromium with 95% deterioration of methyl orange [67]. Using the strontium dopant, TiO_2 could be modified by hindering the electron–hole recombination and, Brilliant green, commercial pigment, can be completely mitigated by it [68].

Immobilization of TiO_2 could be done on hydrophilic polymeric films (e.g. polyvinyl alcohol) that forms a bonding of Ti–O–C group by the dehydrating reactivity during the application of heat. It enables 98% mitigation of methyl orange dye under UV irradiation with enriched recycling potential [69]. Nanocomposite catalytic materials made through the combination of TiO_2/CuO and ZnS/TiO_2 exhibited high porosities with enhanced surface area generation that effectively boosted chemisorption and degradation of Acid Blue 113 (99%) [70], parathion-methyl (95%) and polyoxometalates (97%) [71]. Silica derived from rice husk could be doped with TiO_2 resulting high anatase content which enhanced the surface characteristics for photocatalysis of methylene blue [72]. This was further modified with copper doping that reduced the band gap energy showing 95% mitigation of Rhodamine B [73]. Immobilization of TiO_2 done with PDMS-SiO_2-chitosan on pumice support, $NaYF_4$:(Gd 1%, Si) phosphor, and $NaYF_4$: (Yb, Tm) complex was proved to be efficient in degrading methylene blue and other organic pigments from discarded discharge streams by the reduction of band gap energy while providing strong mechanical support [74, 75]. SnO_2 doped TiO_2 showed lesser recombination of electron and holes, high surface area. Moreover, the use of on clay material palygorskite (Mg, Al)2Si_4O_{10}(OH)·4(H_2O) and Montmorillonite support, the improvisation of nano-structures with surface area showed more than 80–95% dye removal [76–78]. Though the process of hydrothermal production, an excellent adsorbing agent, zeolites could be used as a support for $BiVO_4$ doped TiO_2 which can completely degrade the commercial pigments like methylene blue and Acid red 10 [79]. Formation of composite materials with carbon nanotube, activated carbon can boost up the adsorption efficiency and photoreactivity of TiO_2 while degrading dye chemicals like methyl orange, Acridene, Indigo carmine, Methyl blue, Rhodamine B, and acid orange II and Sunset Yellow along with remarkable reusability of developed photocatalyst [80, 81].

TiO_2 nanoparticles impregnated inside the Graphene oxide framework can take part in efficient treatment of wastewater containing methyl orange and rhodamine B [82], acid navy blue dye. Graphene oxide containing single layer molecular configuration enriches quick transfer of electrons and holes inside the supported catalytic structure which enables quick reactive degradation of target pigments and concentration dyes decrease linearly with concentration of graphene oxide [83]. Reduced graphene oxide supported TiO_2 doped with cobalt(II,III) oxide or carbon nitride composite catalyst manufactured through precipitation method [84] or hydrothermal technology [29], [56] showed more than 95% light induced mitigation of methylene blue, Rhodamine B and crystal violet dyes. Graphene derived component prevents positive–negative charge recombination while reducing the energy between conduction and valence bands [85]. Some of the remarkable researches on dye chemical degradation studies in the recent times has been shown in Table 1.

5 Conclusion

Based on the exhaustive literature review-based analysis, it could be culminated that oxidative degradation of organic dyes through the use of modified TiO_2 nanoparticles would be a preferred pathway for the degradation of polluting effluents discharged from industrial hubs. Such photocatalytic oxidations are having its intrinsic merit of using UV light or sun light for the energy of reactive activation which lessens the cost of energy consumption. Moreover, the target organics could be degraded or transformed into nontoxic materials and the TiO_2 nanoparticles could be easily separated out because of its null aqueous solubility. But, for the operational purpose and to achieve maximum contact area and time, proper design considerations with optimization studies should be surveyed in depth. Wise versatile modification of TiO_2 could be done by metallic and non-metallic doping and co-doping techniques and immobilization. But also, a process suffers due to some of the demerits which could be minimized by thorough scrutinization where the best anticipated techniques should be chosen for the development of catalyst and design of reactor. Employing light eradicated titanium oxide nanocomposite catalysts, dye and organo-phenolic

Table 1 Typical researches on doped TiO_2 photocatalyst on mitigation of organic dyes from industrial effluents. Reproduced with copyright from "Kannappan Panchamoorthy Gopinath, Nagarajan Vikas Madhav, Abhishek Krishnan, Rajagopal Malolan, Goutham Rangarajan, Present applications of titanium dioxide for the photocatalytic removal of pollutants from water: A review, Journal of Environmental Management 270 (2020) 110,906"

Doping agent	Dye	Findings	References
Boron (B)	Acid yellow 1	Complete removal within 120 min by doping of boron. Rate of photoreaction and Effectiveness were twice with respect to application of TiO_2 alone	[86]
Gold (Au)	Methyl orange	Formation of interfacial layers between pure and doped catalysts in heterogeneous doping makes it more efficient than homogeneous doping methodology	[87, 88]
	Rhodamine B Congo red Methylene blue	Remarkable removal efficiency was found under solar radiation than UV light. Lowering of toxic nature confirms biocompatibility of doped TiO_2	[89]
Magnesium (Mg)	Congo red	Incorporation of dopant increased the mitigation efficiency by a factor of 2 due to the reduction of band gap energy through doping process	[90]
Gold (Au)	Methylene blue Auramine O Basic red 5 Basic blue 7	Formation of emerging catalyst which showed high stability and efficiency during dye mitigation. Developed catalyst showed excellent reusability and was able to endure consecutive oxidation and reduction steps because of its exclusive structure and configuration	[91]

(continued)

Table 1 (continued)

Doping agent	Dye	Findings	References
Iron (Fe^{3+})	Acid orange 7	Remarkable degradation efficiency even after 6 cycles of application under any kind of light irradiation	[92]
Cobalt (Co)	Amido black	Spontaneous photo-transformation resulted around 90% dye mitigation efficiency because of improved surface characteristics and amended reactivity between the photocatalyst and target pigments	[93]
Iron (Fe^{3+}) Platinum(Pt^{4+})	Eriochrome black-T	Application of dopants made remarkable improvised the effectiveness of dye mitigation where Pt produced better efficiency than iron by the improvement of active surface area	[94]
Carbon(C) Nitrogen (N)	Violet-3B	The rate of pigment decay was inversely proportional to the pigment concentration. Increase in active specific surface area with adsorption efficiency and colour mitigation of more than 95% was observed by the use of co-dopants	[95]
	Crystal violet	Lesser crystallized structures showed lesser decaying performance because of smaller life period of the evolved pairs of electrons and holes	[96]

components of waste effluents could be completely degraded, and issues of aqueous contamination could be abated. Based on the typical features of TiO_2 nano-catalysts, it could be reckoned as the most promised photo-oxidative agent in which further research will inevitably proceed. In-fact, such techniques accomplish all the aspects of sustainability for the generation of clean water from industrial dye contaminated effluents, which is required to be critically studied, optimized and scaled up for future implementations [97, 98].

References

1. Shakoor S, Nasar A (2016) Removal of methylene blue dye from artificially contaminated water using citrus limetta peel waste as a very low cost adsorbent. J Taiwan Inst Chem Eng 66:154–163. https://doi.org/10.1016/j.jtice.2016.06.009
2. Holkar CR, Jadhav AJ, Pinjari DV, Mahamuni NM, Pandit AB (2016) A critical review on textile wastewater treatments: possible approaches. J Environ Manag 182:351–366. https://doi.org/10.1016/j.jenvman.2016.07.090
3. Hossain L, Sarker SK, Khan MS (2018) Evaluation of present and future wastewater impacts of textile dyeing industries in Bangladesh. Environ Dev 26:23–33. https://doi.org/10.1016/j.envdev.2018.03.005
4. M. Masum, The Bangladesh textile-clothing industry: a demand-supply review. Social Syst Stud 9:109–139. https://core.ac.uk/download/pdf/76063808.pdf
5. Punzi M, Nilsson F, Anbalagan A, Svensson BM, Jönsson K, Mattiasson B, Jonstrup M (2015) Combined anaerobic-ozonation process for treatment of textile wastewater: removal of acute toxicity and mutagenicity. J Hazard Mater 292:52–60. https://doi.org/10.1016/j.jhazmat.2015.03.018
6. Gümüş D, Akbal F (2011) Photocatalytic degradation of textile dye and wastewater. Water Air Soil Pollut Focus 216:117–124. https://doi.org/10.1007/s11270-010-0520-z
7. Mathur N, Bhatnagar P, Nagar P, Bijarnia MK (2005) Mutagenicity assessment of effluents from textile/dye industries of Sanganer, Jaipur (India): a case study. Ecotoxicol Environ Saf 61:105–113. https://doi.org/10.1016/j.ecoenv.2004.08.003
8. Ibhadon AO, Fitzpatrick P (2013) Heterogeneous photocatalysis: recent advances and applications. Catalysts 3:189–218. https://doi.org/10.3390/catal3010189
9. Uddin MK (2017) A review on the adsorption of heavy metals by clay minerals, with special focus on the past decade. Chem Eng J 308:438–462. https://doi.org/10.1016/j.cej.2016.09.029
10. Wang Y (2000) Solar photocatalytic degradation of eight commercial dyes in TiO2 suspension. Water Res 34:990–994. https://doi.org/10.1016/S0043-1354(99)00210-9
11. Neppolian B, Choi HC, Sakthivel S, Arabindoo B, Murugesan V (2002) Solar light induced and TiO2 assisted degradation of textile dye reactive blue 4. Chemosphere 46:1173–1181. https://doi.org/10.1016/s0045-6535(01)00284-3
12. Nakata K, Fujishima A (2012) TiO2 photocatalysis: design and applications. J Photochem Photobiol C Photochem Rev 13:169–189. https://doi.org/10.1016/j.jphotochemrev.2012.06.001
13. Al-Mamun MR, Kader S, Islam MS, Khan MZH (2019) Photocatalytic activity improvement and application of UV-TiO2 photocatalysis in textile wastewater treatment: a review. J Environ Chem Eng 7:103248. https://doi.org/10.1016/j.jece.2019.103248
14. da Silva CG, Faria JL (2003) Photochemical and photocatalytic degradation of an azo dye in aqueous solution by UV irradiation. J Photochem Photobiol A Chem 155:133–143. https://doi.org/10.1016/j.jphotochem.2005.12.013
15. Feizpoor S, Habibi-Yangjeh A (2018) Ternary TiO2/Fe3O4/CoWO4 nanocomposites: novel magnetic visible-light-driven photocatalysts with substantially enhanced activity through pn heterojunction. J Colloid Interface Sci 524:325–336. https://doi.org/10.1016/j.jcis.2018.03.069
16. Yurdakal S, Palmisano G, Loddo V, Augugliaro V, Palmisano L (2008) Nanostructured rutile TiO2 for selective photocatalytic oxidation of aromatic alcohols to aldehydes in water. J Am Chem Soc 130(5):1568–1569. https://doi.org/10.1021/ja709989e
17. Chen D, Cheng Y, Hou N, Chen P, Wang Y, Li K, Huo S, Cheng P, Peng P, Zhang R, Wang L, Liu H, Liu Y, Ruan R (2020) Photocatalytic degradation of organic pollutants using TiO2-based photocatalysts: a review. J Clean Prod 268:121725. https://doi.org/10.1016/j.jclepro.2020.121725
18. Hu X, Li G, Yu JC (2010) Design, fabrication, and modification of nanostructured semiconductor materials for environmental and energy applications. Langmuir 26:3031–3039. https://doi.org/10.1021/la902142b

19. Meng F, Lu F, Sun Z, Lü J (2010) A mechanism for enhanced photocatalytic activity of nano-size silver particle modified titanium dioxide thin films. Sci China Technol Sci 53:3027–3032. https://doi.org/10.1007/s11431-010-4116-z
20. Oller I, Malato S, Sanchez-Perez JA (2011) Combination of advanced oxidation processes and biological treatments for wastewater decontamination—a review. Sci Total Environ 409:4141–4166. https://doi.org/10.1016/J.SCITOTENV.2010.08.061
21. Liu J, Li M, Wang J, Song Y, Jiang L, Murakami T, Fujishima A (2009) Hierarchically macro-/mesoporous Ti-Si oxides photonic crystal with highly efficient photocatalytic capability. Environ Sci Technol 43:9425–9431. https://doi.org/10.1021/es902462c
22. Liu S, Yu J, Jaroniec M (2010) Tunable photocatalytic selectivity of hollow TiO2 microspheres composed of anatase polyhedra with exposed 001 facets. J Am Chem Soc 132:11914–11916. https://doi.org/10.1021/ja105283s
23. Zheng Z, Huang B, Qin X, Zhang X, Dai Y (2010) Strategic synthesis of hierarchical TiO2 microspheres with enhanced photocatalytic activity. Chem Eur J 16:11266–11270. https://doi.org/10.1002/chem.201001280
24. García-Lopez EI, Marcì G, Palmisano L (2019) Photocatalytic and catalytic reactions in gas–solid and in liquid–solid systems. Heterog Photocatal: 153–176. https://doi.org/10.1016/b978-0-444-64015-4.00005-5
25. Prihod'ko RV, Soboleva NM (2013) Photocatalysis: oxidative processes in water treatment. J Chem https://doi.org/10.1155/2013/168701
26. Yu J, Jin J, Cheng B, Jaroniec M (2014) A noble metal-free reduced graphene oxide–CdS nanorod composite for the enhanced visible-light photocatalytic reduction of CO2 to solar fuel. J Mater Chem 2:3407. https://doi.org/10.1039/c3ta14493c
27. Zhou M, Li M, Hou C, Li Z, Wang Y, Xiang K, Guo X (2018) Pt nanocrystallines/ TiO2 with thickness-controlled carbon layers: preparation and activities in CO oxidation. Chin Chem Lett 29:787–790. https://doi.org/10.1016/j.cclet.2018.03.010
28. Gopinath KP, Madhav NV, Krishnan A, Malolan R, Rangarajan (2020) Present applications of titanium dioxide for the photocatalytic removal of pollutants from water: a review. J Environ Manag 270:110906. https://doi.org/10.1016/j.jenvman.2020.110906
29. Konstantinou IK, Albanis TA (2004) TiO2-assisted photocatalytic degradation of azo dyes in aqueous solution: kinetic and mechanistic investigations: a review. Appl Catal B Environ 49:1–14. https://doi.org/10.1016/j.apcatb.2003.11.010
30. Bagwasi S, Tian B, Zhang J, Nasir M (2013) Synthesis, characterization and application of bismuth and boron co-doped TiO2: a visible light active photocatalyst. Chem Eng J 217:108–118. https://doi.org/10.1016/j.cej.2012.11.080
31. Xiang Q, Yu J, Wong PK (2011) Quantitative characterization of hydroxyl radicals produced by various photocatalysts. J Colloid Interface Sci 357:163–167. https://doi.org/10.1016/j.jcis.2011.01.093
32. Alluea LL, Soriaa MTM, Asensioa JS, Salvadorb A, Ferronatoc C, Chovelon JM (2012) Degradation intermediates and reaction pathway of pyraclostrobin with TiO2 photocatalysis. Appl Catal B Environ 115–116:285–293. https://doi.org/10.1016/j.apcatb.2011.12.015
33. Grabowska E, Reszczynska J, Zaleska A (2012) Mechanism of phenol photodegradation in the presence of pure and modified-TiO2: a review. Water Res 46:5453–5471. https://doi.org/10.1016/j.watres.2012.07.048
34. Swarnakar P, Kanel SR, Nepal D, Jiang Y, Jia H, Kerr L, Goltz MN, Levy J, Rakovan J (2013) Silver deposited titanium dioxide thin film for photocatalysis of organic compounds using natural light. Sol Energy 88:242–249. https://doi.org/10.1016/j.solener.2012.10.014
35. Rapsomanikis A, Apostolopoulou A, Stathatos E, Lianos P (2014) Cerium-modified TiO2 nanocrystalline films for visible light photocatalytic activity. J Photochem Photobiol A Chem 280:46–53. https://doi.org/10.1016/j.jphotochem.2014.02.009
36. Umebayashi T, Yamaki T, Sumita T, Yamamoto S, Tanaka S, Asai K (2003) UV-ray photoelectron and ab initio band calculation studies on electronic structures of Cr or Nb-ion implanted titanium dioxide. Nucl Instrum Methods Phys Res Sect B 206:264–267. https://doi.org/10.1016/S0168-583X(03)00740-7

37. Choi J, Park H, Hoffmann MR (2010) Effects of single metal-ion doping on the visible light photoreactivity of TiO_2. J Phys Chem C 114:783–792. https://doi.org/10.1021/jp908088x
38. Zhou W, Fu H (2013) Mesoporous TiO2: preparation, doping, and as a composite for photocatalysis. ChemCatChem 5:885–894. https://doi.org/10.1002/cctc.201200519
39. Bhattacharyya K, Majeed J, Dey KK, Ayyub P, Tyagi AK, Bharadwaj SR (2014) Effect of Mo-incorporation in the TiO_2 lattice: a mechanistic basis for photocatalytic dye degradation. J Phys Chem C 118:15946–15962. https://doi.org/10.1021/jp5054666
40. Peng YH, Huang GF, Huang WQ (2012) Visible-light absorption and photocatalytic activity of Cr-doped TiO2 nanocrystal films. Adv Powder Technol 23:8–12. https://doi.org/10.1016/j.apt.2010.11.006
41. Nasir M, Bagwasi S, Jiao Y, Chen F, Tian B, Zhang J (2014) Characterization and activity of the Ce and N co-doped TiO2 prepared through hydrothermal method. Chem Eng J 236:388–397. https://doi.org/10.1016/j.cej.2013.09.095
42. Wu Y, Xing M, Zhang J (2011) Gel-hydrothermal synthesis of carbon and boron codoped TiO2 and evaluating its photocatalytic activity. J Hazard Mater 192:368–373. https://doi.org/10.1016/j.jhazmat.2011.05.037
43. Cong Y, Tian B, Zhang J (2011) Improving the thermal stability and photocatalytic activity of nanosized titanium dioxide via La3+ and N co-doping. Appl Catal B Environ 101:376–381. https://doi.org/10.1016/j.apcatb.2010.10.006
44. Xing MY, Qi DY, Zhang JL, Chen F (2011) One-step hydrothermal method to prepare carbon and lanthanum co-doped TiO2 nanocrystals with exposed 001 facets and their high UV and visible-light photocatalytic activity. Chem Eur J 17:11432. https://doi.org/10.1002/chem.201101654
45. Gupta S, Tripathi M (2011) A review of TiO2 nanoparticles. Chin Sci Bull 56:1639. https://doi.org/10.1007/s11434-011-4476-1
46. Rauf MA, Meetani MA, Hisaindee S (2011) An overview on the photocatalytic degradation of azo dyes in the presence of TiO2 doped with selective transition metals. Desalination 276:13–27. https://doi.org/10.1016/j.desal.2011.03.071
47. Davis RJ, Gainer JL (1994) Photocatalytic decolorization of waste water dyes. Water Environ Res 66:50–53. https://doi.org/10.2175/WER.66.1.8
48. Hathaisamita K, Sutha W, Kamruang P, Pudwat S, Teekasap S (2012) Decolorization of cationic yellow X-Gl 200% from textile dyes by TiO2 films-coated rotor. Procedia Eng 32:800–806. https://doi.org/10.1016/j.proeng.2012.02.015
49. Chong MN, Jin B, Chow CWK, Saint C (2010) Recent developments in photocatalytic water treatment technology: a review. Water Res 44:2997–3027. https://doi.org/10.1016/j.watres.2010.02.039
50. Xu H, Zheng Z, Zhang LZ, Zhang HL, Deng F (2008) Hierarchical chlorine-doped rutile TiO2 spherical clusters of nanorods: Large-scale synthesis and high photocatalytic activity. J Solid State Chem 181:2516–2522. https://doi.org/10.1016/j.jssc.2008.06.019
51. Ajmal A, Majeed I, Malik RN, Idriss H, Nadeem MA (2014) Principles and mechanisms of photocatalytic dye degradation on TiO2 based photocatalysts: a comparative overview. RSC Adv 4:37003–37026. https://doi.org/10.1039/c4ra06658h
52. Gola D, Malik A, Namburath M, Ahammad SZ (2018) Removal of industrial dyes and heavy metals by Beauveria bassiana: FTIR, SEM, TEM and AFM investigations with Pb(II). Environ Sci Pollut Res 25:20486–20496. https://doi.org/10.1007/s11356-017-0246-1
53. Chung KT (2016) Azo dyes and human health: a review. J Environ Sci Health Part C Environ Carcinog Ecotoxicol Rev 34:233–261. https://doi.org/10.1080/10590501.2016.1236602
54. Khataee AR, Pons MN, Zahraa O (2009) Photocatalytic degradation of three azo dyes using immobilized TiO2 nanoparticles on glass plates activated by UV light irradiation: influence of dye molecular structure. J Hazard Mater 168:451–457. https://doi.org/10.1016/j.jhazmat.2009.02.052
55. Chandra R, Mukhopadhyay S, Nath M (2016) TiO2 @ZIF-8: a novel approach of modifying micro-environment for enhanced photo-catalytic dye degradation and high usability of TiO2 nanoparticles. Mater Lett 164:571–574. https://doi.org/10.1016/j.matlet.2015.11.018

56. Dariani RS, Esmaeili A, Mortezaali A, Dehghanpour S (2016) Photocatalytic reaction and degradation of methylene blue on TiO2 nano-sized particles. Optik 127:7143–7154. https://doi.org/10.1016/j.ijleo.2016.04.026
57. Singh R, Kumari P, Chavan PD, Datta S, Dutta S (2017) Synthesis of solvothermal derived TiO2 nanocrystals supported on ground nano egg shell waste and its utilization for the photocatalytic dye degradation. Opt Mater (Amst) 73:377–383. https://doi.org/10.1016/j.optmat.2017.08.040
58. Zulfiqar M, Chowdhury S, Sufian S, Omar AA (2018) Enhanced photocatalytic activity of Orange II in aqueous solution using solvent-based TiO2 nanotubes: kinetic, equilibrium and thermodynamic studies. J Clean Prod 203:848–859. https://doi.org/10.1016/j.jclepro.2018.08.324
59. Ljubas D, Smoljanic G, Juretic H (2015) Degradation of Methyl Orange and Congo Red dyes by using TiO2 nanoparticles activated by the solar and the solar-like radiation. J Environ Manag 161:83–91. https://doi.org/10.1016/j.jenvman.2015.06.042
60. Pol R, Guerrero M, García-Lecina E, Altube A, Rossinyol E, Garroni S, Baro MD, Pons J, Sort J, Pellicer E (2016) Ni-, Pt- and (Ni/Pt)-doped TiO2 nanophotocatalysts: a smart approach for sustainable degradation of Rhodamine B dye. Appl Catal B Environ 181:270–278. https://doi.org/10.1016/j.apcatb.2015.08.006
61. Gnanasekaran L, Hemamalini R, Saravanan R, Ravichandran K, Gracia F, Gupta VK (2016) Intermediate state created by dopant ions (Mn, Co and Zr) into TiO2 nanoparticles for degradation of dyes under visible light. J Mol Liq 223:652–659. https://doi.org/10.1016/j.molliq.2016.08.105
62. Kerkez-Kuyumcu Ö, Kibar, E., Dayıoğlu K, Gedik F, Akın AN, Özkara-Aydinoğlu S (2015) A comparative study for removal of different dyes over M/TiO2(M= Cu, Ni Co, Fe, Mn and Cr) photocatalysts under visible light irradiation. J Photochem Photobiol A Chem 311:176–185. https://doi.org/10.1016/j.jphotochem.2015.05.037
63. McManamon C, O'Connell J, Delaney P, Rasappa S, Holmes JD, Morris MA (2015) A facile route to synthesis of S-doped TiO2 nanoparticles for photocatalytic activity. J Mol Catal A Chem 406:51–57. https://doi.org/10.1016/j.molcata.2015.05.002
64. Kaur N, Kaur S, Singh V (2016) Preparation, characterization and photocatalytic degradation kinetics of Reactive Red dye 198 using N, Fe codoped TiO2 nanoparticles under visible light. Desalin. Water Treat 57:9237–9246. https://doi.org/10.1080/19443994.2015.1027956
65. Brindha A, Sivakumar T (2017) Visible active N, S co-doped TiO2/graphene photocatalysts for the degradation of hazardous dyes. J Photochem Photobiol A Chem 340:146–156. https://doi.org/10.1016/j.jphotochem.2017.03.010
66. Shao J, Sheng W, Wang M, Li S, Chen J, Zhang Y, Cao S (2017) In situ synthesis of carbon-doped TiO2 single-crystal nanorods with a remarkably photocatalytic efficiency. Appl Catal B Environ 209:311–319. https://doi.org/10.1016/j.apcatb.2017.03.008
67. Zhang Q, Fu Y, Wu Y, Zhang YN, Zuo T (2016) Low-cost Y-doped TiO2 nanosheets film with highly reactive 001 facets from CRT waste and enhanced photocatalytic removal of Cr(VI) and methyl orange. ACS Sustain Chem Eng 4:1794–1803. https://doi.org/10.1021/acssuschemeng.5b01783
68. Sood S, Umar A, Kumar Mehta S, Sinha ASK, Kansal SK (2015) Efficient photocatalytic degradation of brilliant green using Sr-doped TiO2 nanoparticles. Ceram Int 41:3533–3540. https://doi.org/10.1016/j.ceramint.2014.11.010
69. Lei P, Wang F, Gao X, Ding Y, Zhang S, Zhao J, Liu S, Yang M (2012) Immobilization of TiO2 nanoparticles in polymeric substrates by chemical bonding for multi-cycle photodegradation of organic pollutants. J Hazard Mater 227–228:185–194. https://doi.org/10.1016/j.jhazmat.2012.05.029
70. Talebi S, Chaibakhsh N, Moradi-Shoeili Z (2017) Application of nanoscale ZnS/TiO2 composite for optimized photocatalytic decolorization of a textile dye. J Appl Res Technol 15:378–385. https://doi.org/10.1016/j.jart.2017.03.007
71. Xiaodan Y, Qingyin W, Shicheng J, Yihang G (2006) Nanoscale ZnS/TiO2 composites: preparation, characterization, and visible-light photocatalytic activity. Mater Char 57:333–341. https://doi.org/10.1016/j.matchar.2006.02.011

72. Hui C, Lei Z, Xitang W, Shujing L, Zhongxing L (2015) Preparation of nanoporous TiO2/SiO2 composite with rice husk as template and its photocatalytic property. Rare Met Mater Eng 44:1607–1611. https://doi.org/10.1016/s1875-5372(15)30101-6
73. Li J, Zhen D, Sui G, Zhang C, Deng Q, Jia L (2012) Nanocomposite of Cu–TiO2–SiO2 with high photoactive performance for degradation of rhodamine B dye in aqueous wastewater. J Nanosci Nanotechnol 12:6265–6270. https://doi.org/10.1166/jnn.2012.6438
74. Shao L, Liu H, Zeng W, Zhou C, Li D, Wang L, Lan Y, Xu F, Liu G (2019) Immobilized and photocatalytic performances of PDMS-SiO2-chitosan@TiO2 composites on pumice under simulated sunlight irradiation. Appl Surf Sci 478:1017–1026. https://doi.org/10.1016/j.apsusc.2019.02.060
75. Tang Y, Di W, Zhai X, Yang R, Qin W (2013) NIR-responsive photocatalytic activity and mechanism of NaYF 4:Yb, Tm@TiO2 core-shell nanoparticles. ACS Catal 3:405–412. https://doi.org/10.1021/cs300808r
76. Alagarasi A, Rajalakshmi PU, Shanthi K, Selvam P (2019) Solar light photocatalytic activity of mesoporous nanocrystalline TiO2, SnO2, and TiO2-SnO2 composites. Mater Today Sustain. https://doi.org/10.1016/j.mtsust.2019.100016
77. Magdalane CM, Kanimozhi K, Arularasu MV, Ramalingam G, Kaviyarasu K (2019) Self-cleaning mechanism of synthesized SnO2/TiO2 nanostructure for photocatalytic activity application for waste water treatment. Surfaces Interfaces 17:100346. https://doi.org/10.1016/j.surfin.2019.100346
78. Djellabi R, Ghorab MF, Cerrato G, Morandi S, Gatto S, Oldani V, Di Michele A, Bianchi CL (2015) Photoactive TiO2-montmorillonite composite for degradation of organic dyes in water. J Photochem Photobiol A Chem 295:57–63. https://doi.org/10.1016/j.jphotochem.2014.08.017
79. Rahimi B, Jafari N, Abdolahnejad A, Farrokhzadeh H, Ebrahimi A (2019) Application of efficient photocatalytic process using a novel BiVO/TiO2-NaY zeolite composite for removal of acid orange 10 dye in aqueous solutions: modeling by response surface methodology (RSM). J Environ Chem Eng 7:103253. https://doi.org/10.1016/j.jece.2019.103253
80. Saleh TA, Gupta VK (2012) Photo-catalyzed degradation of hazardous dye methyl orange by use of a composite catalyst consisting of multi-walled carbon nanotubes and titanium dioxide. J Colloid Interface Sci 371:101–106. https://doi.org/10.1016/j.jcis.2011.12.038
81. Rajamanickam D, Shanthi M (2014) Photocatalytic degradation of an azo dye Sunset Yellow under UV-A light using TiO2/CAC composite catalysts. Spectrochim Acta Part A Mol Biomol Spectrosc 128:100–108. https://doi.org/10.1016/j.saa.2014.02.126
82. Xu C, Cui A, Xu Y, Fu X (2013) Graphene oxide-TiO2 composite filtration membranes and their potential application for water purification. Carbon N. Y. 62:465–471. https://doi.org/10.1016/j.carbon.2013.06.035
83. Ahmed AS, Ahamad T, Ahmad N, Khan MZ (2019) Removal enhancement of acid navy blue dye by GO—TiO2 nanocomposites synthesized using sonication method. Mater Chem Phys. https://doi.org/10.1016/j.matchemphys.2019.121906,121906
84. Ranjith R, Renganathan V, Chen SM, Selvan NS, Rajam PS (2019) Green synthesis of reduced graphene oxide supported TiO2/Co3O4 nanocomposite for photocatalytic degradation of methylene blue and crystal violet. Ceram Int 45:12926–12933. https://doi.org/10.1016/j.ceramint.2019.03.219
85. Monga D, Basu S (2019) Enhanced photocatalytic degradation of industrial dye by g-C 3 N 4/TiO2 nanocomposite: role of shape of TiO2. Adv Powder Technol. https://doi.org/10.1016/j.apt.2019.03.004
86. Bessegato GG, Cardoso JC, Zanoni MVB (2015) Enhanced photoelectrocatalytic degradation of an acid dye with boron-doped TiO2 nanotube anodes. Catal Today 240:100–106. https://doi.org/10.1016/j.cattod.2014.03.073
87. Zhang Y, Hu H, Chang M, Chen D, Zhang M, Wu L, Li X (2017) Non-uniform doping outperforms uniform doping for enhancing the photocatalytic efficiency of Au-doped TiO2 nanotubes in organic dye degradation. Ceram Int 43:9053–9059. https://doi.org/10.1016/j.ceramint.2017.04.050

88. Zhang Y, Hu Z, Cui X, Yao W, Duan T, Zhu W (2017) Capture of Csþ and Sr2þ from aqueous solutions by using Cr doped TiO2 nanotubes. J Nanosci Nanotechnol 17:3943–3950. https://doi.org/10.1166/jnn.2017.13085
89. Bharati B, Sonkar AK, Singh N, Dash D, Rath C (2017) Enhanced photocatalytic degradation of dyes under sunlight using biocompatible TiO2 nanoparticles. Mater Res Express 4:85503. https://doi.org/10.1088/2053-1591/aa6a36
90. Bhagwat UO, Wu JJ, Asiri AM, Anandan S (2017) Sonochemical synthesis of Mg-TiO2 nanoparticles for persistent Congo red dye degradation. J Photochem Photobiol A Chem 346:559–569. https://doi.org/10.1016/j.jphotochem.2017.06.043
91. Ren ZH, Li HT, Gao Q, Wang H, Han B, Xia KS, Zhou CG (2017) Au nanoparticles embedded on urchin-like TiO2 nanosphere: an efficient catalyst for dyes degradation and 4-nitrophenol reduction. Mater Des 121:167–175. https://doi.org/10.1016/j.matdes.2017.02.064
92. Han F, Kambala VSR, Dharmarajan R, Liu Y, Naidu R (2018) Photocatalytic degradation of azo dye acid orange 7 using different light sources over Fe3þ-doped TiO2 nanocatalysts. Environ Technol Innov 12:27–42. https://doi.org/10.1016/j.eti.2018.07.004
93. Ali I, Alharbi OML, Alothman ZA, Badjah AY (2018) Kinetics, thermodynamics, and modeling of amido black dye photodegradation in water using Co/TiO2 nanoparticles. Photochem Photobiol 94:935–941. https://doi.org/10.1111/php.12937
94. Pal B, Kaur R, Grover IS (2016) Superior adsorption and photodegradation of eriochrome black-T dye by Fe3þ and Pt4þ impregnated TiO2 nanostructures of different shapes. J Ind Eng Chem 33:178–184. https://doi.org/10.1016/j.jiec.2015.09.033
95. Putri RA, Safni S, Jamarun N, Septiani U, Kim M-K, Zoh K-D (2019) Degradation and mineralization of violet-3B dye using C-N-codoped TiO2 photocatalyst. Environ Eng Res. https://doi.org/10.4491/eer.2019.196
96. Amin MT, Alazba AA (2017) Structural study of monoclinic TiO2 nanostructures and photocatalytic applications for degradation of crystal violet dye. Mod Phys Lett B 31:1–11. https://doi.org/10.1142/S0217984917502645
97. Ribeiro MCM, Starling MCVM, Leão MMD, De Amorim CC (2017) Textile wastewater reuse after additional treatment by Fenton's reagent. Environ Sci Pollut Res 24:6165–6175. https://doi.org/10.1007/s11356-016-6921-9
98. Zhang JJ, Fang SS, Mei JY, Zheng GP, Zheng XC, Guan XX (2018) High-efficiency removal of rhodamine B dye in water using g-C3N4 and TiO2 co-hybridized 3D graphene aerogel composites. Separ. Purif. Technol. 194:96–103. https://doi.org/10.1016/j.seppur.2017.11.035

Ozone-Based Processes in Dye Removal

Qomarudin Helmy, I Wayan Koko Suryawan, and Suprihanto Notodarmojo

Abstract The general practice carried out by the textile industry in treating its wastewater is chemically, physically, biologically or a combination of the three. As a case study that occurred in Indonesia, regulations regarding color parameters were not regulated until 2019. To cope with the new and more stringent regulatory threshold values, many textile industries have to modify or even rebuild their existing wastewater treatment plants (WWTPs) by adapting the latest wastewater treatment technologies. One promising alternative that can be added and/or modified to the existing WWTP is advanced oxidation using ozone. The use of ozone in textile wastewater treatment applications has several advantages including having a very large oxidation power so that it only requires a relatively short contact time (CT) to oxidize impurities contained in wastewater in the order of minutes. This chapter will discuss a brief history of ozone use in water and wastewater treatment, its chemistry and generation methods, degradation process, mechanisms, and factors affecting dye removal using ozone, also its practical application and integration with the existing WWTP process.

Keywords Textile · Wastewater · Dye · Ozone · Oxidation · Decolorization · Hydroxyl radical · Degradation · Organic compound · Color removal · Color standard

---

Q. Helmy (✉)
Faculty of Civil and Environmental Engineering, Water and Wastewater Engineering Research Group, Bioscience and Biotechnology Research Center, Institute of Technology Bandung, Ganesa St. 10, Bandung, West Java 40132, Indonesia
e-mail: helmy@tl.itb.ac.id

I W. K. Suryawan
Environmental Engineering Study Program, University of Pertamina, Teuku N.A. St., Simprug, Kebayoran Lama, Jakarta 12220, Indonesia
e-mail: i.suryawan@universitaspertamina.ac.id

S. Notodarmojo
Faculty of Civil and Environmental Engineering, Water and Wastewater Engineering Research Group, Institute of Technology Bandung, Ganesa St. 10, Bandung, West Java 40132, Indonesia

S. S. Muthu and A. Khadir (eds.), *Advanced Oxidation Processes in Dye-Containing Wastewater*, Sustainable Textiles: Production, Processing, Manufacturing & Chemistry,
https://doi.org/10.1007/978-981-19-0987-0_6

1 Introduction

The textile industry is one of the industries that consume water and produces large amounts of wastewater. More than 80% of the water consumed by the textile industry will eventually become wastewater that must be treated before being discharged into the environment. The average amount of water consumption to produce cloth in the wet process of textiles is 150 m^3/ton of product [40]. One of the textile processes that contributes greatly to water consumption and produces the largest wastewater is fabric dyeing. This process requires 80–200 L of water/kg cloth [63, 70, 100] in dyeing cotton cloth using conventional methods. This raises concerns about the availability and need for clean water sources, especially in industrial areas. In general, the textile industry utilizes deep groundwater as a source of raw water. This activity can damage the aquifer and cause land subsidence. Based on land surface measurements carried out between 2000 and 2012, it was found that several locations in the textile industry in Indonesia had experienced land subsidence reaching from 8 to 17 cm/year which was thought to occur due to groundwater extraction by textile industry activities [32].

The wastewater released by the textile industry contains residual dyes that are toxic, mutagenic, and carcinogenic [42]. An example is a dye containing azo chromophore which is the dominant chromophore used in 50–70% of all types of dyes in the textile industry. This azo compound has been reported in many studies to cause cell mutations and has the potential to cause cancer. Another chromophore compound is phthalocyanine with three to four cyanin groups bonded to Cu, Cr, or Co metals making the molecular size of these dyes large, making them very difficult to decompose naturally. In the dyeing process, 10–50% of the dye is wasted as wastewater because it cannot be fixed into the fiber. Around 700,000 tons of dye are consumed per year, thus putting pressure on the textile industry which is considered one of the main polluters [80]. Approximately 280,000 tons of textile dyes are released as textile wastewater effluent [41]. The complex chemical structure of dyes forms cyclic bonds and has a large molecular weight making these compounds non-biodegradable. Some types of dyes also contain azo compounds which are strictly regulated by many countries. For example, the European Union has identified 24 types of aromatic amines that are classified as harmful to humans and prohibits the use of azo dyes that produce 30 mg/kg or more of these aromatic amines. In Asian countries, restrictions on dyes containing azo compounds have been implemented by India starting in 1997 (112 types of dyes), China (in 2005), South Korea (in 2010), Taiwan (in 2011), and Japan (in 2014) [17].

Colored waste is psychologically more frightening to society in addition to its very disturbing aesthetics. The minimum concentration of color that can be seen visually in river flows is around 1–10 mg/L, depending on color, illumination, and the degree of water clarity. In addition to aesthetics, colored substances in water will block the transmission of light through the water so that it can interfere with the photosynthesis process which results in an ecological imbalance. In general, the textile industry in Indonesia uses WWTP with coagulation-flocculation processes, biological processes, and or a combination thereof, with high BOD and COD removal efficiency but not for color removal. Several AOPs methods, such as ozone, Fenton, photo-Fenton, photocatalytic, and UV-based oxidation have been carried out to overcome this problem. The use of ozone is reported to have a good ability to remove color, but its low solubility in water and the use of high energy in producing ozone are some of the drawbacks of this method.

Ozone (O_3) is the allotropic form of oxygen (O_2), comprising of identical atoms, however, they are consolidated in a different structure. The difference is oxygen has just two oxygen atoms, while ozone consists of three oxygen atoms. Ozone has a low molecular weight (MW = 48 g/mol) in which three oxygen atoms are chemically arranged in chains. Ozone is a gaseous compound that naturally exists in the atmosphere and is formed as a result of ultraviolet radiation [43]. Ozone has been used for water disinfection in drinking water treatment plants for centuries and to help remove unpleasant odors and organic/inorganic impurities [54, 72]. Ozone has been used in European countries for a long time and recently the application of ozone in the food and beverages industry has begun to be widely used [33]. The United States Food and Drug Administration (FDA) granted safe status for the use of ozone in bottled water in 1982. Ozone has also been declared Generally Recognized as Safe (GRAS) for use in food processing by the FDA in 1997 [30]. Furthermore, now ozone is also recognized and permitted as a food additive as an antimicrobial agent by the FDA in 2001 [26, 58].

Ozone is formed naturally in the stratosphere in small quantities (0.05 mg/l) by the action of ultraviolet irradiation on oxygen. Small amounts of ozone are also formed in the troposphere as a by-product of photochemical reactions between oxygen, nitrogen, and hydrocarbons, released from industries, oil-fired engine exhaust, forest fires, and volcanic eruptions. However, the gas produced is very unstable and rapidly decomposes in the air [47]. To produce ozone, diatomic oxygen molecules must first be broken down. The resulting oxygen free radical thus freely reacts with other diatomic oxygens to form triatomic ozone molecules. However, to break the O–O bond, it requires a lot of energy [10, 34]. When used in industry, ozone is usually generated at the point of application and in a closed systems. Ultraviolet radiation (wavelength 188 nm) and corona discharge methods can be used to initiate the formation of oxygen free radicals and thereby generate ozone. To generate commercially viable ozone concentration, corona discharge methods are commonly used for large-scale applications [24].

This chapter provides a summary of the physicochemical properties of ozone, the mechanism of ozone formation, and ozone application in water and wastewater treatment. Ozone can be produced through several methods, i.e., quiescent discharge, phosphorus contact, photochemical reactions, and electrochemical reactions, which in principle proceed through the reaction of oxygen atoms with oxygen molecules. However, there are side reactions to ozone formation, which are responsible for ozone depletion including thermal decomposition and cooling reactions by reactive species. The solubility of ozone in water is higher than that of oxygen, indicating that it can be reliably applied in water and wastewater treatment. Based on the resonance structure of ozone, one oxygen atom in the ozone molecule is electron deficient which shows electrophilic properties, while one electron-rich oxygen atom holds nucleophilic properties.

1.1 History of Ozone Use in Water and Wastewater Treatment

Ozone has a long history of research and application. A Dutch chemist called Van Marum in 1785 was probably the first known scientist who noticed the presence of ozone by the specific odor of the air in the neighborhood of his electrostatic generator when subjected to the passage of electric sparks. Another scientist named William Cruickshank, a Scottish chemist in 1801, observed the same specific gas odor that formed near the anode during the electrolysis of water [51, 64, 84]. The history of the ozone discovery has been reported in an excellent articles series by Rubin [9, 49, 68, 76–79, 98, 99]. Table 1 shows the brief history of ozone application and regulation in water and wastewater.

Treating wastewater with ozone, primarily for disinfection, was a major focus to date due to Corona Virus Outbreak since 2019. As per the report published in 2021, it has been indicated that there is a possibility of the virus becoming widespread through the raw water and wastewater network. The risk of exposure via the fecal–oral route, due to its excretion into the sewer, has also been highlighted in areas with inadequate sanitation facilities, especially in developing countries. Although the infectivity of the virus is unknown, the presence of the virus was confirmed in human feces even after 1 month after the patient tested negative for COVID-19. Risk is higher in third-world countries with a high magnitude of open defecation. WHO data in 2010 estimated that 1.1 billion people or 17% of the world's population still defecate in open areas. As many as 81% of the population who practice open defecation are found in 10 countries in the world, where Indonesia is the second-largest country whose people practice defecating in open areas after India [92].

Commercially produced ozone for oxidation reactions is always produced as a gas, from the air at concentrations between 1.0 and 2.0 weight percent, or from liquid oxygen at concentrations greater than 2%. Because ozone is highly reactive and has a short half-life, it cannot be stored as gas and transported. As a result, ozone is always generated on-site for direct use. When ozone is applied as gas for drinking water treatment, it does so primarily because of its oxidative power. This strong oxidizing potential allows ozone to be effective in reducing or eliminating color, residual taste, and odor. More importantly, ozone will effectively destroy dormant bacteria and viruses faster than other disinfectant chemicals. Ozone will also oxidize metals impurities, for instant iron and manganese can be oxidized into iron (III) and manganese (III) form which is easier to remove by simple filtration. This same process is used to liberate organically bound heavy metals, which otherwise are not easily removed. When properly applied in the water treatment process, ozone will not lead to the formation of halogenated compounds such as Trihalomethanes (THMs), which are formed when chlorine is added to raw water containing humic materials. Ozone can be used as an oxidant, which is applied in the final stages of

Table 1 History of ozone, application, and regulation

Year	Achievement	References
1785	Martinus Van Marum was probably the first known scientist who observed the presence of the specific odor which will later be named ozone	[51, 76, 84]
1801	William Cruickshank observed the same specific gas odor that formed near the anode during the electrolysis of water	[73]
1840	Christian Friedrich Schonbein, Professor of Chemistry at the University of Basel, determined that the odor produced during electric sparking was caused by an unknown compound that he called ozone, derived from the Greek word "Ozein" meaning to smell	[68 ,76]
1856	Thomas Andrew, Professor of Chemistry at Queen's University of Belfast, showed that ozone from whatever source derived, is one and the same body, having identical properties and the same constitution, and is not a compound body, but oxygen in an altered or allotropic condition	[77, 94]
1857	Werner von Siemens invented the apparatus of an electric discharge ozone generator, and only this invention made industrial applications of ozone possible at that time	[73]
1865	Jacques Louis Soret determined the molecular formula of ozone and established the relationship between oxygen and ozone, by finding that the three volumes of oxygen produce two volumes of ozone	[77]
1870	The German physician, Lender, published the first study about the biological practice effects related to the application of ozone in the disinfection of water and its antimicrobial properties. This finding eventually revolutionized medical practice during this period, more than half a century before the discovery of penicillin	[84]
1893	The first ozone-based drinking water treatment prototype plant was built in Oudshoorn, Netherlands. After sedimentation and filtration, the water of the Rhine River was purified with ozone. Some French scientists and chemists examined this apparatus and decided to build their own plant, in Nice, France	[74, 75]
1900	Tesla Ozone Co., (the Tesla ozone company) was established and started marketing ozone generators for medical applications	[67]
1906	France commissions its first ozone disinfection unit installation for their water treatment plant in Nice, where ozone was used to disinfect 22,500 m^3/day raw water drawn from the Vesubie River after being filtered by a slow sand unit	[73]
1909	Ozone was used as a preservative for meat cold storage in Germany	[34]
1914	Interest in ozone for water treatment began to decline as many studies led to the production of cheap chlorine gas as a disinfectant	[98]

(continued)

Table 1 (continued)

1916	Around 49 ozone installations were in use throughout Europe (26 of which were located in France)	[49]
1920	The Swiss dentist, Dr. Edwin Parr began using ozone as part of his disinfection system. Dr. Charles S. Neiswanger, Professor of the Chicago Hospital College of Medicine, publishes "Electro Therapeutical Practice." where chapter 32 was entitled "Ozone as a Therapeutic Agent."	[29, 64]
1931	Dr. E. A. Fisch became a pioneer in its dentistry application thanks to his use of ozonated water for dental procedures	[50]
1936	Ozone was used to depurate shellfish in France	[34]
1939	Ozone was used to prevent the growth of yeast and mold during the storage of fruits	[34]
1942	Ozone was used in egg-storage rooms and cheese-storage facilities in the USA	[34]
1957	Ozone is applied for the oxidation of humic substances, undesired odors, taste, iron, and manganese in German drinking water	[49]
1964	Spontaneous flocculation in ozone contact basins led the French to build an ozone plant to improve the removal of particulate matter	[49]
1965	United Kingdom and Ireland started to use ozone in controlling the color of surface water. Switzerland started the use of ozone to oxidize micropollutants such as phenolic compounds and some pesticide residues	[49]
1970	Ozone was used to control algae growth in France	[49]
1982	The US food and drug administration grants GRAS (generally recognized as safe) status for ozone disinfection application in the bottled drinking water industry	[34]
1987	After seven years of trial, an ozonation plant with a 600 MGD (million gallons per day) capacity was commissioned in Los Angeles, USA	[49]
1992	Russia reports the use of ozone in salt water baths for burn treatment applications	[67]
1996	The government of Japan, The Canadian Food Inspection Agency (CFIA), The government of Australia have approved the use of ozone for direct contact with all types of food	[34]
1997	In the United States, ozone has received GRAS (Generally Recognized as Safe) classification	[34]
2001	United State department of agriculture approved the application of ozone as an antimicrobial agent for direct contact with food and can be used in all meat and poultry products	[103]

(continued)

Table 1 (continued)

2004	The International Ozone Association reports the installation of 894 ozone installation projects with a total ozone capacity of 21,246 kg/hour during the 1969–2004 period	[49]
2008	Montreal, Canada, plans to be the first metropolis city in the world using ozonation to disinfect all of its wastewater treatment facilities, with the ozone capacity is expected to be approximately 1,800 kg/h	[49]
2009	More than 50 water treatment plants have installed ozonation as an advanced process for the removal of taste, color, and odor and for the control of trihalomethanes (THMs) formation, with approximately 800 kg/h (42,000 lb/day) ozone capacity in operation in Japan Ozone-treated reclaimed wastewater is being used with a total of final discharge is more than 100,000 m^3/day	[36, 88]
2010	In the US, the installed ozone capacity to treat drinking water exceeds 525,000 lb/day. Meanwhile, wastewater treatment using ozone during 2005–2010 was 7 facilities with a flow capacity of 60 MGD and an ozone capacity of 2,000 lb/day	[49]
2020	Ozone as a potential oxidant for Coronavirus disease (COVID-19) virus inactivation. Ozonated nanobubbles were used in Hospital Wastewater Treatment Plant to eradicate the persistent SARS-CoV-2 residues even though it has been through the final disinfection by chlorine	[3, 8, 46, 52, 89, 97]

water treatment. There are more than 2,000 major installations worldwide that use ozone to treat drinking water, not to mention for the small-scale household drinking water supplier. Ozone is an effective disinfectant for treating municipal and industrial wastewater, effective in dealing with a variety of complex and toxic chemicals.

1.2 *Chemistry of Ozone*

Due to the unstable nature of ozone that spontaneously reverts back into oxygen, ozone production can only be carried out on-site with an ozone generator which can be operated practically and stably. In terms of its application in water and wastewater treatment, it is very important to understand the physical and chemical properties of ozone, with a focus on its solubility and chemical reactivity to the pollutant compounds of concern. Ozone has a structure of 3 oxygen atoms, which are bonded by the same oxygen-oxygen bonds at an angle of 116.8°. The steric hindrance prevents it from forming a triangle with each oxygen atom forming the expected 2 bonds. Instead, each oxygen forms only 1 bond, with the remaining negative charge scattered throughout the molecule. Ozone is a dipole molecule, giving it the characteristic properties that ozone reacts very selectively and is electrophilic (Fig. 1).

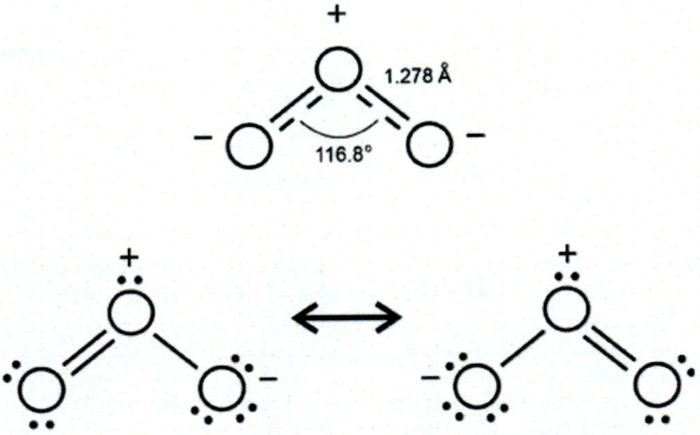

Fig. 1 Schematic of the two resonance structures of ozone with a triatomic molecule with no unpaired electrons and a bent molecular shape, length, and angle formed by three oxygen atoms are shown

Ozone is slightly soluble in water, although it is about 10 times more soluble than oxygen [98]. At 20 °C, the solubility of 100% ozone is about 570 mg/l and decreases to 400 mg/l at 30 °C (Table 2). It is known that ozone solubility is higher than oxygen, however, chlorine solubility is 12 times more than ozone in water. The stability is influenced by the presence of sensitizing impurities, such as metal oxides, heavy metal cations, and by temperature and pressure [37]. Generally, a decrease in the temperature or an increase in pressure will enhance the solubility of ozone in the aqueous phase. Ozone concentrations used in water treatment are generally below 12%, which limits its mass transfer driving force into the water. This results in a low concentration of ozone in the water, only in the range of 0.1–1 mg/l. In addition, due to the limited half-life of ozone in water (±20 min at 20 °C), if we need a certain amount of ozone to react, a larger amount of injection must be given.

Many research reports have shown that ozone decomposes spontaneously during water treatment by a complex mechanism that involves the generation of hydroxyl free radicals. The hydroxyl free radicals are among the most reactive oxidizing agents in water, with reaction rates on the order of 10^{10}–10^{13} $M^{-1}.s^{-1}$, however, the half-life of hydroxyl free radicals is on the order of microseconds, therefore, its concentrations in water can never reach levels above 10^{-12} [93].

$$O_3 + H_2O \rightarrow HO_3 + OH^* \quad (1)$$

$$HO_3 \rightarrow O_3{}^* + H^+ \quad (2)$$

$$HO_3 \leftrightarrow OH^* + O_2 \quad (3)$$

Table 2 Physical and chemical properties of ozone and oxygen

No	Property	Ozone	Oxygen
1	Molecular weight, g/mol	48	32
2	Density (at 101 kPa), kg/m^3		
	Gas (0 °C)	2.144	1.429
	Liquid (−183 °C)	1571	1142
3	Color	Gas: blue-colored Dissolved: purple-blue	Gas: colorless Dissolved: light blue
4	Boiling point (101 kPa), °C	−112	−183
5	Melting point, °C	−192.7	−218.8
6	Solubility in water, mg/l		
	0 °C	1090	14.6
	10 °C	780	11.3
	20 °C	570	9.1
	30 °C	400	7.6
	40 °C	270	6.5
	50 °C	190	5.6
7	Oxydation potential, eV	2.07	1.23
8	Ozone Half-life, Gaseous	Dissolved in water	
	3 months at −50 °C	30 min at 15 °C	
	18 days at −35 °C	20 min at 20 °C	
	8 days at −25 °C	15 min at 25 °C	
	3 days at 20 °C	12 min at 30 °C	
	1.5 s at 250 °C	8 min at 35 °C	

Summarized from [48, 73, 74, 93]

When ozone is in contact with water, several types of oxidizing agents will be produced which will compete for the substrate. In Eq. (1), more HO_3 will be formed than OH*, even though oxidation using dissolved ozone tends to take place more slowly than oxidation using hydroxyl free radicals. On the other hand, at an acidic pH, oxidation using hydroxyl free radicals tends to be small and substrate oxidation is dominated by dissolved ozone as in Eq. (2). On the other hand, at alkaline pH, UV exposure, and the addition of a catalyst, free hydroxyl will dominate as in Eq. (3). In alkaline conditions, the decomposition of ozone in water can also be described in Eqs. (4)–(10) [37].

$$OH^- + O_3 \rightarrow O_2 + HO_2{}^- \overset{H^+}{\leftrightarrow} H_2O_2 \quad (4)$$

$$HO_2^- + O_3 \rightarrow HO_2{}^* + O_3{}^{*-} \quad (5)$$

$$HO_2^* \leftrightarrow H^+ + O_2{}^{*-} \tag{6}$$

$$O_2^{*-} + O_3 \rightarrow O_2 + O_3^{*-} \tag{7}$$

$$O_3^{*-} + H^+ \rightarrow HO_3{}^* \tag{8}$$

$$HO_3^* \rightarrow HO^* + O_2 \tag{9}$$

$$HO^* + O_3 \leftrightarrow HO_2{}^* + O_2 \tag{10}$$

Ozone can react with either and/or both modes in aqueous solution as direct oxidation of compounds by molecular ozone and/or oxidation of compounds by hydroxyl free radicals produced during the decomposition of ozone. Ozone also produces no by-products in the media, very effective in various applications such as oxidizing agents, disinfectants, color removers, odorants, tastes, etc. The biggest drawback of this oxidizing agent is that it must be produced where the reaction takes place so that a production system is needed at the place of use which causes high costs [35, 62]. To carry out oxidation using ozone, the ozone must be dissolved into the water. To get good oxidation results, the ozone level in the water must be kept as high as possible. The solubility of ozone is difficult to predict compared to other gases because the solubility of ozone is influenced by several factors such as temperature, pH, and other solutes. Strategies that can be done to increase the solubility of ozone in water include:

(a) Increase the concentration of ozone in the air
(b) Increase gas pressure
(c) Lowering the temperature of the liquid
(d) Increases pH
(e) Make contact with UV.

Ozone decomposes spontaneously through a series of mechanisms. The exact mechanism and reactions associated have not been established but many researchers proposed several models [6, 12, 19, 38, 98]. Ozone can be decomposed to form free radicals (OH*/hydroxy radicals) which have a very high oxidation potential of 2.8 V, so they act as a stronger oxidizing agent than ozone. Therefore, the ozonation process in water always involves two species, i.e., ozone (direct oxidation) and OH* (indirect oxidation). Hydroxyl free radicals forms are believed as one of the intermediate products that directly react with compounds in water, therefore, ozone demands are associated with the following:

(a) The presence of scavenger compounds, carbonate or bicarbonate ions (usually measured as alkalinity), will react with hydroxyl radicals to form carbonate radicals.

$$HCO_3^- + OH^* \rightarrow CO_3^{*-} + H_2O \quad (11)$$

(b) The presence of natural organic compounds (NOM) in water will trigger an oxidation reaction to form aldehydes, organic acids, and ketoacids. These oxidation by-products are generally more amenable to biological degradation and are one of the strategies applied to increase the biodegradability of compounds (BOD/COD ratio) in wastewater treatment plants. Meanwhile, synthetic organic compounds can be oxidized under favorable conditions. In the case of mineralization of such compounds, hydroxyl free radicals oxidation should be the predominant pathway in the process such as in advanced oxidation processes (AOPs), e.g., O_3/UV, Fenton, and/or O_3/H_2O_2.

(c) Oxidation of bromide ion leads to the formation of bromate ion, hypobromite ion, hypobromous acid, bromamines, and brominated organics.

1.3 Ozone Generation Methods

Probably the first known scientist that notice and report the presence of the specific odor of the air in the neighborhood of his electrostatic generator when subjected to the passage of electric sparks, which will later be named ozone, is a Dutch chemist Martinus van Marum in 1785. The passage of a high voltage, alternating electric discharge through a gas stream containing oxygen will result in the breakdown of the molecular oxygen, to atomic oxygen. Some of the oxygen atoms liberated can react with oxygen to form ozone, while others simply recombine to form oxygen. This process is known as one of the most popular ways to produce ozone gas. The three most popular methods of generating ozone are corona discharge, UV lamp, and cold plasma.

Corona Discharge

Corona discharge in dry process gas containing oxygen is currently the most widely used method of ozone generation for water treatment. Corona occurs due to accelerated ionization events between the two electrodes caused by a high enough electric field. If two electrons are given a high enough voltage, this causes the electric field between the two electrodes to be high enough to be able to move the electrons between the two electrodes. The movement of electrons allows electrons to collide with free molecules. The collision causes the free molecule to have sufficient energy to release its outer electron. The collision event will produce two new electrons, namely electrons that hit the molecule and electrons that come out of the molecule. Because the two electrons are still under the pressure of the electric field, the two electrons will move and collide with other free molecules [27]. In the process of ozone formation, the corona is formed due to the ionization of oxygen. There are two electrodes with different voltages, one is a high-voltage electrode and the other is a low-voltage electrode. These electrodes are connected to a high-voltage source. The two electrodes are separated by a dielectric medium and a narrow discharge gap

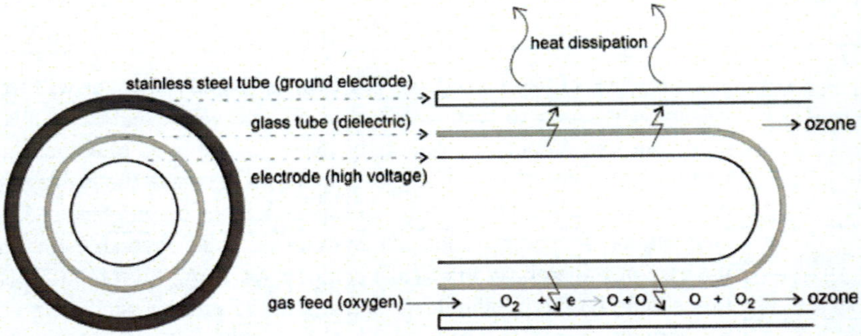

Fig. 2 Ozone generation by electrical discharge method

is provided. Schematic of ozone formation by corona discharge method is shown in Fig. 2.

The presence of a high electric field inside the electrode causes the ionization of air-containing oxygen, which flows in the electrode. The movement of electrons allows the collision of electrons with oxygen molecules and will produce two oxygen atoms (O). Furthermore, these oxygen atoms will naturally collide with the surrounding oxygen molecules to form ozone. The reaction equation for the formation of ozone can be seen in Eqs. (12 and 13).

$$e^- + O_2 \rightarrow 2O + e^- \quad (12)$$

$$O + O_2 \rightarrow O_3 \quad (13)$$

This method is the most widely used method of ozone formation in various industrial activities because it has advantages such as high ozone productivity, does not require complicated maintenance, and is easy to apply. The drawback is that the amount of energy consumed is quite large while the concentration of ozone produced is low, so this technology is considered expensive. Ozone generation by electrical discharge produces heat, where excessive heat can cause decomposition of ozone in the product gas. This makes heat dissipation an important part of the ozone generator unit and must be carried out as quickly and efficiently as possible. Heat can be removed by using heat sinks, water coolers, and/or air coolers.

There are two electrodes in the corona discharge, a low-voltage electrode (ground) and a high-voltage electrode, separated by a dielectric medium in a narrow discharge gap. When the electrons have sufficient energy to separate the oxygen molecules, a certain fraction of these collisions occur and ozone molecules can form from each oxygen atom. The efficiency of ozone production by corona discharge depends on the strength of the micro discharges which is affected by several factors such as the gap width, gas pressure, dielectric and metallic electrode properties, power supply, and the presence of moisture. In a weak discharge, most of the energy is consumed by the ions, while in a stronger discharge, almost all the energy of the release is transferred to the electrons responsible for the formation of ozone. The optimum is a compromise that avoids energy loss for the ions but at the same time obtains a reasonable conversion efficiency of oxygen atoms to ozone. If air is used as the feed gas, it must be dry because the humid air produces nitrogen oxides in the ozone generator which will form nitric acid and will corrode the generator, requiring frequent maintenance. If air is passed through the generator as the feed gas, 1–3% ozone is created; when using high purity oxygen can produce as high as 16% ozone [74].

Ultraviolet Light Lamp

This process of ozone generation is similar to how the ultraviolet from the sun radiation splits O2 to form individual oxygen atoms. Ultraviolet light changes oxygen into ozone when a wavelength at 254 nm hits an oxygen atom. The oxygen molecule splits into two atoms (O) which combine with another oxygen molecule (O_2) to form ozone (O_3). Ultraviolet light occurs naturally through the sun rays, but this process is considered to be less efficient than corona discharge. Ultraviolet lamps have been used for decades to produce ozone. This lamp emits UV light at 185 nanometers (nm). Light is measured on a scale called the electromagnetic spectrum and the increments are referred to as nanometers. When exposed to UV light, oxygen molecules in the ground state absorbs light energy and dissociate to a degree that depends on the specific energy and wavelength of the absorbed light. The oxygen atom then reacts with other oxygen molecules to form ozone (Fig. 3). Because of present technologies with mercury-based UV emission lamps, the 254 nm wavelength is transmitted along with the 185 nm wavelength, and photolysis of ozone is simultaneous with its generation. Moreover, the relative emission intensity is 5–10 times higher at 254 nm compared to the 185 nm wavelength. The advantages of using UV Light include a lower cost than a corona discharge, it is simpler to assemble and use, and the ozone output using UV Light is less affected by humidity.

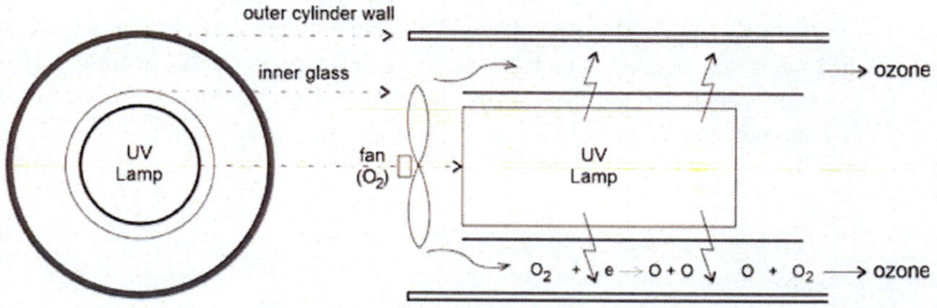

Fig. 3 Photochemical ozone generation using a tubular UV lamp with a cylindrical outer container wall

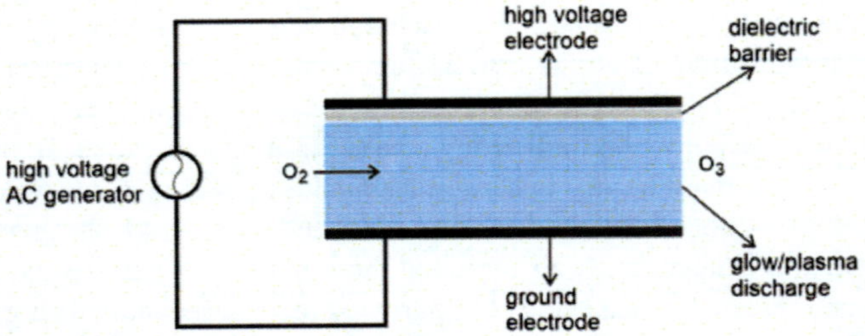

Fig. 4 Schematic of a cold plasma dielectric ozone generator

Cold plasma

Plasma can be called the fourth phase element in nature after solid, liquid, and gas phases. In contrast to the normal gas phase, plasma contains gas where the components of the atomic nucleus (ions) and electrons have been separated due to the energy received and have reactive properties. Plasma can form naturally as happens in the sun or the elements of the stars in space. Plasma can also be formed by providing high energy into the gas medium which makes the gas undergo a dissociation process and an ionization process. Depending on the amount of energy transferred, both processes will result in the transformation of neutral gas into highly reactive negatively and positively charged particles or ions, either partially or completely transformed. Based on the temperature, plasma can be categorized into high-temperature plasma (thermal/equilibrium plasma) and low-temperature plasma (cold plasma/non-equilibrium) [45]. It is a similar design to a corona discharge tube, the difference being that the anode and cathode of this cold plasma are encased in a glass rod filled with noble gases. In this design, the voltage jumps between the anode and cathode rods forming an electrostatic or "plasma" field. The advantage of the cold plasma system is that no heat release is given to the gas as it passes through the electrostatic field. Cold plasma technology has a very long service life due to this design feature. Plasma is a gas that is ionized in an incandescent discharge either partially or completely [44]. This phenomenon can occur when there is a very high potential difference between the two electrodes [61]. Gas ionization that occurs will form free electrons which will collide with gas ions to produce radical active species (OH·, O·, H·), molecules (H_2O_2 and O_3), and UV light [44, 56, 102] (Fig. 4).

2 Ozone for Dye Removal

2.1 *Process and Mechanisms*

Ozone can be formed naturally or human-made. Naturally, ozone can be developed through ultraviolet radiation (UV) rays from sunlight which can decompose oxygen gas in free air. The oxygen molecule breaks down into two oxygen atoms (O*), which then naturally collide with the oxygen gas molecules around them to form ozone (O_3). Artificially, ozone can be developed through some different processes, e.g., through the collision process and light absorption. Ozone can be produced through the collision process through several methods such as Dielectric Barrier Discharge Plasma (DBDP), Corona Discharge, and electrolysis. Meanwhile, through light absorption, ozone can be formed through methods such as ultraviolet radiation. Ozone will experience a decomposition of concentration during the ozone process due to reactions between radicals and non-radical compounds. Ozonation is capable of producing oxidizing hydroxyl radicals, which can decompose organic pollutants in wastewater. The potential for radical oxidation is very high, which means that reactivity can occur with contaminants. Reactive oxygen species (ROS) such as HO•, O_2• –/HO_2•, H_2O_2 are also formed and contribute to redox processes enabling the transformation of pollutants [53]. Radical reactions of organic pollutants depend on chemicals in the waste. Organic components can promote the formation of unstable radical oxidation, which can be easily oxidized to H_2O, CO_2, and acids [53].

Most of the impurities contained in textile wastewater are dyestuffs, especially synthetic dyes. Synthetic dyes are molecules with a delocalized electron system and contain two groups, namely chromophore and auxochrome. Chromophores function as electron acceptors, while auxochromes are electron donors that regulate solubility and color. The important chromophore groups are azo group (−N = N−), carbonyl group (−C = O), ethylene group (−C = C−), and nitro group (−NO_2) which can cause color. While some important auxochrome groups such as –NH_2, −COOH, −SO_3H, and –OH are polar, so they can dissolve in water [69]. Currently, there are various types of synthetic dyes whose use is adjusted to the type of fiber to be dyed, the desired color resistance, other technical and economic factors. The classification of textile dyes based on the method of dyeing is presented in Table 3.

The classification of dyestuffs in Fig. 5 can determine the treatment method selected. For dissolved dyes, physical separation methods are relatively difficult, so chemical, AOPs or biological methods are mostly applied to this type of dye. In contrast to insoluble dyes, absorption or physical methods can be applied with high efficiency. The textile industry is one of the industries that produce wastewater with high non-biodegradable organic content. In general, to reduce degradable organic pollutants, the technology applied is biological wastewater treatment technology, such as activated sludge processes, aerated lagoon, moving bad biofilm reactor, anaerobic–aerobic biofilter, or trickling filter. Meanwhile, wastewater that contains pollutants, long-chain dyes commonly used in textile industries, is such as azo compounds,

Table 3 Type, application, and characteristics of dyes

No.	Type of dyes	Application	Characteristics
1	Direct dyes	Viscose, Cotton	Direct dye having affinity with cellulose fibers, dyeing is carried out directly in solution with suitable additives
2	Mordant dyes	Cotton, Wool, Silk	Mordant dyes have a weak bond with the fiber, the dyeing process is usually done by adding chromium to the dye to form a metal complex
3	Reactive dyes	Viscose, Nylon, Cotton, Wool, Silk	Reactive dyes have a reactive group that can form strong covalent bonds with cellulose, protein, polyamide, and polyester fibers, which can be carried out at low and high temperatures
4	Acid dyes	Nylon, Wool, Silk	Acid dyes have a strong bond with protein and polyamide fibers, dyeing is carried out under acidic conditions and directly added to the fiber
5	Basic dyes	Jute, Acrylic, Paper	Basic/cationic dyes have a strong affinity with protein fibers, dyeing is carried out under alkaline conditions and directly added to the fiber
6	Disperse dyes	Nylon, Acrylic, Polyester, Acetate, and Tri-acetate fiber	Disperse dyes were initially developed to color secondary cellulose acetate fibers, relatively insoluble in water and are prepared for staining by grinding the dye into fine particles in the presence of a dispersing agent
7	Sulfur dye	Viscose, Linen, Cotton	Sulfur dyes have a strong bond with cellulose fibers, the side group contains sulfur which is able to bind strongly to the fiber
8	Vat dyes	Viscose, Rayon, Linen, Cotton, Wool, Silk	Vat dyes are based on the natural indigo dye, which is synthetically produced. Contain a water-insoluble complex polycyclic molecules based on the quinone structure and are known for their characteristic of requiring a reducing agent to get absorbed

(continued)

Table 3 (continued)

9	Azoic dyes	Viscose, Cotton	Azoic dyes are composed of at least one azo group (–N = N–) attached to one or often two aromatic rings. Since azoic dyes are water insoluble, the water fastness of these dyes is excellent. Azoic dyes have high coloring properties, in the range of red to yellow

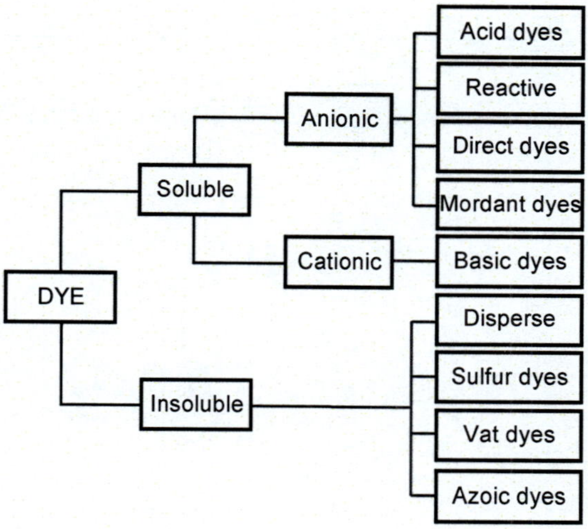

Fig. 5 Textile dyes classifications

which normal biological processes cannot quickly treat non-biodegradable wastewater. The ozonation process can degrade specific dyes well. Previous research has shown that three types of dyeing agents such as reactive black 5, reactive red-239, CI Reactive Blue 19, and Reactive Orange 16, require several stages in the degradation process with ozone (Figs. 6, 7, 8 and 9). The degradation of reactive black 5 dyes is CH_3COOH, H_2C_2O, HCOOH, and CO_2 [101]. Meanwhile, reactive red 239 produces CH_3COOH and CO_2 [22]. CI Reactive Blue 19 can be degraded with ozone to become CO_2, H_2O, and organic acids [25]. Overall, the desired result of the color degradation process with ozone is CO_2. The dye degradation process is quite difficult to explain practically in the field, one of the methods commonly used to determine the dye degradation process using ozonation is by analyzing the biodegradability index (BOD_5/COD).

The increase in the biodegradability index in textile wastewater treatment measures the increased likelihood of dye degradation. The research shows that ozonation as a pre-treatment step could increase the biodegradability index for all types of dyes wastewater or textile wastewater (Table 4). This increase in biodegradability

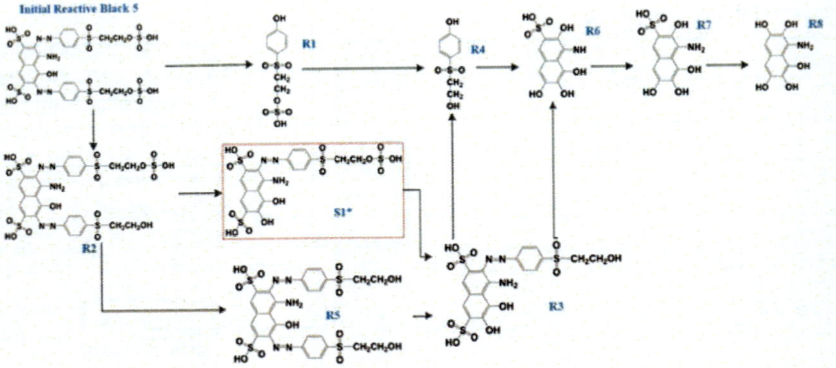

Fig. 6 The process of degradation of reactive black 5 dye with the ozonation process (adapted from [101])

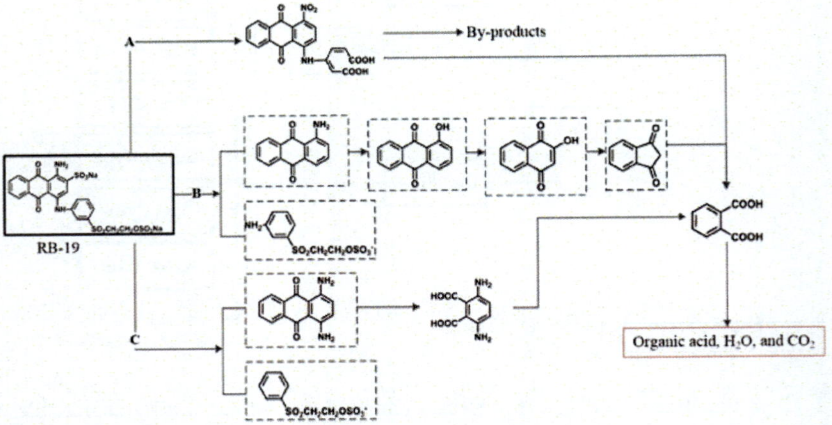

Fig. 7 The process of degradation of ci reactive blue 1 dye with the ozonation process (adapted from [25])

was achieved with moderate COD removal and short ozone times. In addition, the pre-treatment process is considered feasible for a biodegradability index greater than 0.3.

There are two main reactions in the degradation process of RB5 dye with the ozonation process, namely the detachment of the sulfonate group and the cleavage of the azo bond which occurs simultaneously with the increase in the OH group formed during the RB5 oxidation process. Specifically, the release of the sulfonate moiety from the parent compound leads to the formation of P2 (when one –O–SO3H group is cut off) and P5 (when both –O–SO3H groups are eliminated or degraded). Meanwhile, the cleavage of the azo group in the initial RB5 dye molecule will produce S1 and R1 compounds. S1 compounds can also be produced due to the

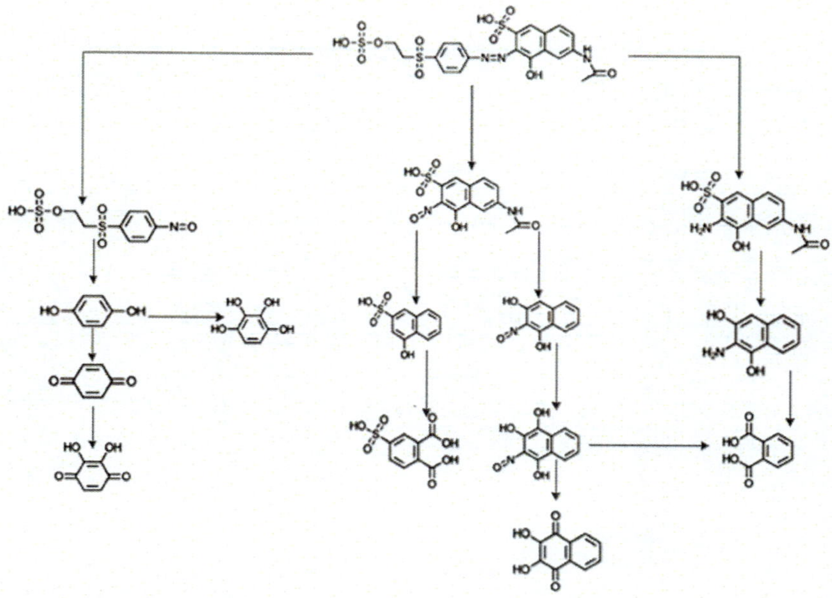

Fig. 8 The process of degradation of reactive orange 16 with the ozonation process (adapted from [15])

process of breaking the azo bond in R2. Compound R3 is formed from compound S1 through the removal/elimination of sulfonic groups or formed from compound R5 through breaking azo bonds and subsequent hydroxylation reactions. The same azo bond cleavage reaction can also occur at R3 with the formation of hydroxylated naphthalene compounds (R6) and alkylsulfonyl phenolic compounds (R4). Then, desulphonation and subsequent hydroxyl addition reactions to the R6 molecule result in the formation of R7, which can be converted more quickly to R8. Furthermore, R1 is converted to R4 through the release of a sulfonate group. In the end, further oxidation of R8 will produce organic acids with smaller molecular weights such as formic, acetic, and oxalic acids, as well as mineralization to carbon dioxide and water [100, 101].

The possible mechanism for the degradation of dyestuffs through the oxidation process with ozone as described in the Figure above generally consists of two main mechanisms, namely the mechanism of reductive severing of double bonds in dyes which causes color loss in the visible color absorbance spectrum. Degradation is continued by further oxidation mechanism of aromatic amine intermediate compounds which produce simpler degradation products in the form of aromatic and aliphatic acids which will eventually be mineralized. Possible mechanisms of dye degradation through oxidation by ozone are as follows [85]:

1. Cleavage of azo bonds (–N = N–)

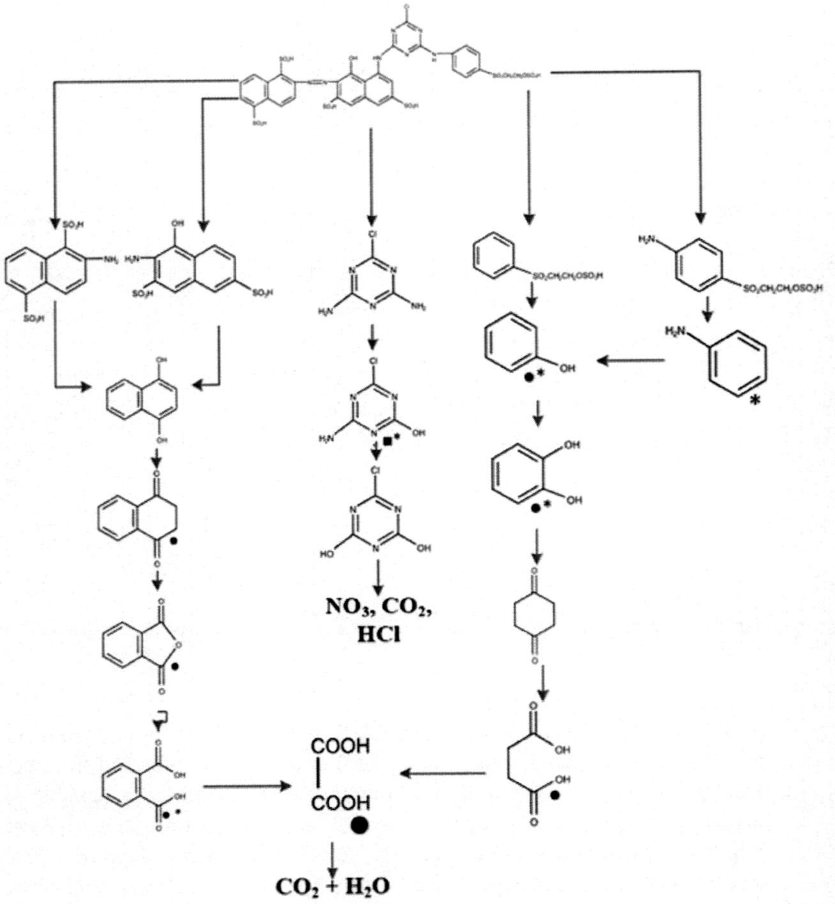

Fig. 9 The process of degradation of reactive red 239 dye with the ozonation process (adapted from [22])

The mechanism of reductive breaking of azo bonds in dye molecules appears to be the initial reaction to produce intermediate compounds in the oxidation of azo dyes.

2. Cleavage of C–C, C–N, and C–S bonds

The C–C and C–N bonds of the color chromophore groups and the C–S bonds between the aromatic ring and the sulfonate groups are broken by hydroxy radicals. This leads to further breakdown of the intermediate product.

3. Naphthalene ring termination

The intermediate compound formed from the breaking of the adsorbed azo bond undergoes a structural change.

Table 4 Results of increased biodegradability of dyes by ozonation

No.	Dye	Initial biodegradability	Biodegradability after ozone pre-treatment	Dosage	Contact time	References
1	Acid Red 14	–	±0.45	–	25 min	[95]
2	C.I. Reactive Blue 19	0.15	0.33	88.8 mg/min	10 min	[25]
3	Polypropylene and polyester yarn dyeing industry	0.18	0.32	26 mg/L	75 min	[91]
4	Reactive Black 5	0.056	0.46	2 g/h	270 min	[23]
5	Reactive Black 5	–	0.27	15 mg/L	60 min	[4]

4. Benzene ring termination

With further reaction on azo dye, benzene derivatives are degraded into smaller molecular organic acids. Hydroxy radicals also attack COOH compounds to produce simpler organic acids such as maleic acid, acetic acid, oxalic acid, and formic acid.

5. Mineralization toward the final product in the form of carbon dioxide and water.

2.2 *Factors Affecting Dye Removal*

The ozone process as a pre-treatment shows an increase in the biological degradation of organic pollutants. The pre-treatment process can improve the quality of waste to meet existing quality standards [5, 14]. The main drawback of the ozonation process is its short half-life, and it can be increased if the dye is present in acidic conditions, so adjusting the pH of the textile waste is necessary [2]. In the ozone operation, the short half-life can be caused by a diffuser that is too large, resulting in low ozone solubility. This can be done utilizing porous glass or metal armor, solid catalysts, stirring, contact, and increased retention time by large bubble columns or diffusers to overcome low solubility [11]. The wastewater that has just come out of the textile industry has a relatively high temperature, around 70–80. This high temperature is also an obstacle in the textile industry because this wastewater must be cooled first in the cooling tower before being processed further. Theoretically, the solubility of ozone in water is getting lower due to the increase in temperature. The solubility of ozone in the liquid phase decreases at a temperature of 43 °C. In chemical reactions such as that occurring in the color decomposition reaction using ozone, an increase in temperature every 10 °C will speed up the reaction speed twice as fast. Thus there is an optimum temperature where the solubility of ozone and the reaction rate of decomposition is high. The effect of temperature on ozonization must be observed for the consequences of both products. However, the presence of other factors such as homogeneous and heterogeneous catalysts, lighting, initiators, and inhibitors in the reaction flux will determine the temperature dependence in very complex modes [55].

The ozone decomposition process is faster in alkaline conditions, pH > 8.5, so continuous monitoring of the textile waste pH is required. Increasing the pH value will increase the ozone decomposition in the water. For example, at a solution pH of 10, the half-life of ozone in water is less than 60 s. Oxidation of organic compounds can occur due to a combination of the reaction of the ozone molecule and the reaction with OH radicals [55]. The reaction between ozone and hydroxide ions can trigger the formation of superoxide anions of O_2 radicals and hydroperoxyl HO radicals [35]. Bicarbonate and carbonate play an essential role in forming OH radicals in the system. The reaction between OH radical and the carbonate or bicarbonate is the passive carbonate or bicarbonate radical, which has no further interaction with ozone or organic compounds. OH radicals reaction rate is usually 106–109 times faster than the appropriate reaction rate for ozone molecules. In an acidic environment, protons H^+ react with O_3 to form O_2 and H_2O, preventing the direct reaction of O_3 with pollutants as described in the equation [11].

Most textile dyes use long-chain organic compounds. Dyestuff is a complex compound that can be retained in a network of molecules. The dye is a combination of organic substances that are not far away, so the dye must consist of chromogen as the color carrier and Auxochrome as the binder between the color and the fiber. The ozonation mechanism for color removal is very complex, and its interpretation based on the composition of the intermediates and the final product is difficult to determine. The reaction rate constant is calculated using a simple mathematical model; the rate constant for ozone reaction with dyes is also calculated using the same model. The variation of ozone in the reactor output is significant for studying dye decomposition dynamics and ozonization kinetics. The first pseudo-order constant for dye decomposition shows a strong dependence on the hydrogen peroxide concentration for O_3/H_2O_2 and the O_3/H_2O_2 /UV process [7]. Decomposition of color by the ozonation process can reach the optimum condition influenced by the type of dye, initial concentration, the dose of ozone given, and detention time. Table 5 shows the results of the exclusion of various kinds of research. Generally, the color removal cannot reach 100%, and it seems that some studies only produce around 70% removal at 10–14 min of contact time.

3 Practical Application

3.1 Ozone System Configuration

System for generating and applying ozone to water and/or wastewater typically consist of five components:

Table 5 References to previous research regarding colour removal efficiency with ozonation process

No.	Dye	Initial dye concentration (mg/L)	Ozone dose	Dye removal (%)	Detention time (min)	Source
1	Acid Red 14	1500	5 g/h	93	25	[22]
2	C.I. Reactive Black 5 (RB5)	200	3.2 g/h	70.21	10	[21]
3	Direct Red 28	1500	5 g/h	92	25	[100, 101]
4	Methylene blue	10	46.32 mg/L	73.01	13	[31]
5	Methylene orange	10	0.83 mg/s	100	15–30	[31]
6	Methylene blue	1500	5 g/h	(93.5)	25	[82]
7	Reactive Black 5	1500	5 g/h	94	25	[95]
8	Reactive Red 239	50	40 and 20 mg/L	More than 95	4–12	[95]
9	Reactive Red 239	50	20 mg/L	100	20	[95]
10	Reactive Red 239	500	16.6 mg/min	90	90	[60]
11	Reactive Red X-3B	100	0.66 L/h	92	6	[1]
12	Reactive Yellow 176	500	16.6 mg/min	90	90	[96]
13	Reactive Black 5	100	40.88 mg/min	96.9	60–300	[86]
14	Acid Red 14	1500	–	93	25	[94]

(a) Electrical power generation,
(b) Ozone generation method,
(c) Feed gas preparation,
(d) Contacting of liquid with ozone, and
(e) Excess ozone destruction method.

The five basic components of an ozonation system, all of which must be taken into account when designing ozone to ensure effectiveness and safety at the same time. Instrumentation and controls systems can be added to ensure the effective and safe operation of the entire ozonation system. Electrical power generation is closely related to the type of ozone generation method that is usually categorized by the frequency of the power applied to the ozone generator. Low-frequency (50 or 60 Hz), medium-frequency (60–1000 Hz), and high-frequency (>1000 Hz)

Table 6 Comparison of low, medium, and high frequency ozone generators

Characteristics	Low frequency	Medium frequency	High frequency
Hertz (Hz)	50–60	60–1000	More than 1000
Level of electronics sophistication	Low	High	High
Turndown ratio	5:1	10:1	10:1
Cooling water required (L/kg ozone produced)	4.1–8.2	4.1–12.5	2.1– 8.2
Optimum cooling water differential	8°–10 °F	5°–8 °F	5°–8 °F
Typical application range	<225 kg/day	To 900 kg/day	To 900 kg/day
Optimum O_3 Production (%)	60–75	90–95	90–95
Operating concentrations			
– Air as feed gas (wt-%) – Oxygen as feed gas (wt-%)	0.5–1.5 2.0–5.0	1.0–2.5 2.0–12	1.0–2.5 2.0–12
Power requirements for air feed system, (kW-h/kg O_3)	11–15	11–15	11–15
Power required, (kW-h/kg O_3)			
– Air as feed gas – Oxygen as feed gas	17–26 8–13	17–26 8–13	17–26 8–13

Adapted from [93], with modifications

ozone generators can be found in the market this day. Low-frequency generators generally produced less heat compared to medium and high-frequency generators, although using medium or high-frequency generators tend to effectively produce ozone more than low generators. Table 6 summarized a comparison of the three types of generators.

Ozone can be generated on-site, either by corona discharge or ultraviolet (UV) radiation methods. With the UV radiation technique, usually low concentrations of ozone (below 0.1 wt%) can be generated, whereas, with corona discharge, ozone concentrations in the range of 0.5–2.5 wt% can be generated when dry air is fed to an ozone generator. When pure oxygen is used as the feed gas, ozone concentration of 2–5 wt % is produced typical for low output generators and can reach 10–12 wt % in general for large-scale applications by ozone generator manufacturers. Oxygen concentrators often replace air drying units to supply oxygen-enriched air to ozone generators to provide higher output and gas-phase ozone concentrations, thus avoiding the need for on-site oxygen production or storage facilities. In all cases, ozone is only partially soluble in water and must be in contact with water and/or wastewater to be treated in such a way as to maximize the transfer of ozone to a solution. For this purpose, many types of ozone contactors have been developed; all of them are effective for the designed water treatment purpose. However, as higher ozone concentrations are used, the contact system design becomes more critical due to the lower ozone gas-to-liquid ratio. Figure 10 shows the simplified ozone system configuration. In addition, the use of pure oxygen as feed gas can lead to oxygen saturation in the treated water which causes operational problems in following the

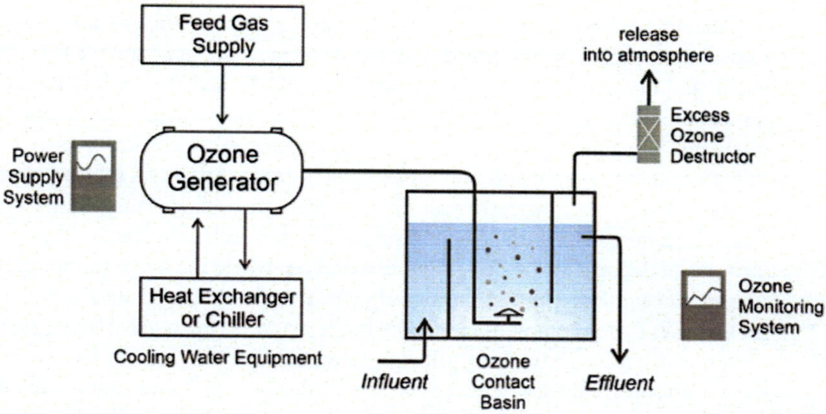

Fig. 10 Simplified typical ozone system configuration

treatment process and aesthetics in the distribution system [71]. Table 7 summarized comparison of the advantages and disadvantages of either air or oxygen as feed systems.

Table 7 Comparison of air and oxygen feed system in ozone generation

Feed gas	Advantages	Disadvantages
Air	Proven technology, simple and uncomplicated mechanical equipment, suitable both for small and large-scale installation systems	Low efficiency, used more energy per ozone volume produced, required gas handling and adjustment
High purity oxygen	Produce higher ozone concentration for approximately double for the same generator that uses air as the feed gas, suitable both for small and large-scale installation systems	Required oxygen concentrators, oxygen resistant material, and safety concerns
Liquid oxygen (LOX)	Simple to operate and to maintain, less equipment required, can store excess oxygen to meet peak demands, and suitable for small to medium scale installation systems	Higher cost for liquid oxygen, on-site storage including safety concerns, and potential loss of LOX in storage when not in use
Cryogenic O_2 generation	Equipment is similar to air preparation systems, can store excess oxygen to meet peak demands, and is feasible for large-scale installation systems	Higher capital cost, extensive gas handling equipment that is more complex than the LOX system, and sophisticated in its operation and maintenance

Adapted from [93], with modifications

The air feed system for an ozone generator is quite complicated because the air must be properly conditioned to prevent damage to the generator. The air must be clean and dry, with a dew point of −80° to −40° F) and free of impurities. An air preparation system usually consists of an air compressor, filter, dryer, and pressure regulator. For this purpose, an air dryer unit is needed to reduce or even eliminate the content of moisture and impurities in the free air. Air dryers that are often found in the industry are compressed air dryers, where the main unit is an air compressor and a compressed air storage tank. The compressed air that is dried will experience a process of decreasing the dew point temperature, where the water vapor contained in the air can condense so that the air output of the air dryer is relatively dry. One type of air dryer that is commonly available in the market is the regenerative desiccant air dryer, which uses the working principle of an adsorption system to remove moisture. Generally, desiccant dryers are supplied with dual towers/tubes containing moisture-absorbing substances such as silica gel, activated alumina, zeolite, or other water-absorbing materials. Water vapor is removed from the dryer by using an external heat source or by passing a fraction of dry air (between 10 and 30%) through a saturated tower/tube at reduced pressure. Another type of air dryer available in the market is refrigeration dryers and membrane air dryers.

Compressed air usually contains water in the form of liquids and vapors which are affected by local ambient air conditions. The 100 cfm compressor capacity and dryer-cooler combination, operating for 4000 h in typical climatic conditions produces approximately 8300 L of liquid condensate per year and will increase significantly with larger compressors or in warmer and humid climates. Water, in any form, must be removed for the system to run properly and efficiently. That's why dryers are so important to produce clean, dry air.

Air drying operations can range from trapping condensed water and preventing additional condensation of moisture to remove nearly all of the water present. The more water removed, the higher the cost. However, if too much water is left in the compressed air supply, more costs will be incurred in the future due to downtime, higher maintenance, corrosion, product failure, and premature equipment failure. A typical simplified air preparation system with several optional unit operations available is shown in Fig. 11. Aftercoolers reduce the temperature and water content of compressed air. Bulk liquid separators remove liquid condensed in the distribution system. Particulate filter removes solid-particle impurities down to 1–5 microns in size and separates bulk liquid from the air stream. Coalescing filters remove liquid aerosols and particle impurities down to 0.01 micron in size [13].

Ozone contact system options include pressurized gas-to-liquid mass transfer processes, the use of ceramic or stainless steel fine bubble diffusers, static mixers, or venturi injectors that can be used to mix ozone gas with the water and/or wastewater to be treated (Table 8, Fig. 12). In small systems, small in-line injectors and pressure reaction vessels replace large concrete, cost-effective 20 foot deep bubble diffuser tanks on a large scale. Once dissolved in water, ozone is now available to act on water contaminants to achieve the purpose of disinfection and/or oxidation of the intended pollutant. The final component required in ozone treatment is a unit for the destruction of excess ozone which is always present in the off-gas contactor

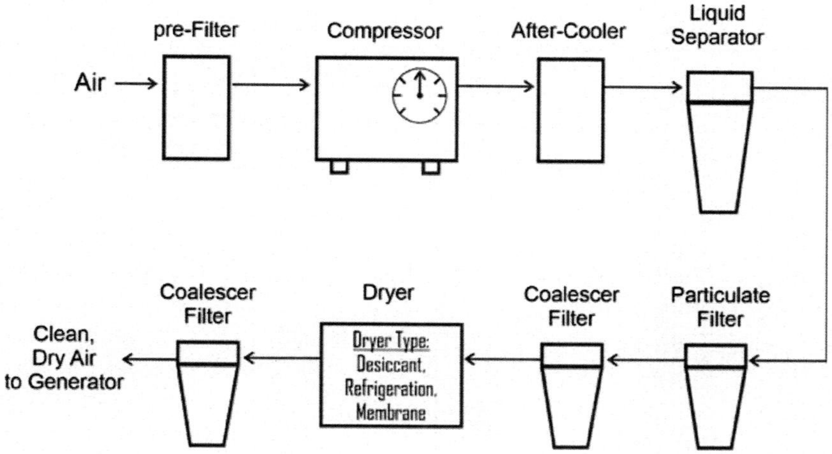

Fig. 11 Simplified typical air preparation system

Table 8 Type of ozone contact system and apparatus

Diffuser type	Characteristics
Stainless steel fine bubble	• Material: SS304, SS316 • Size Pore: 2.5–160 μm • Airflow per band: 15–63 Nm^3/h • Withstand Pressure: 0.5–3 Mpa • Temperature Resistance: 600–1000 °C
Ceramic fine bubble	• Airflow per unit: 1.2–4 Nm^3/h • Effective Surface Area per unit: 0.5–1 m^2 • Standard Gas Transfer Rate: 0.2–0.6 kg O_2/h. unit
Static mixer (Teflon/Fiberglass/Resin)	A static mixer based on the turbulent mixing principle, the mixing effect is mainly achieved by internal rotation and the shearing of the layers of fluid at the point where the direction of rotation is reversed. When small rates of shearing are required, a pitch between two rotations can be programmed. This pitch delays the rotation process and causes an additional mixing effect. Common advantages: no moving parts, no mechanical seals, less to no maintenance, no leakage, predictable homogeneity, low energy dissipation, easy scale-up, and in-line processing
Jet mixer/Aerator	Jet Mixer uses a low shear rotor–stator mixing head to create a unique tank flow pattern efficiently and uniformly. Common advantages: eliminates vortexing-stratification and dead mixing area. Standard Gas Transfer Rate: 1.5–40 kg O_2/h

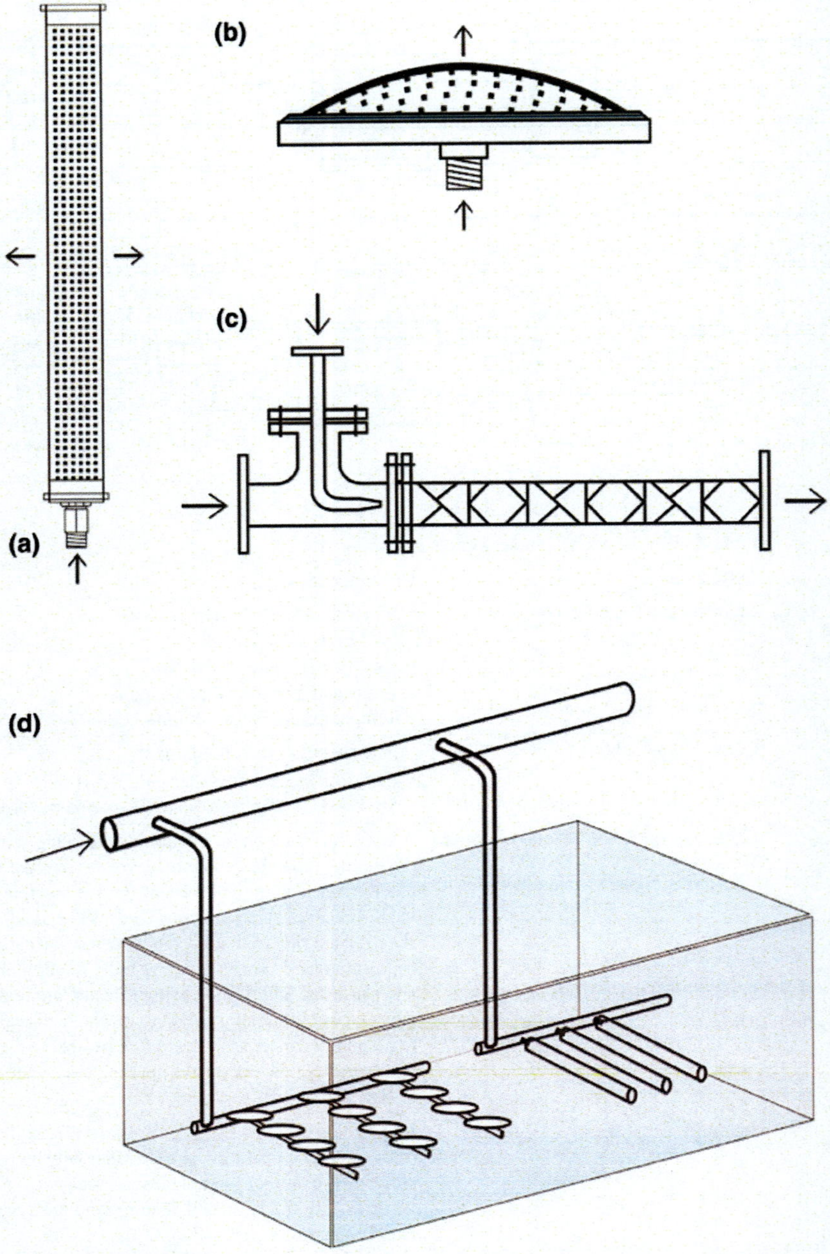

Fig. 12 Schematic of various ozone diffusers **a** stainless steel fine bubble, **b** ceramic fine bubble, **c** static mixer, and **d** unit placement perspective

when generated by corona discharges. Without an effective off-gas ozone-destroying unit, this excess ozone will expose people and surrounding materials to a hazardous oxidizing environment. Destruction of the ozone off-gas contactor is easily accomplished thermally (370 °C), thermally catalytically, or by passing through the catalyst medium alone.

3.2 *Integration with Existing WWTP Process*

The textile industry usually consists of 4 processes that produce wastewater: desizing, bleaching, dyeing, and mercerization. Based on [39], the effluent of the bleaching and dyeing process needs to be given more attention related to the biodegradability value. At first, the raw material is subjected to a singeing process to burn the hairs on the fabric surface, and then a desizing process is carried out to remove starch. Starch must be removed from the fabric to not interfere with the following process because it will block the absorption of the substances used in the process. The starch removal process aims to convert water-insoluble starch into water-soluble glucose and maltose compounds. Hydrolysis can occur in hot water, acidic solutions, and alkaline solutions. For starch that cannot be hydrolyzed but is easily oxidized, an oxidizing agent can be used. In addition, it can also use enzymes that function as catalysts and convert starch into water-soluble sugars [57]. Usually, this process is wastewater with good biodegradability for the biological treatment of wastewater [39].

In the bleaching process, auxiliary chemicals often used are sodium hypochlorite, sodium silicate, hydrogen peroxide, and organic stabilizers such as enzymes as bleaching agents. The use of chloride or peroxides causes inhibition problems, and these problems cause acidic pH. The dyeing process is the process of giving color evenly to the fabric. Before dyeing, several stages of dirt removal were performed, including desizing (removing starch that was still attached to the fabric), scouring (removing dirt adhering to the fabric), singeing (removing projecting fiber), and washing. After washing, the preparation stage is carried out, namely the selection of the dye, which is then carried out by dissolving and entering the textile materials into the solution [57]. The coloring process requires a substantial volume of water compared to other processes [81]. Not only immersion, but rinsing also involves a lot of water. In the staining process, many chemicals such as metals, salts, surfactants, organic matter, sulfides are added to the fiber. Wastewater resulting from the bleaching and dyeing processes must be re-analyzed to see the biodegradability value. If it meets the biodegradability index above 0.3, it can be continued with biological treatment. Meanwhile, if it does not meet the biodegradability index, it is necessary to do pre-treatment with ozonation (Fig. 13).

Dye wastewater that is disposed of into the environment will cause environmental disturbances such as eutrophication, production of hazardous, and toxic by-products through chemical reactions such as oxidation, hydrolysis, and adverse ecological,

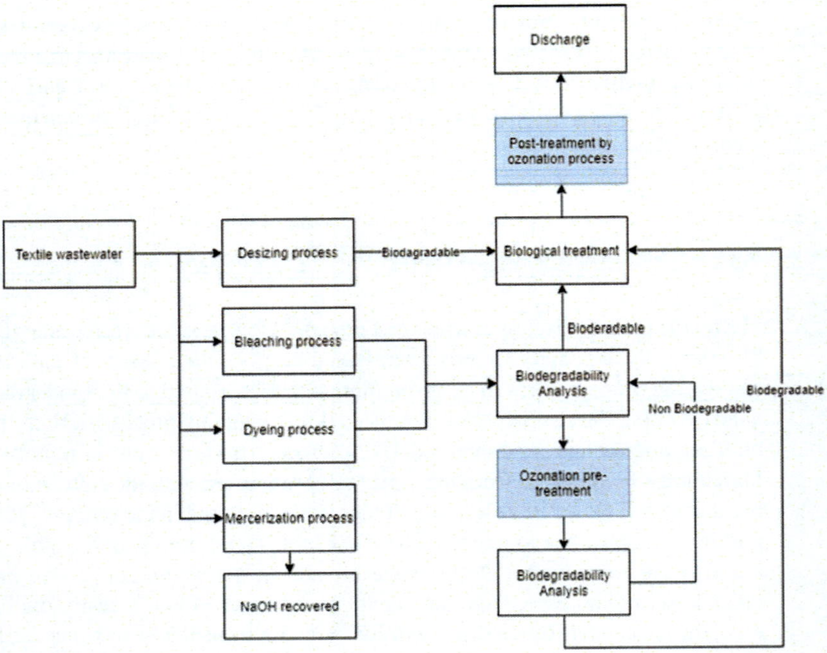

Fig. 13 Recommended textile wastewater management scheme

aesthetic effects, water quality degradation [59]. Therefore, the treatment of contaminated wastewater with color has an essential role in environmental pollution and the practical application of technology. These colors usually will not be treated by aerobic biodegradation and cannot quickly be processed by conventional biological treatment processes [20, 28]. The reason for causing the color to be non-biodegradable waste is the lack of enzymes needed for color degradation in the environment. Physical and chemical treatment methods such as adsorption, coagulation, sedimentation, and oxidation can treat colored wastewater. However, the application of this method requires advanced processing, high costs, and a lack of complete conversion to inorganic compounds [20]. Reactive colors containing poly-aromatic molecules are highly soluble in water, making absorption more difficult by absorbent agents. In several studies, activated sludge is often used as a sorbent to remove reactive colors, where the absorption capacity depends on environmental conditions, pH, type, and color concentration. High operational costs such as restoring the sorbent or treating the resulting solid waste are considered the most important limitations.

The results of previous studies showed color removal with ozone pre-treatment was higher than without pre-treatment (Table 9, Fig. 14). The finding confirms that color removal by biological treatment alone is not effective enough compared to biological treatment with ozonation pre-treatment regardless of the type of biological treatment used.

Table 9 Biological treatment types integrated with ozone pre-treatment process for wastewater containing dye

Dye	Biological treatment	Color removal with ozone pre-treatment (%)	Color removal without ozone pre-treatment (%)	References
Reactive Black 5	Upflow anaerobic sludge blanket (UASB)	94	–	[96]
Reactive Orange 16	Moving-bed biofilm reactor (MBBR)	More than 97	0	[15]
Reactive Orange 16	Anaerobic moving-bed biofilm reactor (MBBR)	61 ± 18	–	[16]
Real Textile Wastewater	Anoxic-aerobic activated sludge	76.60	30	[87]
Remazol Black 5	Moving-bed biofilm reactor (MBBR)	86.74	68.60	[65]
Remazol Black B	Activated sludge (biomass from a municipal wastewater treatment plant)	91	59.30	[90]

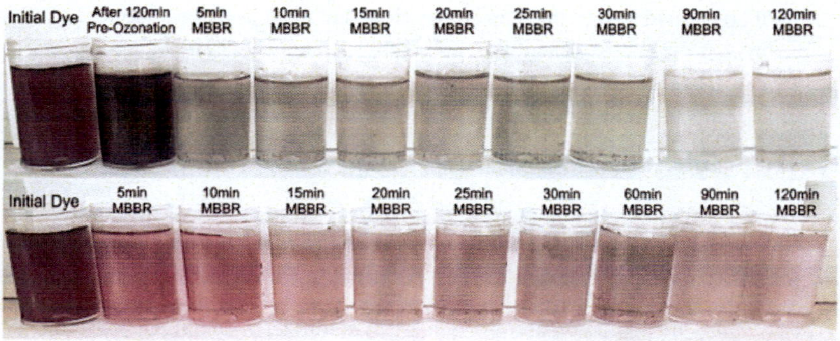

Fig. 14 Visual image of reactive black 5 color gradation with an initial concentration of 50 mg dye/l after 5–120 min degradation processes in the MBBR reactor. The upper image shows effluent color of MBBR reactor with pre-treatment using ozonation for 120 min with 3.81 mg O_3/min (color removal eff. 86.74%), and the lower image shows effluent color of MBBR reactor without pre-treatment (color removal eff. 68.60%). *Source* modified from [66]

The mechanism of color removal by ozone can occur based on two mechanisms, i.e., directly, where ozone molecule O_3 oxidizes dyes, or indirectly, where ozone decomposes into OH radicals which have a higher oxidation potential of 2.8 eV compared with 2.07 eV of ozone. The direct mechanism mostly occurs at the pH of the solution which is acidic and neutral. Meanwhile, the indirect reaction occurs at the pH of the alkaline solution because ozone will decompose into hydroxyl radicals (OH radicals) with a higher reaction rate [18, 100, 101].

The first stage of the color degradation process in the ozonation reaction that may occur in most dyes is the ozone reaction with a single chromophore group such as an azo group or a carbon double bond (C = C) in an aromatic ring [18]. Ozone can cleave the conjugate bonds of azo dyes, and convert dye compounds with high molecular weight to compound with smaller molecular weight such as organic acids, and cause the color to deteriorate. Ozone and OH radicals, both in the dissolved phase, are able to break aromatic rings and oxidize inorganic and organic compounds [83].

The initial design of textile WWTPs in Indonesia was not specifically designed to eliminate color because before 2019, color parameters were not regulated by national quality standards. Knowing the ineffectiveness of the current WWTP in removing color, the textile industry must modify and improve the WWTP system by adding a decolorization unit combined with the previously existing WWTP unit. Based on the poor performance in decolorization, an appropriate solution is needed to improve the WWTP so that the effluent released complies with quality standards by evaluating the existing WWTP unit and planning new alternative designs that are effective and able to provide benefits to the textile industry from both technical and financial aspects. Alternative textile wastewater treatment technology that can be used is the addition of an ozonation unit that can be positioned in front as a pre-treatment of a biological process or post-treatment as a polishing system (Fig. 15).

4 Concluding Remarks

Textile wastewater treatment can be done chemically, physically, biologically or a combination of the three. Chemical treatment is carried out by coagulation, flocculation, and neutralization. The coagulation and flocculation processes are carried out by adding coagulants and flocculants to stabilize colloidal particles and suspended solids to form flocs that can settle under gravity. The formed flocs are able to absorb the color and then settle as sludge. Physical wastewater treatment can be done by adsorption, filtration, and sedimentation. Adsorption is done by adding adsorbent, activated carbon or the like. Filtration is a solid–liquid separation process through a filter. Sedimentation is a solid–liquid separation process by depositing suspended particles in the presence of gravity. Biological wastewater treatment is the utilization of microorganism activity to decompose organic materials contained in wastewater. Each of the three processing methods above has advantages and disadvantages. Chemical wastewater treatment will produce large amounts of sludge, thus creating new problems for the handling of the sludge. The use of activated carbon in the

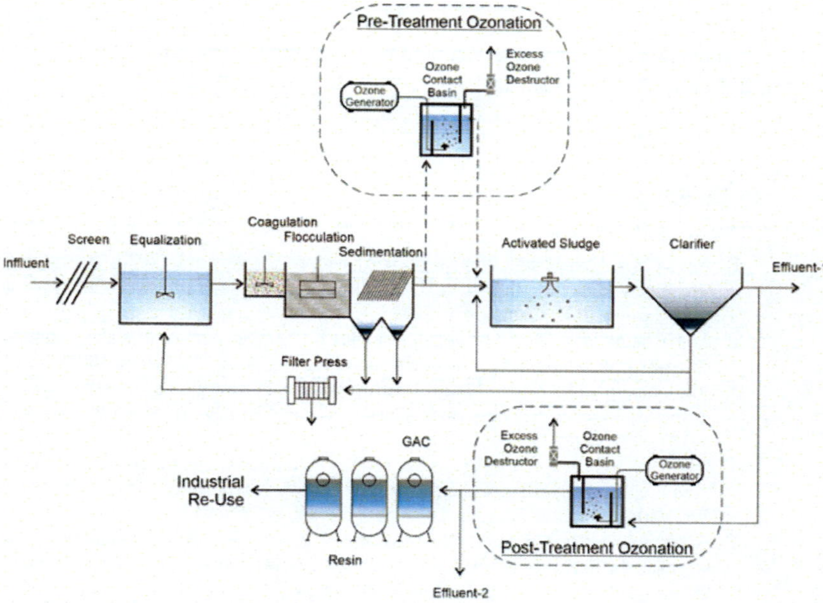

Fig. 15 Modification of the existing textile WWTP option with the addition of an ozone system as pre- and/or post-treatment

treatment of waste-containing dyes results in a high percentage of dye reduction, but the price of activated carbon is relatively expensive and will also increase the cost of equipment for the regeneration of the activated carbon. The use of ozone in textile wastewater treatment applications has several advantages. For starters, it has a very high oxidizing power, so it only takes a few minutes of contact time (CT) to oxidize the impurities in the liquid waste. Second, ozone makes oxygen more soluble in water. Third, by using ozone, the WWTP almost eliminates the need for chemicals, except for those absolutely necessary in the sedimentation process (coagulation and flocculation). Fourth, ozone can react with organic compounds dissolved in water (TOC, total organic compounds) and remove them very effectively, thus improving the performance of biological processes. Fifth, because ozone decomposes quickly in water (on the order of 1–15 minutes), the residual effects of ozone are easily overcome. On the other hand, some of the weaknesses or obstacles of the application of this ozone technology include, ozone is a toxic gas (TLV from OSHA is 0.1 ppm), whose toxicity is directly proportional to concentration and exposure time. The installation cost of ozonation is high due to its energy consumption and is relatively more complex than other oxidation processes. An ozone destruction catalyst is highly recommended to be installed at the outlet, to prevent ozone poisoning and fires hazard. Can produce unwanted carbonyl compounds (aldehydes and ketones), especially if the contact time is too short. Recently, the application of ozone to both air, water, and wastewater treatment as a disinfection agent has received great attention worldwide. New technological research related to the development of disinfection chambers using ozone as a decontamination agent is growing, largely driven by the 2019 COVID-19 pandemic.

References

1. Adelin MA, Gunawan G, Nur M, Haris A, Widodo DS, Suyati L (2020) Ozonation of methylene blue and its fate study using LC-MS/MS. In: Journal of physics: conference series, vol 1524, no 1. IOP Publishing, p 012079
2. Ahmet B, Ayfer Y, Doris L, Nese N, Antonius K (2003) Ozonation of high strength segregated effluents from a woollen textile dyeing and finishing plant. Dyes Pigm 58:93–98
3. Alimohammadi M, Naderi M (2021) Effectiveness of ozone gas on airborne virus inactivation in enclosed spaces: a review study. Ozone Sci Eng 43(1):21–31. https://doi.org/10.1080/01919512.2020.1822149
4. Alvares ABC, Diaper C, Parson SA (2001) Partial oxidation of hydrolysed and unhydrolysed textile azo dyes by ozone and the effect on biodegradability. Process Saf Environ Prot 79(2):103–108. https://doi.org/10.1205/09575820151095184
5. Baban A, Yediler A, Lienert D, Kemerdere N, Kettrup A (2003) Ozonation of high strength segregated effluents from a woollen textile dyeing and finishing plant. Dyes Pigment 58(2):93–98. https://doi.org/10.1016/S0143-7208(03)00047-0
6. Bailey PS (1972) Organic groupings reactive toward ozone mechanisms in aqueous media. In: Evans FL (ed) Ozone in water and wastewater treatment. Ann Arbor Science, Ann Arbor, Michigan, pp 29–59
7. Bilinska L, Gmurek M, dan Ledakowicz S (2016) comparison between industrial and simulated textile wastewater treatment By AOPs—biodegradability, toxicity and cost assessment. Chem Eng J 306:550–559
8. Blanco A, Ojembarrena FB, Clavo B, Negro C (2021) Environ Sci Pollut Res 28:16517–16531
9. Braslavsky S, Rubin MB (2011) The history of ozone Part VIII. Photochemical formation of ozone. Photochem Photobiol Sci 10(10):1515–20
10. Bocci V (2006) Is it true that ozone is always toxic? The end of a dogma. Toxicol Appl Pharmacol 216(3):493–504
11. Boczkaj G, Fernandes A (2017) Wastewater treatment by means of advanced oxidation processes at basic pH conditions: a review. Chem Eng J 320:608–633. https://doi.org/10.1016/j.cej.2017.03.084
12. Bühler RE, Staehelin J, Hoigné J (1984) Ozone decomposition in water studied by pulse radiolysis. 1. HO2/O2—and HO3/O3—as intermediates. J Phys Chem 88:2560–2564
13. CAGI (2016) Compressed air & gas handbook, 7th edn. Compressed Air and Gas Institute
14. Carmen Z, Daniela S (2012) Textile organic dyes—characteristics, polluting effects and separation/elimination procedures from industrial effluents—a critical overview. In: Puzyn T (ed) Organic pollutants ten years after the stockholm convention—environmental and analytical update. InTech. ISBN: 978-953-307-917-2
15. Castro FD, Bassin JP, Dezotti M (2017) Treatment of a simulated textile wastewater containing the Reactive Orange 16 azo dye by a combination of ozonation and moving-bed biofilm reactor: evaluating the performance, toxicity, and oxidation by-products. Environ Sci Pollut Res 24(7):6307–6316
16. Castro FD, Bassin JP, Alves TLM, Sant'Anna GL, Dezotti M (2020) Reactive Orange 16 dye degradation in anaerobic and aerobic MBBR coupled with ozonation: addressing pathways and performance. Int J Environ Sci Technol 1–20.
17. Cattoor T (2007) European legislation relating to textile dyeing. In: Environmental aspects of textile dyeing. Elsevier Inc. , pp 1–29

18. Colindres P, Yee-Madeira RR (2010) Removal of Reactive Black 5 from aqueous solution by ozone for water reuse in textile dyeing processes. Desalination 258:154–158
19. Criegee R (1975) Mechanismus der Ozonolyse (Mechanisms of the ozonolysis). Angew Chem 87:765–771
20. Dehghani M, Nasseri S, Mahdavi P, Mahvi A, Naddafi K, dan Jahed G (2015) Evaluation of acid 4092 dye solution toxicity after uv/zno mediated nanophotocatalysis process using daphnia magna bioassay. Colour Sci. Technol 5:285–292
21. Dias NC, Alves TL, Azevedo DA, Bassin JP, Dezotti M (2020) Metabolization of by-products formed by ozonation of the azo dye Reactive Red 239 in moving-bed biofilm reactors in series. Braz J Chem Eng 37(3):495–504
22. Dias NC, Bassin JP, Sant'Anna Jr GL, Dezotti M (2019) Ozonation of the dye Reactive Red 239 and biodegradation of ozonation products in a moving-bed biofilm reactor: Revealing reaction products and degradation pathways. Int Biodeterior Biodegrad 144:104742
23. Dinçer AR (2020) Increasing BOD5/COD ratio of non-biodegradable compound (reactive black 5) with ozone and catalase enzyme combination. SN Appl Sci 2(4):736. https://doi.org/10.1007/s42452-020-2557-y
24. Duguet JP (2004) Basic concepts of industrial engineering for the design of new ozonation processes. Ozone News 32(6):15–19
25. Fanchiang JM, Tseng DH (2009) Degradation of anthraquinone dye CI Reactive Blue 19 in aqueous solution by ozonation. Chemosphere 77(2):214–221
26. FDA (2001) Secondary direct food additives permitted in food for human consumption. Federal Regist 66(123):33 829–30
27. Garniwa I, Sudiarto B, Gaol EHL (2006) Study of partially discharge and flash off corona waves in a simulated cubicle covered with insulating material (Studi gelombang korona peluahan sebagian dan lepas denyar dalam kubikel simulasi dengan dilapisi bahan isolasi). Jurnal Teknologi 1:24–31
28. Girish K (2019) Microbial decolourization of textile dyes and biodegradation of textile industry effluent. In: Kumar S (ed) Advances in biotechnology and bioscience. AkiNik Pub. ISBN: 978-93-5335-855-6
29. Giunta R, Coppola A, Luongo C, Sammartino A, S Guastafierro, A Grassia, L Giunta, L Mascolo, A Tirelli, L Coppola (2001) Ozonized autohemotransfusion improves hemorheological parameters and oxygen delivery to tissuesin patients with peripheral occlusive arterial disease. Ann Hematol 80(12):745–748. https://doi.org/10.1007/s002770100377
30. Graham DM (1997) Use of ozone for food processing. Food Technol 51:72–75
31. Güneş Y, Atav R, Namırtı O (2012) Effectiveness of ozone in decolourization of reactive dye effluents depending on the dye chromophore. Text Res J 82(10):994–1000
32. Gumilar I, Abidin HZ, Hutasoit LM, Hakim DM, Sidiq TP, Andreas H (2015) ISEDM 3rd international symposium on earthquake and disaster mitigation land subsidence in Bandung Basin and its possible caused factors. Procedia Earth and Planetary Science 12:47–62
33. Guzel-Seydim ZB, Greene AK and Seydim AC (2004) Use of ozone in the food industry. Lebensm-Wiss U Technologie 37:453–460
34. Goncalves AA (2009) Ozone—an emerging technology for the seafood industry. Braz Arch Biol Technol 52(6):1527–1539
35. Gottschalk C, Libra A, dan Saupe A (2010) Ozonation of water and wastewater: a practical guide to understanding ozone and its applications. Wiley-VCH, Weinheim
36. Hashimoto T, Nazazawa H, Murakimi T (2009) State of ozonation to municipal wastewater treatment in Japan. In: Proceedings of 19th ozone world congress, Tokyo, Japan, Paper 4-K-1
37. Hoigne J, dan Bader H (1978) Ozonation of water: kinetics of oxidation of ammonia by ozone and hydroxyl radicals. Swiss Federal Institute for Water Resources and Water Pollution Control. Dubendort, Switzerland
38. Hoigné J, Bader H (1975) Ozonation of water: role of hydroxyl radicals as oxidizing intermediates. Science 190:782–783
39. Holkar C, Jadhav A, Pinjari D, Mahamuni N, Pandit, dan AB, et al (2016) A critical review on textile wastewater treatments: possible approaches. J Environ Manage 182:351–366

40. Hussain T, Wahab A (2018) A critical review of the current water conservation practices in textile wet processing. J Clean Prod 198:806–819. Elsevier Ltd.
41. Hussain Z, Arslan M, Malik MH, Mohsin M, Iqbal S, Afzal M (2018) Treatment of the textile industry effluent in a pilot-scale vertical flow constructed wetland system augmented with bacterial endophytes. Sci Total Environ 645:966–973. https://doi.org/10.1016/j.scitotenv.2018.07.163
42. Iervolino G, Vaiano V, Palma V (2020). Enhanced azo dye removal in aqueous solution by H2O2 assisted non-thermal plasma technology. Environ Technol Innov 19:100969
43. Jakob SJ, Hansen F (2005) New chemical and biochemical hurdles. In: Sun D (ed) Emerging technologies for food technology. Elsivier Ltd., pp 387–418
44. Jiang B, Zheng J, Qiu S, Wu M, Zhang Q, Yan Z, Xue Q (2014) Review on electrical discharge plasma technology for wastewater remediation. Chem Eng J 236:348–368
45. Kasih TP, Nasution J (2016) Development of cold plasma technology for modification of surface characteristics of material without changing the basic properties of the material (Pengembangan Teknologi Plasma Dingin untuk Modifikasi Karakteristik Permukaan Material tanpa Mengubah Sifat Dasar Material). J Res Appl Indus Syst Appl X(3):373–379
46. Kataki S, Chatterjee S, Vairale MG, Sharma S, Dwivedi SK (2021) Resour Conserv Recycl 164:105156. https://doi.org/10.1016/j.resconrec.2020.105156
47. Khadre MA, Yousef AE (2001) Decontamination of a multilarninated aseptic food packaging material and stainless steel by ozone. J Food Safety 21:1–13
48. Kirschner MJ (2000) Ozone. Ullmann's encyclopedia of industrial chemistry. Wiley-VCH Verlag. https://doi.org/10.1002/14356007.a18_349
49. Loeb BL, Thompson CM, Drago J, Takahara H, Baig S (2012) Worldwide ozone capacity for treatment of drinking waterand wastewater: a review. Ozone Sci Eng 34:64–77
50. Makkar S, Makkar M (2011) Ozone-treating dental infections. Indian J Stomatol 2(4):256–259
51. Manjunath RGS, Singla D, Singh A (2015) Ozone revisited. J Adv Oral Res 6(2):5–9
52. Morrison C, Atkinson A, Zamyadi A, Kibuye F, McKie M, Hogard S, Mollica P, Jasim S, Wert EC (2021) Critical review and research needs of ozone applications related to virus inactivation: potential implications for SARS-CoV-2. Ozone Sci Eng 43:1, 2–20. https://doi.org/10.1080/01919512.2020.1839739
53. Mondal S, Bhagchandani C (2016) Textile wastewater treatment by advanced oxidation processes. J Adv Eng Technol Sci 2:2455–3131
54. Muthukumarappan K, Halaweish F and Naidu AS (2000) Ozone. In: Naidu AS (ed) Natural food antimicrobial systems.CRC Press, Boca Raton, FL, pp 783–800
55. Munter R (2001) Advanced oxidation processes—current status and prospects. Proc Estonian Acad Sci Chem 50(2):59–80
56. Murugesan PV, Monica VE, Moses JA, Anandharamakrishnan C (2020) Water decontamination using non-thermal plasma: concepts, applications, and prospects. J Environ Chem Eng 8(5):104377
57. Moertinah S (2008) Clean production opportunities in the textile finishing bleaching industry (Peluang-peluang produksi bersih pada industri tekstil finishing bleaching). Diponegoro University, Thesis
58. Nath A, Mukhim K, Swer T, Dutta D, Verma N, Deka BC, Gangwar B (2014) A review on application of ozone in the food processing and packaging. J Food Product Dev Packag 1:7–21
59. Nakhjirgan P, dan Dehghani M (2015) The evaluation of the toxicity of reactive red 120 dye by daphina magna bioassay. J Res Environ Health 1:1–9
60. Nashmi OA, Mohammed AA, Abdulrazzaq NN (2020) Investigation of ozone microbubbles for the degradation of methylene orange contaminated wastewater. Iraqi J Chem Pet Eng 21(2):25–35
61. Nur M (2011) Plasma physics and its applications (Fisika Plasma dan Aplikasinya). Diponegoro University Press. ISBN: 978-979-097-093-9
62. Parsons S (2004) Advanced oxidation process for water and wastewater treatment. Gray Publishing, Turnbridge Wells, UK

63. Petek J, Glavič P (1996) An integral approach to waste minimization in process industries. Resour Conserv Recycl 17(3):169–188
64. Pulga A (2018) Oxygen-ozone therapy in dentistry: current applications and future prospects. Ozone Ther 3;7968:37–42
65. Pratiwi R, Notodarmojo S, Helmy Q (2018) Decolourization of remazol black-5 textile dyes using moving bed bio-film reactor. In: IOP conference series: earth and environmental science, vol 106, no 1. IOP Publishing, p 012089
66. Pratiwi R (2018) Reactive Black 5 (RB 5) dye treatment using Moving Bed Biofilm Reactor (MBBR). Thesis, Environmental Engineering Department, Institute of Technology Bandung
67. Pressman S. 2007. The Story of Ozone. Plasmafire International, Canada.
68. Preis S (2008) History of ozone synthesis and use for water treatment. In: Munter R (ed) Encyclopedia of Life Support Systems (EOLSS). EOLSS Publishers, pp 6–192
69. Ramachandran P, Sundharam R, Palaniyappan J, Munusamy AP (2013) Adv Appl Sci Res 4(1):131–145
70. Raja ASM, Arputharaj A, Saxena S, Patil PG (2019) Water requirement and sustainability of textile processing industries. Water Text Fash: 155–173
71. Rice RG, Overbeck PK, Larson K (1998) Ozone treatment of small water systems. International Ozone Association Pan American Group, Vancouver, British Columbia, Canada
72. Rice RG (1999) Ozone in the United States of America-State-of-the-art. Ozone Sci Eng 21:99–118
73. Rice RG, Browning ME (1980) Ozone for industrial water and wastewater treatment, a literature survey. USEPA Research and Development, EPA-600/2-80-060
74. Rice RG, Robson CM, Miller GW, Hill AG (1981) Uses of ozone in drinking water treatment. J Am Water Works Assoc 73:44–57
75. Remondino M, Valdenassi L (2018) Different uses of ozone: environmental and corporate sustainability. Literature review and case study. Sustainability 10:4783. https://doi.org/10.3390/su10124783
76. Rubin MB. 2001. The History of Ozone. The Schonbein Period, 1839–1868. Bull. Hist. Chem., 26(1):40–56
77. Rubin MB (2002) The history of ozone. II. 1869–1899 (1). Bull Hist Chem 27(2):81–105
78. Rubin MB (2003) The history of ozone. Part III. C.D. Harries and the introduction of ozone into organic chemistry. Helv Chim Acta 86:930–940
79. Rubin MB (2004) The history of ozone. IV. The isolation of pure ozone and determination of its physical properties (1). Bull Hist Chem 29(2):99–106
80. Samsami S, Mohamadi M, Sarrafzadeh MH, Rene ER, Firoozbahr M (2020) Recent advances in the treatment of dye-containing wastewater from textile industries: overview and perspectives. Process Safety Environ Prot 143: 138–163
81. Sarayu K, dan Sandhya S (2012) Current technologies for biological treatment of textile wastewater—a review. Appl Biochem Biotechnol 167:646–661
82. Shen Y, Xu Q, Wei R, Ma J, Wang Y (2017) Mechanism and dynamic study of reactive red X-3B dye degradation by ultrasonic-assisted ozone oxidation process. Ultrason Sonochem 38:681–692
83. Shimizu A, Takuma Y, Kato S, Yamasaki A, Kojima T, Urasaki K, Satokawa S (2013) Degradation kinetics of azo dye by ozonation in water. J Fac Sci Tech Seikei Univ 50:1–4
84. Srikanth A, Sathish M, Harsha AVS (2013) Application of ozone in the treatment of periodontal disease. J Pharm Bioallied Sci 5(Suppl 1):S89–S94. https://doi.org/10.4103/0975-7406.113304
85. Sugiyana D, Soenoko B (2016) Identification of photocatalytic mechanism in the degradation of azo reactive black 5 dye using TiO2 microparticle catalyst (Identifikasi mekanisme fotokatalitik pada degradasi zat warna azo reactive black 5 menggunakan katalis mikropartikel TiO2). Arena Tekstil 31(2):115–124. https://doi.org/10.31266/at.v31i2.1939
86. Suryawan IWK, Helmy Q, Notodarmojo S (2018) Textile wastewater treatment: colour and COD removal of reactive black-5 by ozonation. In: IOP conference series: earth and environmental science, vol 106, no 1. IOP Publishing, p 012102

87. Suryawan I, Siregar MJ, Prajati G, Afifah AS (2019). Integrated ozone and anoxic-aerobic activated sludge reactor for endek (Balinese textile) wastewater treatment. J Ecol Eng 20(7)
88. Takahara H, Kato Y, Nakayama S, Kobayashi Y, Kudo Y, Tsuno H, Somiya I (2009) Estimation of an appropriate ozonation system in water purification plant. In: Proceedings of 19th ozone world congress, Tokyo, Japan, Paper 17–2
89. Tizaoui C (2020) Ozone: a potential oxidant for COVID-19 virus (SARS-CoV-2). Ozone Sci Eng 42(5):378–385. https://doi.org/10.1080/01919512.2020.1795614
90. Ulson SMDAG, Bonilla KAS, de Souza AAU (2010) Removal of COD and colour from hydrolyzed textile azo dye by combined ozonation and biological treatment. J Hazard Mater 179(1–3):35–42
91. Ulucan-Altuntas K, Ilhan F (2018) Enhancing biodegradability of textile wastewater by ozonation processes: optimization with response surface methodology. Ozone Sci Eng 40(7):1–8. https://doi.org/10.1080/01919512.2018.1474339
92. UNICEF-WHO (2015) Progress on sanitation and drinking water—2015 update and MDG assessment. WHO Press, Geneva, Switzerland
93. USEPA (1999) Alternative disinfectants and oxidants-guidance manual. Office of Water (4607), EPA 815-R-99-014
94. Venkatesh S, Venkatesh K (2020) Ozonation for degradation of acid red 14: effect of buffer solution. Proc Natl Acad Sci India Sect A 90(2):209–212
95. Venkatesh S, Quaff AR, Pandey ND, Venkatesh K (2015) Impact of ozonation on decolourization and mineralization of azo dyes: biodegradability enhancement, by-products formation, required energy and cost. Ozone Sci Eng 37(5):420–430
96. Venkatesh S, Venkatesh K, Quaff AR (2017) Dye decomposition by combined ozonation and anaerobic treatment: cost effective technology. J Appl Res Technol 15(4):340–345
97. Verinda SB, Yulianto E et al (2021) Ozonated nanobubbles—a potential hospital waste water treatment during the COVID-19 outbreak in Indonesia to eradicate the persistent SARS-CoV-2 in HWWs. Ann Trop Med Public Health 24(1):197. https://doi.org/10.36295/ASRO.2021.24197
98. Von Sonntag G, von Gunten U (2012) Chemistry of ozone in water and wastewater treatment from basic principles to applications. IWA Publishing
99. Wisniak J (2008) Thomas Andrews. Revista CENIC Ciencias Químicas 39(2):98–108
100. Zheng H, Zhang J, Yan J, Zheng L (2016a) An industrial scale multiple supercritical carbon dioxide apparatus and its eco-friendly dyeing production. J CO2 Utilization 16:272–281
101. Zheng Q, Dai Y, Han X (2016b) Decolourization of azo dye CI Reactive Black 5 by ozonation in aqueous solution: influencing factors, degradation products, reaction pathway and toxicity assessment. Water Sci Technol 73(7):1500–1510
102. Zhang C, Sun Y, Yu Z, Zhang G, Feng J (2018) Simultaneous removal of Cr(VI) and acid orange 7 from water solution by dielectric barrier discharge plasma. Chemosphere 191:527–536
103. Ziyaina M, Rasco B (2021) Inactivation of microbes by ozone in the food industry. Afr J Food Sci 15(3):113–120. https://doi.org/10.5897/AJFS2020.2074

Nanomaterials in Advanced Oxidation Processes (AOPs) in Anionic Dye Removal

Aiswarya Thekkedath, Samuel Sugaraj, and Karthiyayini Sridharan

Abstract Water, popularly known as Universal Solvent, plays a vital role for surviving in this environment. Over a decade, water bodies are being polluted in several ways (wastewaters, chemicals, dyes, papers, etc.). Out of these pollutants, dyes are relatively more toxic. Dyes are the pigment used in textile industries. The dye waste from the industries is likely to pollute the water. For the degradation of dyes from water, different methods were introduced. Advanced oxidation processes are one of the prominent methods. In this process, there are physical, chemical as well as biological methods involved. This chapter coveys a detailed description of water pollution, how dyes are affecting the quality of the water, advanced oxidation process, and the nanomaterials used in the advanced oxidation process.

1 Water Pollution—Introduction

In this era, the most challenging situation is freshwater access, a major issue faced by every region. In rural parts the major cause for drinking water are pollution and scarcity. Interpreting the causes of pollution, we can say the major reason will be over population, and besides the increasing demand in agriculture and industry. When the strange particles enter into the water, the quality is being disturbed resulting in a change in the environment and human health hazard [1].

Major issues for human health as well as surroundings is the sewage or effluents disposed to the water as well as environment. It will destroy the common natural resources, thereby resulting in severe diseases like typhoid like fever, diarrhea, vomiting, etc. There is another possibility for the quickest destruction of animals and plants. The pollutants from the water will kill marine life and our ecosystem also. In our normal food processes some insecticides also influencing more [2].

A. Thekkedath · K. Sridharan (✉)
Department of General Sciences, BITS Pilani Dubai Campus, Dubai, UAE
e-mail: rajkar6761@gmail.com

S. Sugaraj
Department of Physics, The New College, Chennai, India

S. S. Muthu and A. Khadir (eds.), *Advanced Oxidation Processes in Dye-Containing Wastewater*, Sustainable Textiles: Production, Processing, Manufacturing & Chemistry,
https://doi.org/10.1007/978-981-19-0987-0_7

2 Hazards of Water Pollution

Factories discharge a huge amount of toxic waste, colors, and organic chemicals. Heavy metals like copper, arsenic, lead, cadmium, mercury, nickel, and cobalt and certain auxiliary chemicals and acids, dyes, soaps, chromium compounds, etc., make the water highly toxic. Some formaldehyde-based dyes, hydrocarbons, non-biodegradable dyeing chemicals are other kinds of toxic materials. Some factories dispose of water having high temperature and pH, which may affect the system. Other forms which make water appear bad with a foul smell are the colloidal matter present along with colors and oily scum. This prevents the light penetration into the water that is necessary for photosynthesis, resulting in refraining the oxygen transfer mechanism at the air interface. Another issue is the depletion of dissolved oxygen in water causing hindrance to marine life which will stop the self-purification of water.

In addition, when this effluent is allowed to flow in the fields, it clogs the pores of the soil resulting in loss of soil productivity. The texture of soil gets hardened and penetration of roots is prevented. The wastewater that flows in the drains corrodes and incrustates the sewerage pipes. If allowed to flow in drains and rivers, it affects the quality of drinking water in hand pumps making it unfit for human consumption. It also leads to leakage in drains increasing their maintenance cost. Such polluted water can be a breeding ground for bacteria and viruses. Impurities in water affect textile processing in many ways. In scouring and bleaching, they impart a yellow tinge to the white fabric. In the dyeing stage, metallic ions present in water sometimes combine with the dyes causing dullness in shades. Textile effluent is the cause of a significant amount of environmental degradation and human illnesses. About 40% of globally used colorants contain organically bound chlorine a known carcinogen. All the organic materials present in the wastewater from the textile industry are of great concern in water treatment because they react with many disinfectants, especially chlorine. Chemicals evaporate into the air we breathe or are absorbed through our skin and show up as allergic reactions and may cause harm to children even before birth.

2.1 Effects of Water Pollution on Human Health

Bonding between human health and pollution is much higher. The spreading will be through microorganisms commonly known as pathogens and they will spread diseases easily [2]. Most of the diseases which we commonly know are waterborne diseases [3]. Floods and heavy rainfall also favor the spreading in developing as well as developed nations [4]. Only a few percent of the whole population, say 10%, depends on veggies and others grown from this wastewater [5]. Diseases caused through water were linked with waste excreted toward the water sources and their resultants [6]. Human health became high risk with this contaminated water as it results in cancer, nervous disorder, etc. [7]. Chemicals excreted into the water will cause blue baby syndrome and cancer [8]. Both urban and rural areas' mortality rates are increased due to this sewage water. People who are poor will be more at risk of unhygienic and improper water supply [9]. Women are also affected because of effluents, as it will cause a lower fetal birth rate [10]. Lower the quality of water, lesser the production of crops and will affect the marine as well as human life [11]. Whenever the pollutants are heavier, it will enter into fish's body which will be directly eaten by the humans thereby results the entering the pollutants inside the human body as the pollutants effects the respiratory system of fishes [12]. Water contamination due to metals will result in liver cirrhosis, hair fall, nervous system failure [13] http://www.alliedacademies.org/environmental-risk-assessment-and-remediation/. ISSN: 2529–8046.

2.2 Categories of Water Pollution

The two main sources for water pollution can be generally concluded as point sources and non-point sources.

Point sources

The directly identifiable sources that are causing pollution are popularly known as point sources. For example, pipe attached to a factory, oil spill from a tanker, effluents coming out from industries. Point sources of pollution include storm sewer discharge and wastewater effluent (both municipal and industrial) and affect mostly the area near it.

Every type of source related to pollution causes its own consequences and health issues. The substances which are considered as major water pollutants are thus classified into several groups which are organic compounds (pesticides, hydrocarbons, dyes, and oil), inorganic substances (salts, acids, phosphate, sulfates, fluorides, and toxic heavy metals) or microbial (bacteria, protozoa, and viruses). If any substance exceeds their sill, then it will cause a series of issues which will be harmful to human as well as aquatic life. Apart from these, the waste from textiles and paint industries have colored components (persistent color, organics, toxicants, surfactants, chlorinated, and inhibitory compounds), that are resultants from dyeing also one of the major issues.

Water pollutants may be (i) Organic and (ii) Inorganic water.

(1) Organic water pollutants:

Contaminants include organic wastes such as detergents, disinfection by-products found in food processing waste, insecticides, chemically disinfected drinking water, herbicides, chlorinated solvents, petroleum hydrocarbons, volatile organic compounds, etc.

(2) Inorganic water pollutants:

Inorganic pollutants include acidity causing substances from industrial discharges, ammonia from food processing waste, chemical waste as industry by-products, fertilizers containing nutrients such as nitrates and phosphates, heavy metals form motor vehicles, secretion of creosote preservative into the aquatic ecosystem, slit sediment (adapted from [14]).

Non-point sources

The non-identifiable sources which are appeared from numerous sources of origin and number of ways by which contaminants that may enter into groundwater or surface water and occur in the environment. Water drained from agricultural lands and waste from cities are some examples. Causing severe pollution in one place is known as transboundary pollution. The best example of this pollution is radioactive waste. This waste will flow to the ocean and different water bodies, thereby resulting in disaster. The main non-point sources of pollution are the following:

- Industrial Wastes: Waste coming out from factories and mines will come under this category. Industrial wastes are of different types, some may be chemical, scrap metal, solvents, food waste from hotels. These wastes may be in liquid, solid or semi-solid form.
- Agro-chemical Wastes: This waste includes a wide area of pesticides which includes herbicides, fungicides, nematicides, and insecticides. Agro waste also includes some hormones synthetic fertilizers.
- Nutrient enrichment: Heavy production of nutrient will damage the algal growth and affect finally with a high toxicity. It manages rich productivity in water which will either be natural or artificial.
- Thermal pollution: Unexpected rise in the temperature of a water body is thermal pollution. When a plant put back in water with change in temperature will cause this pollution.
- Sediment pollution: Land erosion is the main reason for sediment pollution.
- Acid Rain Pollution: Chemicals that were excreted into the surroundings will cause acid rain. It may be because of oxides of nitrogen and sulfur dioxide. Exhausts from trucks, buses, and cars are also one of the reasons.
- Radioactive waste: This is one of the most dangerous waste which is the after effect of nuclear research, nuclear medicine, etc.

  [https://www.researchgate.net/publication/321289637]

3 Different Sets of Pollution

Water can be polluted by different factors. Some are described briefly as follows (Fig. 1).

Water pollution	
Point source	**Non-point source**
occurs through many diffuse sources	occurs from single identifiable source
effect is less	effect is high
treatment plant not needed	treatment plant needed
example: agriculture, domestic, etc.	example: industries, sewage, etc.

Fig. 1 Sources of water pollution

3.1 Chemical Pollution

The cause for chemical pollution may be either through organic or inorganic species. Dyes and pesticides include in organic chemicals and also their derivatives such as nitrophenols, trihalomethanes, etc. [15, 16]. Another category such as hydrocarbon, namely phenols, benzene, toluene, xylene, in which an amount of 10–59% gasoline is the regular contaminant of natural waters. Most newborns have been observed to be at potential risk of drinking elevated levels of sodium ions [17]. These chemicals are highly toxic for kidney, liver, and nervous system. Toxic heavy metals having inorganic ions cause cancer, hypertension, poisoning, and infantile cyanosis [18].

3.2 Heavy Metal Pollution

Zinc, arsenic, cobalt, nickel-like heavy metals released into the environment of textile industries is great consternation all over the world as these pose a high risk to aquatic life, human life, and nature. Also these are damaging non-biodegradable nature, biological half-lives, and potential to accumulate in the body. Due to solubility also, some metals are highly toxic. Metals like Cu, Mn, Mo, Zn having lower concentration, plays a vital role in physiological functions in marine as well as human life. These metals are even more dangerous in their free state and also in combined form. Heavy metals are one of the old toxicants for humans [18].

3.3 Water Pollution Due to Organic Dyes

Diverse dyes are used in various applications and they are dispersed to nature in the form of sewage or effluents. Out of several types of dyes, azo dyes are commonly used because of their stability and versatility, it is used in textiles, printing cosmetics, and tattooing. However, their non-biodegradability and durability cause pollution once these dyes are released into the water bodies. Some of the azo dyes are very toxic, carcinogenic, and mutagenic [19]. One of the malignant pollutants is wastewater discharge. This discharge consists of almost all types of contaminants that causes a severe threat to humanity [18].

3.4 Dyes-The Major Pollutant and Its Types

One of the most beautiful words which might give the human a visual pleasure is none other than color!!! At the same time, it kills us as a sweet poison!!! Color enhances its self-appearance. Nowadays, pigments and dyes are commonly used in mills, and various coloring compounds are being utilized for the dyeing process. The most dangerous and toxic nature of coloring compounds became a threat to aquatic life. Thus, various methods have been undergoing for the removal of these dyes from water such as filtration, oxidation, etc. [20].

The major class of dyes is reactive dyes. They are highly distinguished because of their excellent binding capability due to the covalent bond formation with the reactive groups of dyes and surface groups seen on textile fibers. Dyes when discharged into the environment, cause a disturbance to the ecosystem, thereby increasing the level of toxicity and threat to humanity. Most common dyes can be seen in fabric materials (Fig. 2).

Fig. 2 Types of dyes

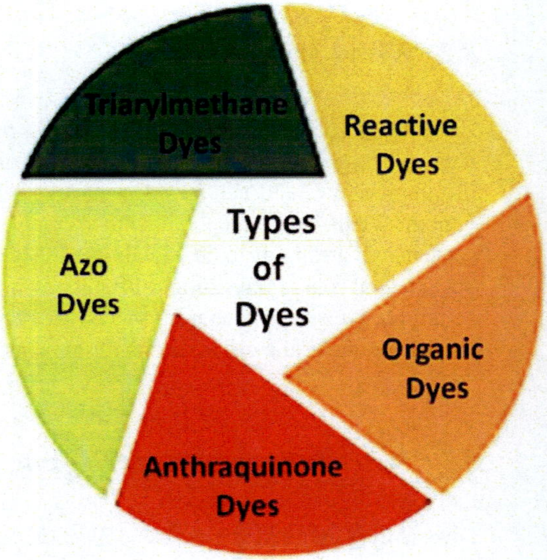

3.4.1 Azo Dyes

Azo dyes (mono, dia, tria, poly) account for the majority of synthetic aromatic dye, which are composed of one, two, three or more (N=N) groups and connected to benzene and naphthalene rings that are substituted frequently with some functional groups, e.g., triazine amine, chloro, hydroxyl, methyl, nitro, and sulphonate. They are basically used in cosmetics, paper printing, pharmaceutical, textile dyeing, and food. However, 80% of azo dyes are used in textiles because of high stability, low cost, and color variation. Studies estimated that 10–15% of dyes from textiles are released into water bodies. For example, the dyestuff of 30–60 g and water of 70–150 L were needed for dyeing 1 kg of cotton, and about half percent of the dyes which are used ones were thrown by either direct or indirect sources [21].

3.4.2 Reactive Dyes

Dyes are formed by the formation of covalent bonds between diflurochloropyrimidine, chlorotriazine, vinyl sulfone, trichloropyrimidine. In contrast to other classes of dyes, reactive dyes have many advantages in textiles because of their simple application techniques with low energy consumption bright colors, which are water-fast. Thus, reactive dyes are usually used at present. During the dyeing process, covalent bond bonds faster with reactive dyes. These are widely used in wool an cotton dyeing processes. Coloring cellulose fibers is done through reactive dyes. Second case these dyes also be put in nylon, wool but under weaker conditions. In reactive dyes, the functional groups create hydrolysis by bonding with water resulting in low utilization.

Reactive dyes are the most problematic dyes among others, as they tend to pass unaffected through conventional treatment systems [Vol. 4, No. 1, 22–26 (2012) Natural Science http://dx.doi.org/10.4236/ns.2012.41004].

However, reactive dyes were found to be mutagenic and carcinogenic and even the intermediate products, i.e., mineralization, aromatic amine, and arylamine, also have severe adverse effects on human beings, including damages caused to the brain, liver, kidney, central nervous system, and reproductive system [22].

3.4.3 Anthraquinone Dyes

After azo dyes, another set of dyes that are commonly used in textiles were Anthraquinone dyes and have been widely used in the garments owing to their different colors, fastness properties and ease of application, and low energy consumption. Another major advantage of anthraquinone dyes applied in dyeing processes is their high affinity to silk and wool. However, due to their fused aromatic structure, anthraquinone dyes exhibit a lower rate of decolorization. Most of these dyes are mutagenic, carcinogenic, and toxic, therefore, the removal of anthraquinone dyes has attracted an enormous recognition.

3.4.4 Triarylmethane Dyes

Triphenylmethane dyes are aromatic xenobiotic compounds that were formed by a carbon atom placed in center connected with two benzene rings and one p-quinoid group-chromophore. The common auxochromes are –NH2, NR2, and –OH. These dyes have brilliant colors with typical shades of green, violet, red, and blue, such as Malachite green (MG), Crystal violet (CV), Bromophenol blue (BB), etc. These dyes are widely applied in the garments for coloring the substrates like modified nylon, polyacrylonitrile, cotton, silk, and wool.

Another important application of triphenylmethane dyes is their use in paper and leather industry. Their relatively inexpensive dyeing, high tinctorial strength, and low light fastness, especially in the slow washing and light are considered apt for dyeing a wide range of garment substrates. In addition, triphenylmethane dyes have been extensively employed in medicine as medical disinfectants, because these dyes exhibit antibacterial, antifungal, and antiprotozoal properties, they can be used for the disinfection of post-operative wounds, and also employed in controlling diabetes. Some of these dyes (phenolphthalein, fuchsine, and fluorescein) are used as indicator dyes because of their pH sensitivity.

3.4.5 Other Dyes

One of the most popular azo dyes is Congo red (CR). Congo Red is an anionic dye (diazo benzidine) that contains –N=N– linkage (double azo). This is highly dangerous, carcinogenic too. Janus Green B (JGB) is incorporated by conjugating dimethyl aniline to diethyl safranine through double azo linkage. It has been used for reduction–oxidation capability. The color of JGB is blue. The dyes which are highly used in the manufactural and industrial processes such as textile dyeing, cosmetics, and food includes triphenylmethane dyes. These dyes are carcinogenic for microbes and mammals. This will result in abnormalities in the reproduction of aquatic animals. An aromatic amine and an important element of polymers, drugs are o-Phenylenediamine (OPD). It has been found that this dye is highly dangerous in case of eye contact and inhalation. A synthetic dye that consists of a large amount of organic compounds is Brilliant Cresyl Blue (BCB). Some of the cationic dyes are BCB, JGB, etc., are some organic dyes that have different industrial as well as scientific applications. Rhodamine 6G (R6G) is widely used in biotechnological applications. This dye is one among the rhodamine family of dyes. This dye shows unfortunate effects on humans and microorganisms. R6G is water-soluble and shows photostability with better absorption coefficient.

Another water-soluble dye which is commonly used as a coloring agent in garments and food industries is Rhodamine B (RhB). It is basically N-[9-(ortho-carboxyphenyl)-6- (diethylamino)-3Hxanthen-3-ylidene. RhB is an irritant to eyes, skin, and highly toxic.

A water-soluble anthraquinone dye, Alizarin Red S (ARS). It is basically salt known as 1,2-dihydroxy-9,10-anthraquinonesulfonic acid, sodium. ARS is meant to be a biologically active dye, which is highly toxic toward human life.

Methylene blue (MB) is a cationic dye that is yet another dye used widely for colouring in paper, leather, and wool industries.

Crystal Violet (CV) is commonly used in textiles and also used as detection of structures or tissues in veterinary and human medicine. This causes tumors in fishes and is also a poisoning agent.

A benzidine-based dye, namely Malachite Green (MG) has been used preferably in fish farms to stop infections caused by fungi. The combination of MG and leucamalachite green causes severe health hazards toward humans as well as the environment. Direct contact of MG will affect our eyes and irritate the skin, which results in redness [18, 23].

4 Detection of Dyes in Water Samples

In water samples, the detection of dyes is a big issue. As we know the dyes are colored and spectrophotometry is not possible as the interference of spectral lines in a mixture may contain a number of dyes including both anionic as well as cationic dyes, namely BCB, R6G, MB, CV < MO, AO, etc. The quickest analysis of dyes in water effluents is strenuous, excessive, and requires particular techniques and procedures. Several dyes may undergo biological, chemical, and photochemical degradation in water. The final products formed from degradation were tenacious to the environment and more injurious to water bodies. Studies reported that under direct sunlight the photodegradation green dye is transformed into more toxic chemicals. Due to this degradation, some dyes form intermediates like benzidine, benzene, etc., which makes the detection even more difficult [18, 24].

5 Removal of Dyes

One of the huge water consumption industries is the textile industry. The resultant coming from these industries include different chemicals and coloring compounds that should be properly treated before disposal. But this treatment is difficult as the sewage contains various compounds for which the differentiation is tougher. Currently, the usage of a combination of different methods of treatment so as to remove the contaminants seen in wastewater treatment (Holkar 2016). Thus, adsorption became the most effective method for the removal of dye in textile water.

In adsorption, the affordable source of adsorbents is sludge (plant biosolids), magnetically modified brewer's yeast, cassava peel activated carbon, tapioca peel activated carbon, soil, fly ash, jack fruit peel activated carbon, groundnut shell activated carbon activated with Zinc chloride solution, neem leaf powder, kaolinite, montmorillonite, hazelnut activated carbon, bagasse pith, natural clay, maize cob, rice bran based activated carbon, guava seeds activated with Zinc chloride solution followed by pyrolysis, etc. These sources are commonly used for the removal of colors from dye-house by various research studies.

5.1 Dye Removal Techniques

The major removal techniques involve physical, chemical, and biological method. The limitations that are faced under these techniques are huge capital costs, lower efficiency, and formation of excess sludge. Out of these methods, some are adaptable and better, and appropriate for the removal of dyes [25].

5.1.1 Physical Methods

The physical method involves, first, membrane filtration, which includes reverse osmosis, nanofiltration, and electrodialysis, Second, it involves ion exchange, and third is the adsorption. In membrane filtration, the demand for periodic replacement have a severe problem in membrane fouling, which is the major limitation. Because of low cost, easy operative, simple design, high efficiency, the adsorption technique became attractive, and also it serves as a better alternative compared to other treatments. A detailed description of every method is given below.

5.1.2 Membrane Filtration

A membrane is a thin layer material that separates the substance when a force is applied. This method involves microfiltration (MF), ultrafiltration (UF), nanofiltration (NF), and reverse osmosis (RO), an advanced treatment technology to remove color, biochemical oxygen demand (BOD) and chemical oxygen demand (COD). Usually, the pore size is determined to check whether the particle is passing or retaining within the membrane, which is the core principle for membrane filtration accompanied by different membrane techniques. When particles are larger than the pore size of the membrane, particles will be hindered to pass through.

Membranes with a pore size of 0.03–10 microns having molecular mass greater than 1,000,000 daltons and low feed water pressure of 100–400 kPa are separated using microfiltration process.

For treatment of organic colloids or inorganic particles having a separation of 0.05–0.15 μm, is generally used in ultrafiltration method (UF). The studies were carried out with Reactive Black 5 [26]. The rejection remained as 70% for dye concentration of 500 mg/L. Thus, UF is considered less suitable for dye removal.

Nanofiltration involves the size of the pore ranging from 1 to 10 angstronm. Nanofiltration is a combination of ultrafiltration and reverse osmosis. Compared to

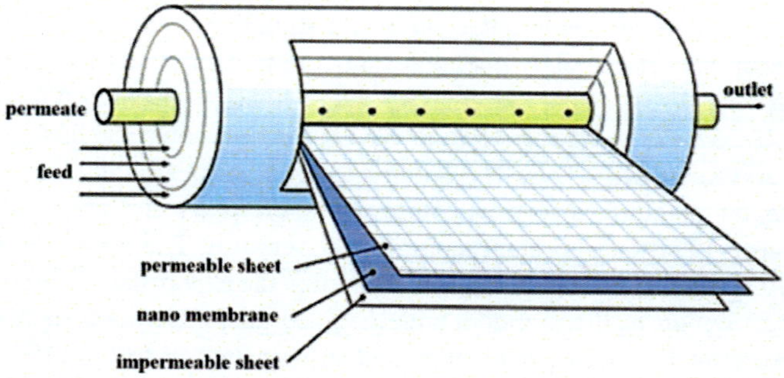

Fig. 3 Nanofiltration process

UF and RO, the maintenance costs, low operation, low investment, higher diffusion of flux, detention of salts were lower. So nanofiltration is considered superior to UF and RO (Fig. 3).

Usage of cellulose nanocrystals proved as an effective filtration process for removal of dye (Methylene blue (MB), Victoria blue (VB), and Methyl violet (MV)). Later Reactive black 5 was studied by Amin et al. using nanofiltration technique from sewage under different conditions such as flow rate, pressure, temperature, etc. The results show as the rejection of dye exceeding 99% is possible with a pressure of 15 bar, temperature of 35 °C, and flow rate of 480 L/h. From this study, nanofiltration is a highly efficient technique in the removal of reactive dyes. Inorganic contaminants present in water are removed by reverse osmosis (RO). RO removes natural organic substances, radium. The advantages of reverse osmosis are low waste concentration is possible, removes nearly all ions contaminated in water, bacteria, and particles are removed.

5.1.3 Ion Exchange

An attractive feature with high efficiency, low cost for the purification of water is the Ion exchange method. The strong interaction of functional groups and charged dyes on the exchange of resins is possible in the removal of dyes in aqueous solution. Resins are classified into four parts, namely cation exchange membrane, anion exchange membrane, cross-linkage membrane, and other ion exchange membrane. Out of these four categories, anion and cation are mostly used for the removal process. Anion exchange removal represents positively charged groups such as $-NH^{3+}$, $-NRH^{2+}$, $-NR2H^{+}$, which pass anions and restrict cations. Whereas cation exchange membrane contains negatively charged groups, namely-SO^{3-}, $-COO^{-}$, etc. which allows only cations and prevents anions.

Some removal examples are depicted below.

Congo Red (CR) an anionic dye is removed from water bodies using an anion exchange membrane studied by Ismail et al. [18]. The study showed that the membrane dosage and contact time is a significant influence on CR removal efficiency. The effect of temperature and ionic strength was compared to be very less.

Removal of Acid dye Orange (AO7) through anion exchange was carried out by Akazdam and Chafi. The studies include various parameters like pH, contact time, temperature, adsorbent dosage, agitation speed, etc. The results indicated that solution pH is an important factor for the removal. The resin almost removed 75% of AO7 within 100 min. Results indicate that the experimental data was fitted well to the Langmuir isotherm model with the maximum adsorption capacity of 200 mg/g. Adsorption kinetics showed that the adsorption process followed the pseudo-second-order model and thermodynamic studies further confirmed that the adsorption of AO7 on the macroporous basic anion exchange resin under the investigated conditions was a spontaneous and exothermic process.

5.1.4 Adsorption

Sticking of ions or atoms or molecules of dissolved solid or liquid or gas to a surface is known as adsorption. On the surface of the adsorbent, a thin film of the adsorbate will be created during the process. Adsorption is different from absorption as the absorption phenomenon involves the volume of the material and adsorption involves only surface.

The adsorption process is categorized into two, namely: physical adsorption and chemical adsorption depending upon how the adsorbate is adsorbed on the adsorbent surface.

Physical adsorption involves the sticking of adsorbate to adsorbent through Van der Waals force (weak force), hydrogen bonding, hydrophobicity, static interactions, polarity, dipole interactions, and π-π interactions. Whereas in chemical adsorption, adsorbates are chemically bound to the surface of an adsorbent by the force of the electrons exchange.

For environmental remediation purposes, this adsorption method is widely used as it has been considered superior to other dye removal methods. Moreover, adsorption does not cause hazardous substances and avoids secondary pollution. Adsorption by eco-friendly and cost-effective adsorbent was selected for the adsorption process depending on the adsorption capacity, surface area, and potential for reuse cost of the materials (Fig. 4).

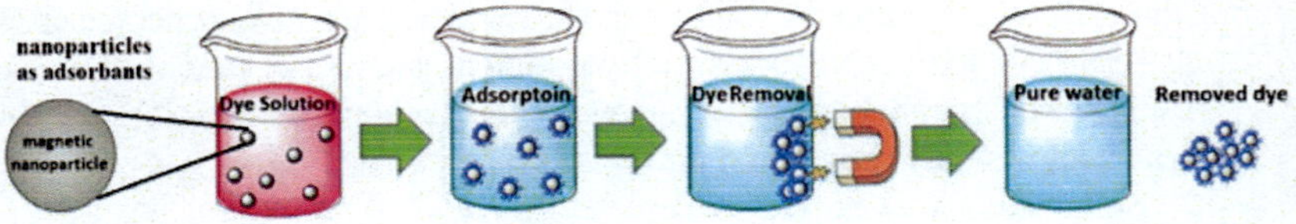

Fig. 4 Nanoparticles as adsorbents

5.1.5 Chemical Methods

Chemical methods include ozonation, coagulation-flocculation, precipitation-flocculation with $Fe^{2+}/Ca(OH)_2$, photocatalytic degradation, electrochemical treatment. These chemical techniques are efficient for the removal of dyes from wastewater, but the excessive use of chemicals causes the difficulty of sludge disposal and the possibility of a secondary pollution problem. In addition, high electrical energy and high cost are needed for these techniques resulting in limited use.

5.1.6 Ozonation

The introduction of ozone into water is known as ozonation or ozonization. It is basically chemical water treatment. One of the strong oxidants is Ozone (contains 3 oxygen atoms).

They can easily decompose into oxygen and free radicals. These free radicals then combine with dyes causing color destruction. Ozone can strike on pollutants by the indirect or direct reactive path.

The direct path involves the ozone molecular activity and the indirect path results in highly oxidative free radicals. The rate of decomposition of ozone is truly affected by pH and dye concentration. For a low pH, the ozone will be present as molecules to react with dye as an electrophile.

At high pH, ozone can decay suddenly into hydroxyl radicals in solution, which are a powerful, effective, and nonselective oxidizing agent. The decolorization efficiencies of Reactive orange 7 (RO 7), Reactive blue 19 (RB 19), and Reactive black 5 (RB 5) were almost 100% are reported in Shaikh et al. For pH 4, 7, 9, and 11 the percentage of color removal was studied in RO7 as 19, 50, 60, and 75%. The homogenous trend can be noted in the decolorization of RB 5 when the pH reached 11. O^3 decomposition and OH radicals production occurred at higher pH values which hold the performance of decolorization. The research investigated by Manali and coauthors demonstrated that the degradation of Methylene blue (MB) was up to 94.6% after 26 min [27].

5.1.7 Coagulation–Flocculation

The treatment of drinking water, textile wastewater, and reducing cop partly were carried out using coagulation–flocculation technique. It is considered an essential and most efficient method. The usefulness is being reported in Sanchez-Martin et al. [28].

By adding coagulant to neutralize the negative charges is accompanied with the destabilization of colloids is happening in the coagulation process.

In the flocculation process, the huge particles are collected to form microflocs that will be taken out in subsequent sedimentation stages or flotation stages. An agent that causes a liquid or sol to coagulate is called a coagulant.

Coagulants are divided into two main types coagulants and coagulant aid. Coagulants are either metallic salts, polymers or naturally occurring materials. The removal of dyes by aluminum sulpfate and polyaluminium chloride has been reported. Khayet et al. stated that the effectiveness of aluminum sulpfate for the removal of Acid black 210 (AB 210) is more than 90% under the condition of 40 mg/L of dosage, 4–8 range of pH, and 35 mg/L of dye concentration [29].

To modify or to assist a process, coagulant aid is always used. It is basically not a coagulant. It can be a material or can be a chemical. The use of a coagulant aid will increase the density of slow-settling flocs and increase the toughness of the flocs so that they do not break up during the mixing and settling processes.

5.1.8 Electrochemical Methods

In recent years, the most attractive method due to complete decolorization, low final temperature, operating under moderate pH range, reduced BOD and COD with no sludge formation is the electrochemical method. Electrochemical method is used in the preparation of compounds or retrieval of metals, huge growth of decolorization of dye applications. In this process, highly toxic pollutants are destroyed either directly or indirectly by oxidation process.

In anode surface, the pollutants are adsorbed first and then destroyed by an anodic transfer reaction, which is a direct anodic oxidation process.

Strong oxidants are released during electrolysis and then destroyed in the oxidant solution during its oxidation process, which is known as the indirect anodic oxidation process [30].

5.1.9 Biological Methods

The most regular and universal technique used in wastewater dye treatment is biological treatment [31]. Microbial technique implementation is shown to be highly economical, less intensive to treat industrial waste or sewage. These can be treated better through microbial biosorption and biodegradation. Transformation of dye molecules to be harmless is done through microorganisms. Functional groups like amino, hydroxyl, phosphate, carboxyl present in cell wall components are accredited with the above property of microorganisms. This will result in a strong attractive force between the cell wall and azo dye.

Different species were used for the mineralization and decolorization of various dyes. This method is advantageous with inexpensive, low run cost, nontoxic resultants. Biological methods are of two types: aerobic and anaerobic, but sometimes a combination of aerobic and anaerobic is also predicted which is depicted below in detail [32–36].

5.1.10 Aerobic Treatment

The potentiality to treat dye wastewaters was carried out by bacteria and fungi worldwide. Bacteria present in wastewater secretes enzymes which will give out organic compounds. Aerobic bacteria isolation has been still persisting [37]. Straining *Kurthia* sp. Triphenylmethane dyes such as crystal violet, pararosaniline brilliant green, malachite green have decolorized upto 92–100%. In this study, after biotransformation, the COD reduction of cell extracts triphenylmethane more than 88% except ethyl violet (70%). Synthetic dyes does not allow to decompose by activated sewage in aerobic process [38]. Anerobic strain of bacteria through decolorization showed outstanding ability on the structure of dye [39].

Research is still going on which is based on the study of decolorization of azo dyes and triphenylmethane dyes with the help of fungal strains [40]. Different studies were performed with different fungi, namely Phanerochaete chrysosporium for the decolorization of dyes. For the same purpose, several microorganisms were also included. For example, Cyathus bulleri, Funalia trogii, etc. They concluded during various tests that concentration, temperature, pH like factors are affecting the decolorization process.

5.1.11 Anaerobic Treatment

Degradation of synthetic dye was demonstrated [41]. Under aerobic condition, dye decolorization was succeeded but in conventional aerobic systems perception of non biodegradability still persists [37]. Mordant Orange 1 and azodisalicylate were decolorized through methanogenic granular sludge [42]. Zee Van der demonstrates the possibility of the application of anaerobic granular sludge for the decolorization of 20 azo dyes [43].

The problem of bulk sludge can be avoided through anaerobic pretreatment. It was also possible for exclusion of expensive aeration. With the removal of BOD, decolorization of dyes can be done easily. And also foaming problems can be rectified with the surfactants, through sulfate reduction heavy metals can be maintained, effluents' high pH can be acidified, refractory organics degradation can be initiated. The main disadvantage was insufficient on BOD removal, dyes were loving, nutrients were not removed [41].

5.1.12 Combined Aerobic–Anaerobic Treatment

A combined form of aerobic and anaerobic treatment is advised to get better results of decolorization of dyes from industries. A major advantage of this treatment is the full mineralization through the collaboration of microorganisms. Enhancing the controlled anaerobic bioreactors resulting in the reduction of azo bond, colorless aromatic amines. Hence, it will make azo dye treatment fetching. Thus, the combination is now advised for dye treatment in wastewaters [44–46].

The factors influencing this treatment are the concentration and pH of the dye and the temperature of sewage. The main disadvantage of the biological method is low biodegradability of dyes, less flexibility in design and operation, larger land area requirement, and longer times required for decolorisation–fermentation processes, thereby making it incapable of removing dyes from effluent on a continuous basis in liquid state fermentations [47].

5.1.13 Bioremediation of Dyes

Biodegradation or biological degradation means degradation of the dyes to a less toxic compound, which is economically feasible, eco-friendly, and eventually aids color removal [48]. The removal mechanism of azo dyes involves two steps. In the first step, azo bonds were broken down to form amines and in the second step, the amines were metabolized into nontoxic molecules under aerobic condition [49]. Under aerobic and anaerobic environments, the bacteria have to survive so that the degradation will be completely fulfilled for the formation of azo bonds within the dyes themselves.

The wastewater treatment with using different fungi in dying industries has been reported [50]. With the combination of aerobic and anaerobic treatment by different microbes, biological degradation of textile dyes has been noticed a better improvement [51]. Replacement of physical and chemical with the fungal based dye decolorization have conducted [52]. An important tool for the effluent treatment from textiles, paper and pulp industries that comprises PAH-polycyclic aromatic hydrocarbon is Phanerocheate chrysosporium [53]. This develops lignin and manganese peroxidases like enzymes. Another study involves Trichoderma harzianum, introduced for cleaning the sludge from different industries [54]. Combo of bromophenol blue dyes and Congo red were completely degrades using semi-solid PDA as a medium with fungal mycelium process. The growth of Trichoderma harzianum was inhibited by the medium which contains Basic blue, Congo red, Acid red, Direct green, and Bromophenol blue.

Bromophenol blue was considered as a good inhibitor compared to other dyes. Fungi is distinct as it is fast-growing with larger biomass and the hyphal spectra are comparatively higher than bacteria. Fungi have a high surface to cell ratio, thus it is considered as a string degrader [55]. Filamented fungi can be developed as a strong degrader as it has high surface to cell ratio. The degradation of textile dyes like Orange 3R was studied based on the combination of Aspergillus Strain (MMF3) with NaCl having different concentrations [54].

The high decolouration even in a naïve condition, the fungal strain is considered in the biological treatment of the wastes from the dyeing mills. The anthraquinone dye-Reactive Blue4 dye can be removed by marine derived fungus as bioremediation process [56] involves two steps for degradation. First is the partial treatment of purified laccase and the second is with compounds (with low molecular weight) as the final enzymatic degradation process. It is found that 29% reduction is confirmed in total carbon and in toxicity, a twofold reduction is found. The azo dyes found in wastewater by the biosorption of paramorphogenic form of Aspergillus oryzae are reported [57]. The laccase enzyme isolated from Trametes polyzona strain WR710-1is examined for the decoloration and deprivation of bisphenol-A and other synthetic dye. About 61% decoloration was observed by Anthraquinone dye (Reactive Blue 4) and was found that the COD was reduced by twofold in 12 h by laccase treatment. The whole process of metabolites synthesis was monitored by mass spectrometry, UPLC, and UV–vis spectroscopy to analyze the Characterization. Recently, for eliminating tannins along with dyes, Biodegradation combined with ozonation has been investigated to minimize pollutant load in the effluent. The biodegradation of wattle extract and various synthetic dyes was extensively studied using various fungal cultures, like Penicillium sp. and Aspergillus niger. The removal process of tannins and dyes was successful with the usage of a hybrid model of ozonation and biodegradation. As the degradation rate is higher in ozonation, this method is highly beneficial compared with biodegradation. The pollution level of different physical parameters like BOD, COD, TOC, TSS, and TDS was found to be at a minimum by this combined model [58].

5.1.14 Bacterial Degradation of Dyes

In bioremediation, microorganisms capable of catabolizing the organic pollutants to either recognize the toxic pollutants or to degrade the dyes in wastewater is to be thoroughly studied [59, 60]. The study of the dye degradation ability of bacteria must be done vigorously because it is easy to culture and grow them. Further, it is found that chlorinated and aromatic hydrocarbon-based organic pollutants are catabolized by bacteria. Carbon source will decompose the bacteria and the same bacteria have the ability to oxidize the sulfur-based textile dyes to sulphuric acid. Many bacteria were identified to degrade different azo-based dyes at a faster rate. By the reduction of azo dyes, different bacterial group under traditional aerobic, anaerobic and under extreme oxygen deficient conditions are employed for decolorization. The azo dyes start with the breaking of –N=N– linkage is the first chemical reaction during the reduction under anaerobic environment by the azoreductase enzyme which results in a colorless solution of aromatic amines. It is also reported that after dye reduction, the metabolites formed [61] can further be catabolized using aerobic or anaerobic processes. Also during dye decolorization, the intermediate products can also be reduced by hydroxylase and oxygenase and other enzymes produced by the bacteria [62, 63]. The aerobic bacteria that propagates in azo compounds are responsible for the formation of intermediate sulfonated amines which are aerobically degraded. Bacillus subtilis, Clostridium perfringens, Proteus sp., Pseudomonas aeruginosa, Pseudomonas putida are some Gram-positive bacterial strains that were found to decolorize various structurally different textile azo dyes effectively. Similarly, Klebsiella pneumonia, Enterococcus sp., and Escherichia coli, being Gram-negative bacterial strains, exhibit decolorizing efficacy on various dyes. Thus, some naturally occurring bacteria in the natural water and soil systems are used to biodegrade textile wastewater.

6 Advanced Oxidation Processes

Advanced Oxidation processes (AOPs) are the processes that involves the powerful generation of hydroxyl radical (•OH). The hydroxyl radical, •OH, is the neutral form of the hydroxide ion. Hydroxyl radicals are short-lived and highly reactive. Hydroxyl radicals are produced from different kinds of energy (electrical, sound, etc.), hydrogen peroxide, ozone, catalyst. AOPs play a vital role in removing organic waste from the water. Numerous studies were out regarding the AOPs in water treatment. Generally, in AOPs •OH radical will be released for the degradation of pollutants. They attack the molecules quickly and unselectively. AOPs depend on the production of highly reactive hydroxyl radicals (•OH). Hydroxyl radicals are strong oxidants, can oxidize anything present in water. It also reacts indiscriminately and can easily fragmented results in the conversion of small inorganic molecules. These radicals are formed with the help of hydrogen peroxide, ozone or UV light. Two types of systems are possible in AOPs. Homogenous system and Heterogeneous system.

Nonbiodegradable materials such as pesticides, aromatics, some volatile compounds present in water can be cleansed up using AOP process. The secondary treated water with sewage can be treated using AOPs known as tertiary treatment. The contaminants will undergo mineralization. The ultimate goal of AOPs is the purification of water from chemical contaminants there by reducing the toxicity of water (Fig. 5).

For the generation of hydroxyl radicals, the AOPs are classified into different types.

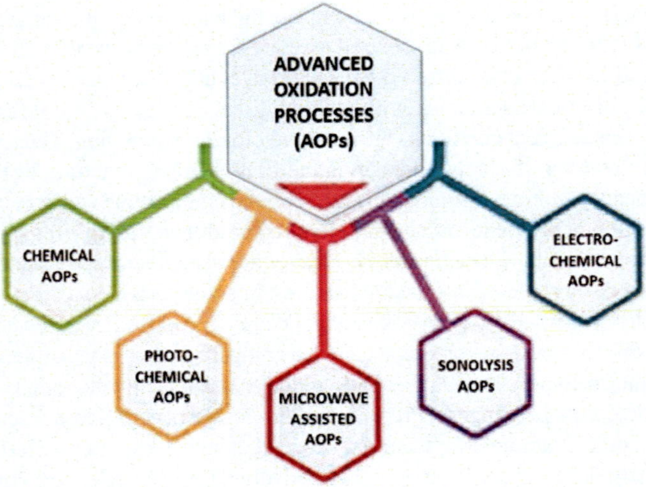

Fig. 5 Advanced oxidation process-brief

6.1 Chemical AOPs

Various combinations of chemical agents produce some temporary products. Usually, hydrogen peroxide is used as an oxidant, sometimes ozone and air are also used.

Fentons' Reagent (Fe^{2+}/H_2O_2)

A combination of soluble ion (II) salt and H_2O_2 known as Fenton's reagent is used for the degradation process. As it was developed by Fenton, the reagent was later known as Fenton's reagent. Fenton tried a mixture of Fe^{2+} and H_2O_2 for the tartaric acid destruction and oxidation.

$$Fe^{2+} + H_2O_2 \rightarrow Fe^{3+} + HO\bullet + OH^- \text{ (initiation of Fenton's reagent)}$$

By reduction of oxygen on different electrodes such as gas diffusion electrodes, graphite hydrogen peroxide can be formed.

$$O_2 + 2H^+ + 2e^- \rightarrow H_2O_2 \quad (1)$$

With the addition of Fe^{2+} ions as a catalyst, very strong, active oxidant hydroxyl radical can be generated which initiate the oxidation of pollutant (RH) molecules

$$Fe^{2+} + H_2O_2 + H^+ \rightarrow Fe^{3+} + \bullet OH + H_2O \quad (2)$$

$$Fe^{3+} + H_2O_2 \rightarrow Fe^{2+} + \bullet OOH + H^+ \text{ (regeneration of } Fe^{2+} \text{ catalyst)} \quad (3)$$

$$\bullet OH + RH \rightarrow R\bullet + H_2O \quad (4)$$

$$H_2O_2 \rightarrow HO\bullet + HOO\bullet + H_2O \text{ (Self scavenging and decomposition of } H_2O_2\text{)}$$

Controlled production of radicals is the advantage of the Fenton process. Fenton process has been studied in vast contaminants such as pesticides, organic molecules, also degradation of benzene ring compounds, azo dyes. As this process is easy to handle, simple inexpensive chemicals and energy output truancy paved the advantages toward the water treatment.

Heterogeneous Fenton Process (Fenton Like Reactions)

Homogeneous Fenton process has shortcomings for Fenton reactions, heterogeneous Fenton process taken into account. In this system, the Fe (III) species immobilized within the structure of pores of catalyst generates hydroxyl radicals from H_2O_2. In this method, the strain of iron ions into water results in the precipitation of iron hydroxide is prevented.

This method is easily separable, recovery of the catalysts and extendable during continuous operations. In this process, nano zerovalent iron (NZVI) is highly acceptable. Oxidation of NZVI delivers an alternative solution for the production of Fenton's reagent.

$$O_2 + Fe^0 + 2H^+ \rightarrow Fe^{2+} + H_2O_2$$

Nano zerovalent iron commonly influences Carbon-based materials, palygorskite composites, chitosan, etc.

Catalytic Wet Peroxide Oxidation (H_2O_2/Catalyst)

The heterogeneous Fenton process using water-soluble catalyst is known as the Catalytic wet peroxide oxidation process. It involves Fe^{2+}/Fe^{3+} ions for degradation process. Some variable valency materials such as cobalt, copper, gold, nickel, and carbon-based materials will decompose hydrogen peroxide to hydroxyl radical.

Perozonation (O_3/H_2O_2)

A fast reaction of O_3 and H_2O_2 which produces hydroxyl radical is known as perozonation.

$$O_3 + HO_2^- \rightarrow O_2 + \bullet OH + O_2^-$$

This process is simple and has good bacterial activity. For this reason, perozonation is highly recommended for the disinfection of potable waters. However, the drawbacks of this process are sensitivity to pH, temperature, and low water solubility.

Catalytic Ozonation (Ozone/Catalyst)

Due to the efficient oxidation capacity of ozone, it plays a vital role in water treatment. Mainly the removal of contaminants from drinking water is carried out in the catalytic ozonation process. Heterogeneous catalysts are used to improve the reactivity through milder conditions, reduces the cost. A complex will be formed in between the catalyst and organic compound which will dispose into ozone and metal particles.

$$Fe^{2+} + O^3 \rightarrow Fe^{3+} + O_3^-$$

$$O_3^- + H^+ \leftrightarrow HO_3 \rightarrow O_2 + \bullet OH$$

Materials like $NiFe_2O_4$, MgO, Co_3O_4 have been tried as a catalyst in ozonation.

Catalytic Wet Air Oxidation (O_2/Catalyst)

Molecular oxygen is used as an oxidizing agent for the treatment of water. This process needs high temperature (473–593 K) and pressure (20–200 bar) for the generation of hydroxyl radicals. This method is frequently used for the treatment of sewage having organic matter, COD, defiant to biological purification. The compounds will break down in the form of easily treatable, simple before their release.

$$RH + O_2 \rightarrow R_{\bullet} + HO_{2\bullet}$$

$$RH + HO_{2\bullet} \rightarrow R_{\bullet} \rightarrow H_2O_2$$

$$H_2O_2 + M \longrightarrow 2\bullet\ OH + þzM$$

The main disadvantage of the wet air oxidation process is that it's unfit for mineralization. Mixed oxides, ruthenium, Ru/nitrogen-doped carbon nano fibers have been used for the oxidation process.

6.2 Photochemical Advanced Oxidation Processes

Light is generally used in photochemical advanced oxidation processes as it is clean, inexpensive, more efficient than chemical AOPs, can disinfect waters. In this method, UV radiations are coupled with ozone and hydrogen peroxide with the catalyst either Fe^{3+} or TiO_2.

This method can be degraded under three processes, namely photooxidation in presence of H_2O_2, photodecomposition in presence of O_3 on UV irradiation, and oxidation by photocatalysis.

Photo Peroxidation (H_2O_2/UV)

H_2O_2 is ionized by absorption of UV radiations (200–300 nm). The decomposition of uncharged radicals of O–O bond of hydrogen peroxide molecule leads to form hydroxyl radicals through the following reactions

$$H_2O_2 + h\nu \rightarrow 2\bullet OH$$

$$\bullet OH + H_2O_2 \rightarrow H_2O + HO_2\bullet$$

$$HO_2\bullet + H_2O_2 \rightarrow \bullet OH + H_2O + O_2$$

$$\bullet OH + HO_2^- \rightarrow HO_2\bullet + OH^-$$

$$2\ HO_2\bullet \rightarrow H_2O_2 + O_2$$

$$\bullet OH + HO_2\bullet \rightarrow H_2O + O_2$$

$$2\bullet OH \rightarrow H_2O_2$$

For higher efficiency degradation it is advisable to use a higher concentration of hydrogen peroxide as the molar absorption of hydrogen peroxide is weaker in UV radiation.

Photo Ozonation (O_3/UV)

Ozone present in water absorbs UV region releases hydroxyl radicals which are proven highly efficient for sewage removal. The series of reactions are as follows:

$$O_3 + H_2O + h\nu \rightarrow 2\bullet OH + O_2$$

$$O_3 + \bullet OH \rightarrow HO_2\bullet + O_2$$

$$O_3 + HO_2\bullet \rightarrow \bullet OH + 2\,O_2$$

$$\bullet OH + HO_2\bullet \rightarrow H_2O + O_2$$

Heterogeneous Photocatalysis

When light is exposed to a semiconductor having bandgap, valence electron is excited to conduction band leaving a hole in the valence band. In photocatalysis method, the electron hole pair will move to the surface of the semiconductor leading to redox reactions, and as a result, hydroxyl radicals with ion radicals are formed. The hole thus formed in a semiconductor serves as an oxidant which degrades the contaminants directly and hence purifies the water. Titania nanoparticles under this photocatalysis method are exclusively studied and now emerged as second widest method for purification of water.

Photo-Fenton/Fenton Like Systems

UV radiation dissociates hydrogen peroxide directly into hydroxyl radicals with a range of 250–400 nm. The intensity of UV radiation is significant in the degradation process. Various materials like ferric hydroxide, Fe/ZSM-5 zeolite Fe- supported bentonite, nanostructures like nickel ferrite have been experimented under this process.

6.3 *Microwave Enhanced AOPs*

Microwave enhanced AOPs are categorized into two processes namely homogenous and heterogeneous processes. In a homogeneous process, the chemical oxidants are activated primarily under the influence of microwaves and result in the production of hydroxyl and superoxide anion free radicals which will help in the degradation of pollutant species. In a heterogeneous process, primarily the formation of solid substrates (including solid catalyst) and hot spots on the edges occur. These are also under the influence of microwaves. The hotspots make the formation of hydroxyl and superoxide anionic radicals which will prompt the degradation of organic pollutants.

In this process, microwaves are generated at 2.5 GHz frequency and electrodeless discharge lamps were used as a source of ultraviolet radiation. Under continuous microwave exposure, it will give an ultraviolet–visible electromagnetic spectrum. There will be a glass tube containing electron–ion plasma material that acts as a plasma chamber. When the microwave enters the chamber, the electrons get excited and collide with the gas atoms and ionize them to produce more excited electrons.

In the microwave coupled ultraviolet process, microwaves cause the dye molecules to get stabilized at high energy excited states which promote the generation of hydroxyl radicals and result in rapid decay of dyes. In microwave-assisted ultraviolet/H_2O_2 process, H_2O_2 yields a huge quantity of additional hydroxyl radicals under ultraviolet radiation. A combined microwave/ultraviolet/H_2O_2 process is highly recommended for the removal of dyes as coupling of microwaves will rise the temperature of the aqueous pollutant matrix, thereby resulting in enormous production of hydroxyl radicals and increasing the degradation of microwave/ultraviolet/H_2O_2 process.

The microwave/Fenton process improves the efficiency of dye removal. The priorities in this process involve pH-insensitive, less floc formation, better settling, and no need for pretreatment. This method is efficient for the degradation of dyes/mineralization due to synergistic effect of ultraviolet/microwave radiation. The limitations of ultraviolet/H_2O_2 and ultraviolet Fenton processes are encountered by the investigation of microwave coupled photocatalysis. In this method, secondary pollutants are produced and offer low mineralization.

Acid Orange 7 is degraded through microwave radiation by using polyaniline as a catalyst in the absence of light source. Bromothymol blue dye in an aqueous medium is degraded through microwave/ultraviolet/TiO_2/O_3/H_2O_2 hybrid process [64].

6.4 *Sonolysis/Sonochemical AOPs*

In sonolysis, ultrasonic irradiation is used to produce hydroxyl radicals through the pyrolysis of water. For removing the toxic pollutants, this process won't produce any exterior chemicals in the water system. Accordingly, sonolysis is referred to as the "green method". Initially, acoustic bubbles were formed during the ultrasound irradiation. The bubble formation is controlled by changing the frequency and power that were given in the form of ultrasound serves as a parameter for controlling the degradation rate of pollutants. Sonochemical treatment efficiency is controlled by cavitational activity which is yet another crucial parameter in this process. Usage of intense ultrasonic waves increases the temperature and pressure of the aqueous matrix resulting in the destruction/oxidation of organic compounds or pollutants into simpler products due to the formation of highly free reactive radicals.

The major drawback of this process is that it does not require any additional chemicals for the production of hydroxyl radicals. Later it requires high energy and extra chemicals for mineralization. With this drawback, the sonolysis process is limited with lab scale or low initial COD value pollutants. The final products of this process are carbon dioxide, water, inorganic salts, hydrogen, etc. The combination of ultrasonic, photocatalytic, Fenton, ultrasonic radiation process has more influence on the degradation of dyes by the release of a huge amount of hydroxyl radicals.

6.5 Electrochemical Advanced Oxidation Processes (EAOPs)

The electrochemical AOPs involve electrochemical treatments which are used to remove toxic pollutants. This method is based on the generation of the most powerful oxidizing agent known as hydroxyl radicals, able to destroy the organic compounds present in the water. Electrochemical AOPs basically involves three methods

1. Anodic Oxidation: One of the popular EAOP is Anodic oxidation, where organic compounds are directly oxidized at the anodic surface by electron transfer or indirectly oxidized by •OH weakly physisorbed at anodic surface or agents at the bulk solution (active chlorine species, O_3, persulfates, and H_2O_2). When anodic oxidation is combined with cathodic electrogeneration of H_2O_2, then the process can be named anodic oxidation with electrogenerated H_2O_2 (AO-H_2O_2). The oxidation process in anodic oxidation is carried out by the following processes:

   - Direct electron transfer to anode surface "X"
   - During the oxidation of water to oxygen, heterogeneous reactive species are formed as intermediates including physisorbed •OH at anode surface [X(•OH)], H_2O_2 (weaker oxidants produced from [X(•OH)]) and ozone formed from water discharge at the anode surface
   - Other weaker oxidant agents produced electrochemically formed from ions existing in the bulk.

   The equations for anodic oxidation:

$$X + H_2O \rightarrow X(\bullet OH) + H^+ + e^-$$

$$2X(\bullet OH) \rightarrow 2MO + H_2O_2$$

$$3H_2O \rightarrow O_3 + 6H^+ + 6e^-$$

The efficiency of anodic oxidation is purely dependent on the mass transfer of pollutants from the bulk solution to the anode surface. Anodes are usually made up of Ti-based alloys, TiO_2, graphite, Ru or Ir oxides, boron-doped diamond (BDD). Studies arose interest in the mineralization of organic pollutants by direct electrochemical oxidation using boron-doped diamond. These studies then conclude that BDD exhibits good chemical and electrochemical stability, a wide potential for water discharge, and longer durability.

2. Electro-Fenton (EF) Process
Electro-Fenton (EF) process is the most known method in EAOP, which involves an indirect electrochemical way to produce/generate hydroxyl radical in aqueous solutions. This method was developed to overcome the drawbacks of the chemical Fenton process. It can also be called as electrochemically assisted Fenton process. •OH is generated through Fenton's reagent (please refer chemical AOPs) in which Fenton's reagent is electrochemically produced in situ preventing the use of high quantities of H_2O_2 and Iron (II) salt. The production of hydrogen peroxide is one of the crucial parameters as H_2O_2 controls the rate of Fenton's reagent. Hydrogen peroxide is continuously supplied to wastewater in an electrochemical reactor from the two-electron cathodic reduction of oxygen gas, directly injected as compressed air.

$$O_2(g) + 2H^+ + 2e^- \rightarrow H_2O_2$$

The above H_2O_2 production efficiency is not high and depends on operating conditions such as solubility, temperature, pH, O_2).

The second component of Fenton's reagent is Fe^{2+} ion, which is formerly started in a catalytic amount (0.1 nM) in the form of ferrous salts and is electrocatalytically regenerated from the reduction of Fe^{3+} formed by Fenton's reaction. Now, this Fenton's reagent is produced continuously producing hydroxyl radicals through Fenton's reaction, thereby destroying the organic pollutants. The hydroxyl radicals react in the bulk solution with organic compounds resulting in mineralization/oxidation [65].

The main advantages of the EF process over the classical Fenton process are:

- Elimination of parasitic reactions that waste hydroxyl radicals.
- Probability of controlling the degradation kinetics and performing mechanistic studies.
- In situ controlled generation of Fenton's reagent.
- Full control of processing by current/potential control.
- Total mineralization of organic compounds including the intermediates.

The selection of the electrochemical process is purely dependent on the structure and nature of the electrode material, electrolyte composition, and experimental conditions which prevent electrical fouling. Due to versatility, efficiency, easy handling, safer toward environment compatibility, electrochemical AOPs are highly recommended.

7 Nanomaterials in AOPs

The metal, metal oxide, and inorganic nanomaterials like nano zerovalent zinc, magnetic Fe_3O_4, magnesium oxide (MgO), titanium dioxide (TiO_2), zinc oxide (ZnO) are broadly used for the removal of dyes. It is mainly due to their provide high surface area and specific affinity properties. Further, it is found that nano metal oxides have a low impact on the environment, low solubility, and no secondary pollution (Fig. 6).

7.1 *Nano Zerovalent Iron*

Nano zerovalent iron (nZVI) is powerful in eliminating pollutants like nitrates, organochlorine pesticides, chlorinated compounds, heavy metals, and dyes. It is assumed that the oxidation of the Fe^0 is responsible for the reactivity of core–shell nanoparticles of ZVI with their microscale counterparts. The high density and high intrinsic reactivity of their reactive surface sites also lead to their higher reactivity.

nZVI is effectively found to remove three azo dyes like Sunset yellow, Acid blue A, and Methyl orange [66]. Increasing the dosage of nZVI, increased the decolonization of the dyes. But the degradation decreased with an increase in the concentration of dyes. The acidic condition was favorable in pH effect to the adsorption which may be due to reduction of H^+ ions to atoms by the electrons released from iron particles. A cleavage of formed by the chromophore group and conjugated system in dyes by these

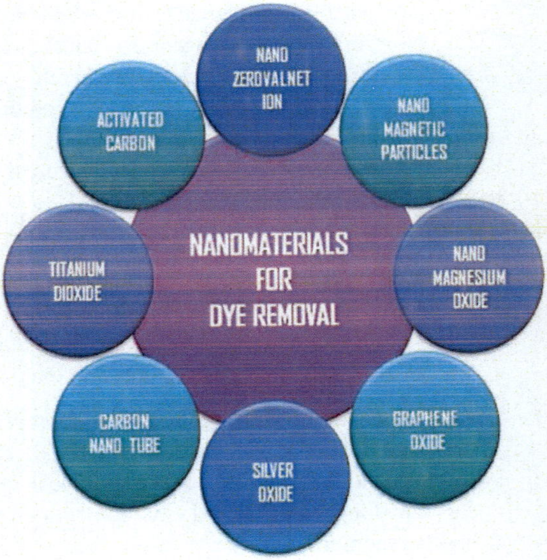

Fig. 6 Nanomaterials for dye removal

atoms with the decrease of pH. The results in the highly efficient decolorization of Sunset yellow, Acid blue A, and Methyl orange by nZVI. Reactive black 5, Reactive red 198 and Light green are removed. 500 mg of nZVI with concentration of 100 ppm removed Reactive black 5 and Reactive red 198 completely and effectively removed 97% of light green dye.

7.2 *Nanomaterials with Magnetic Properties*

Iron oxide like magnetite (Fe_3O_4), hematite (γ-Fe_2O_3), and maghemite (α-Fe_2O_3) [67] due to their high ratio of surface area to volume, high magnetic susceptibility, and excellent biocompatibility, have the tendency to oxidation by air and aggregation in aqueous systems.

Fe_3O_4 nanoparticles coated with humic acid (HA), which enhances the stability by reducing oxidation of Fe_3O_4 are used to remove Rhodamine B (Rh B). The adsorption of Rh B by Fe_3O_4/HA reached equilibrium within 15 min, with a maximum adsorption capacity of 161.8 mg/g. It removed over 98.5% of Rh B at an optimized pH [68]. Fe_3O_4 nanomaterials were fabricated from cetyltrimethylammonium bromide by chemical precipitation method. This was used as a surfactant to remove Acridine orange (AO), Coomassie brilliant blue R-250 (CBB), and Congo red (CR) at pH of 4 for CBB, 6 for AO and CR, and at the dosage amount of 0.02 g. Methylene blue (MB) was removed by impregnating magnetic nanoparticles onto maize cobs with 99.9% efficiency. Reactive red 120, Rhodamine 6G, and Direct blue 15 were removed by surface modification of IONPs like ionic liquids, polyacrylic acid, and silica-based cyclodextrin (Al-CD-MNPs) [69].

7.3 *Nano Magnesium Oxide*

Nano magnesium oxide (nano-MgO), a nanosized alkaline earth metal oxide with high surface reactivity and adsorption capacity, is used as a destructive adsorbent for the removal of many toxic chemicals. Besides, it has excellent optical, electrical, thermodynamic, mechanical, electronic, and special chemical properties. Hence, it is utilized as catalyst support, for toxic-waste remediation, refractory as an additive, in paints and superconducting products, and as bactericide and adsorbent [70]. It is reported that MgO nanoparticles act as an effective sorbent for Reactive black 5 (RB 5) and Reactive orange 122 (RO 122) with maximum adsorption capacity of 500 mg/g and 333.34 mg/g, for RB 5 and RO 122, respectively [71].

Removal of toxic dyes such as Congo red and malachite green has been studied with different morphologies of MgO such as nanorods, nanoflakes, etc. Hence, synthesis of MgO with enhanced surface area with varying morphologies such as rods, wires, belts, tubes were recently attempted [72]. In a comparative study of MgO as nanorods, hierarchical nanostructures and nanoflakes for removal of dyes were done. It exhibited excellent adsorption with maximum sorption capacities of 1205.23 mg/g and 1050.81 mg/g, respectively, for the removal of both the dyes which was relatively considered much higher than other absorbents, maybe due to the high surface area and hierarchical structures.

An advanced composite material fabricated by the modification of nano-MgO form was used as the adsorbent for the removal of Rhodamine B (Rh B) [73]. The adsorption process was explained by Langmuir isotherm, where the maximum adsorption capacity of 16.2 mg/g at pH of 6.75, dosage of 100 mg, and contact time of 2 h was observed. By using modified co-precipitation method, Rice straw charcoal/MgO nanocomposites were prepared to investigate the ability to remove Reactive blue 221 (RB 221) from aqueous solutions by varying the amount of shaking time of 90 min, shaking speed of 200 rpm, pH of 7.0, temperature of 26 °C and adsorbent dosage of 250 mg at an initial adsorbate concentration of 30 mg/L. The pseudo first-order kinetics and the Freundlich isotherm were attributed for the adsorption of this dye. The Langmuir isotherm was calculated to be 27.78 mg/g as the maximum adsorption capacity. Thus, adsorption of RB-221on nanocomposite of rice straw charcoal/MgO nanocomposite from aqueous solutions was shown possible.

7.4 *Graphene Oxide and Graphene Oxide Based Nanomaterials*

A two-dimensional nanomaterial with tightly packed six-membered rings of sp^2 carbon atoms is known as graphene. Graphene has strong mechanical, electrical, and optical properties. It is found in applications like nano-electronic devices, sensors, and nanocomposite materials. Further, it to have strong π-π stacking due to the graphitized basal plane structure, found to strongly interact with the aromatic moieties present in various dyes [73]. Graphene oxide (GO) containing oxygen-rich functional groups on its surface, (i.e., carboxyl, carbonyl, hydroxyl groups) is found to disperse in water and enhance the electrostatic interactions with cationic dye molecules. Hence, it is considered as a promising substrate for the preparation of various graphene-based nanocomposites.

Graphene oxide is found to have high efficiency in removal of lead to the increase of solution pH. Thus, the magnetite/reduced graphene oxide nanocomposites (MRGO) are used to remove Rh B (over 91%) and MG (over 94%) [74]. And the removal efficiencies were over 80% even after five cycles. Hence, MRGO may effectively be used for the removal of dye pollutants. Reduced graphene oxide-supported nanoscale zerovalent iron (nZVI/rGO) was used in the removal of Rh B using artificial intelligence tools [75]. To optimize and predict the optimum conditions for the maximum removal efficiency Response surface methodology (RSM) and artificial neural network hybridized with genetic algorithm (ANN-GA) were used. The results predicted and experimental value by the ANN-GA model were found to be

(90.0%) and (86.4%), respectively. Freundlich isotherm was used to fit the experimental data, and the maximum adsorption capacity based on the Langmuir isotherm was 87.72 mg/gm [76].

7.5 Silver Nanoparticles.

Antibacterial mechanism and microbial inactivity in water is investigated in silver. Their lower toxicity nature is reported [77]. Silver nanoparticles are derived from silver chloride and silver nitrate. The effectiveness of silver nanoparticles in dye degradation is well studied [78].

Water treatment with silver and medicine with silver nanoparticles has been studied. Anti-infective efficiency is more in silver nanoparticles. Compared to bulk material, the antimicrobial characteristics of Ag NPs are improved with mechanical, physical, and chemical properties. For diabetic patients, antimicrobial bandages with Ag NPs is the most promising application. According to the synthesis of Ag NPs, the efficiency varies. Some of such applications are the type of metal precursors, solvents, and reducing agents. Silver nanoparticles are synthesized as a reducing agent in which water is the primary solvent, thereby efficiency is determined.

Methyl Orange (MO) can be degraded using silver as a nanocatalyst under photodegradation/visible light. In general, Ag Nanoparticles have been used as a catalyst for the degradation of dyes namely CR, MO through chemical reduction by NaBH4 method [18]. Silver nanoparticles incorporated with the novel nanocomposite adsorbent, namely the graft copolymer of Poly (AA)/GG have been studied for the removal of methylene blue from the water [79].

Silver nanoparticle with activated carbon was examined for the removal of methylene blue [80]. Silver nanoparticles with nano silica powder combination show the removal of dyes such as Eosin yellow, Bromophenol blue 2, Congo Red, Brilliant blue on adsorption. Novel composite of silver nanoparticles with ploy (styrene-N-isopropyl-methacrylic acid) reduces adsorption of methylene blue (Sivasankari Marimuthu 2020).

7.6 TiO_2 Nanoparticles

The most promising nanoparticle used for water purification is TiO_2 nanoparticles. It has been used as a photocatalyst because of its nontoxic nature, highly stable, and reactive nature.

The removal of Direct black 38 (DB 38) with UV/TiO_2 shows the decolorization process which is sensitive to pH, which arrives to the conclusion that pH is a controlling factor in the removal process [81]. Ni-doped TiO_2 thin films developed through chemical bath deposition under UV light irradiation method are prominent for removal of PonceauS dye [82]. RB 5 a prominent azo dyes are removed by the synthesized TiO_2-NPs through adsorption process. The behavior of the adsorption process can be explained with pseudo second-order kinetic model. The initial concentration of RB5 increases with the adsorption of RB 5 on TiO_2. The synthesized Titanium oxide nanoparticles thus proved to remove azo dyes from aqueous solution in better proportion. Further, the adsorbate and adsorbent ratio provides an economical way to produce expensive semiconductor material support which is convenient for the detoxification of pollutants.

7.7 Carbon Nanotubes (CNT)

Carbon nanotubes are one of the remarkable materials in nanotechnology which is used for different applications. Though it possess in single walled and multi walled, both were used for water remediation. Multi walled Carbon nanotube was studied for the removal of Ismate violet 2R dye from contaminated water [83]. The effects were studied using a batch process with influencing factors such as pH, adsorbent dosage, dye concentration, etc. Multi walled CNT, magnetically modified, was used for the removal of cationic dyes such as Crystal violet (CV), Thiionine (Th), Janus green B (JG B), and Methylene blue (MB) from contaminated water. This was easily adsorbed in water and separated magnetically. The influencing factors like pH, concentration, adsorbent were studied and concluded by Madrakian et al. [84]. Fe_3O_4 nanoparticles with multi walled CNT through chemical methods were used for removing dyes such as methylene blue and neutral red [85].

7.8 Activated Carbons

Water is polluted by synthetic dyes that are released from textile wastewater. Dyes and their degradation products are highly venomous. Activated Carbon (ACs) are known as good absorbents. ACs are used in many processes to remove the contaminants as they readily adsorb color and odor from wastewater as well as drinking water. The particle size, pH, surface properties of activated carbon affect the adsorption of dye. Activated carbon has a distinct molecular structure and it is considered to have high affinity to a variety of dyes. Methods like physical and chemical methods of dye wastewater treatment can be boosted by the addition of activated carbon [86].

AC prepared from waste cassava peel employing physical and chemical methods were tested in the removal of dyes and metal ions from an aqueous solution. The material impregnated with H_3PO_4 was reported more efficient than the heat-treated materials although both efficient as adsorbents for dyes and metal ions. The removal of Rhodamine-B was removed by using tapioca peel activated carbon as an adsorbent was reported [87]. The dye wastewater mainly consists of dying ingredients, sodium sulpfate anhydride (Na_2SO_4), and PVA (polyvinyl alcohol). Granular activated carbon (GAC) and zeolite was used as the adsorbent for dye wastewater [88].

Activated carbon obtained from peel, crown, and core of pineapple with phosphoric acid can be used for the adsorption of methylene blue and malachite green by the chemical activation method. The abundance of pineapple waste problem can be solved through this process [89].

Activated carbon with ground nut shell powder and Zinc Chloride as an adsorbent was used for the dye removal process. The results concluded with the high adsorption and high removal. Activated carbon with neem leaf powder was used for the removal of dyes, namely congo red, methylene blue, and brilliant green. The influence factors were tested and their interactions were studied [90].

The removal of acid red 183 from an aqueous solution was studied by activated carbon, raw kaolinite, and montmorillonite using an agitated batch adsorber [91]. Rice bran-based activated carbon and guava seeds activated carbon, followed by pyrolysis were also used as adsorbents to remove dyes from aqueous solutions.

8 Summary

Contamination of water through disposal of dyes was a major threat for human beings and the environment. Various aspects of dyes, their origin, effects toward human health and hazards, detection and degradation were explained in detail. Advanced oxidation processes, a vital process for dye removal, offer a promising solution for the degradation or removal of dyes (both organic and inorganic) from contaminated water. In future, we can expect more studies in this area for making mankind conservative (Table 1).

Table 1 Nanomaterials and the dyes removed

Nanomaterials	Dyes removed	References
Nano zerovalent iron (nZVI)	Sunset Yellow Acid blue A Methyl Orange	Rahman [66]
nZVI	Reactive red 198 Light Green Reactive black 5	Ruan [76]
Humic Acid coated with Fe3O4	Rhodamine B (Rh B)	Chaudhary [68]
Fabricated Fe3O4	Acridine Orange (AO) Coomassie brilliant blue R-250 (CBB) Congo Red (CR)	Liang [92]
Ionic nanoparticles (ionic liquids, polyacrylic acid and silica-based cyclodextrin (Al-CD-MNPs)	Reactive Red 120, Rhodamine 6G, Direct Blue 15	Absalan [69]
Nano-MgO	Reactive Black 5 Reactive Orange 122 (RO122)	Jamil [71]
MgO nanoflakes, hierarchical nanostructures, nanoflakes	Malachite Green Congo Red	Dhal [72]
Rice straw charcoal/MgO nanocomposite	Reactive Blue 221 (RB 221)	Moazzam [73]
nZVI/rGO	Rh B	Shi [75]
Magnetite/reduced graphene oxide nanocomposites	Rh B MG	Sun [74]
Silver NP loaded Activated Carbon	Methylene Blue	Ghaedi [80]
Nano silica powder fabricated with Ag NPs	Congo Red Eosin Yellow Bromophenol blue 2 Brilliant Blue	Sivasankari Marimuthu (2020)
UV/TiO2	Direct Black 38 (DB 38)	Seyyedi [81]
Ni-doped TiO2	PonceauS Dye	Marathe [82]
TiO2	RB 5	Hussein (2014)
Multi walled CNT	Ismate violet 2R	Abualnaja [83]
Multi walled CNT with magnetically modified	Crystal violet (CV), Thiionine (Th), Janus green B (JG), and methylene blue (MB)	Madrakian [84]
Fe3O4 nanoparticles with multi walled CNT	Methylene blue and neutral red	Qu [85]

(continued)

Table 1 (continued)

Nanomaterials	Dyes removed	References
Tapioca peel with activated carbon	Rh B	Fathima [87]
Activated carbon with neem leaf powder	congo red, methylene blue, and brilliant green	Sharma [90]
Activated carbon from pineapple with phosphoric acid	Malachite Green, methylene blue	Selvanathan [89]

References

1. Briggs D (2003) Environmental pollution and the global burden of disease. Br Med Bull 68:1–24
2. Kamble SM (2014) Water pollution and public health issues in Kolhapur city in Maharashtra. Int J Sci Res Publ 4(1):1–6
3. Halder JN, Islam MN (2015) Water pollution and its impact on the human health. J Environ Human 2(1):36–46
4. Ahmad SM, Yusafzai F, Bari T et al (2014) Assessment of heavy metals in surface water of River Panjkora Dir Lower, KPK Pakistan. J Bio Env Sci 5:144–152
5. Corcoran E, Nellemann C, Baker E et al (2010) Sick water the central role of wastewater management in sustainable development. A Rapid Response Assessment. United Nations Environment Programme
6. Nel LH, Markotter W (2009) New and emerging waterborne infectious diseases. Encycl Life Support Syst 1:1–10
7. Ullah S, Javed MW, Shafique M et al (2014) An integrated approach for quality assessment of drinking water using GIS: a case study of Lower Dir. J Himal Earth Sci 47(2):163–174
8. Krishnan S, Indu R (2006) Groundwater contamination in India: discussing physical processes, health and sociobehavioral dimensions. IWMI-Tata, Water Policy Research Programmes, Anand, India
9. Jabeen SQ, Mehmood S, Tariq B et al (2011) Health impact caused by poor water and sanitation in district Abbottabad. J Ayub Med Coll Abbottabad 23(1):47–50
10. Currie J, Joshua GZ, Katherine M et al (2013) Something in the water: contaminated drinking water and infant health. Can J Econ 46(3):791–810
11. Khan MA, Ghouri AM (2011) Environmental pollution: its effects on life and its remedies. J Arts Sci Commer 2(2):276–285
12. Ahmed T, Scholz F, Al-Faraj W et al (2013) Water-related impacts of climate change on agriculture and subsequently on public health: a review for generalists with particular reference to Pakistan. Int J Environ Res Public Health 13:1–16
13. Chowdhury S, Annabelle K, Klaus FZ (2015) Arsenic contamination of drinking water and mental health 1–28
14. Capenter SR, Caraco NF, Correll DL, Howarth RW, Sharpley AN, Smith VH (1998) Non-point pollution of surface waters with phosphorus and nitrogen. Ecol Appl 8(3):559–568
15. Chauhan N, Singh V, Kumar S, Sirohi K, Siwatch S (2019) Synthesis of nitrogen- and cobalt-doped rod-like mesoporous ZnO nanostructures to study their photocatalytic activity. J Sol-Gel Sci Technol 91:567–577. https://doi.org/10.1007/s10971-019-05059-3
16. Chen GS, Haase H, Mahltig B (2019) Chitosan-modified silica sol applications for the treatment of textile fabrics: a view on hydrophilic, antistatic and antimicrobial properties. J Sol-Gel Sci Technol 91:461–470. https://doi.org/10.1007/s10971-019-05046-8
17. Ojo O, Bakare S, Babatunde A (2007) Microbial and chemical analysis of potable water in public-water supply within Lagos University, Ojo. Afr J Infect Dis 1:30–35

18. Ismail M, Akhtar K, Khan MI, Kamal T, Khan MA, Asiri AM, Seo J, Khan SB (2019) Current pharmaceutical design 25:3653–3671
19. Jadhav JP, Parshetti GK, Kalme SD, Govindwar SP (2007) Decolourization of azo dye methyl red by Saccharomyces cerevisiae MTCC 463. Chemosphere 68(2):394–400. https://doi.org/10.1016/j.chemosphere.2006.12.087 [PMID: 17292452]
20. Muthirulan P, Nirmala Devi C, Meenakshi SM (2017) Synchronous role of coupled adsorption and photocatalytic degradation on CAC-TiO_2 composite generating excellent mineralization of alizarin cyanine green dye in aqueous solution. Arabian J Chem 10(Suppl. 1):S1477–S1483
21. Aksu Z, Karabayur G (2008) Comparison of biosorption properties of different kinds of fungi for the removal of Gryfalan Black RL metalcomplex dye. Bioresour Technol 99:7730–7741
22. Aspland JR (1997) Textile dyeing and coloration, Res. Triangle Park, NC, USA
23. Kushwaha AK, Gupta N, Chattopadhyaya MC (2014) Removal of cationic methylene blue and malachite green dyes from aqueous solution by waste materials of Daucus carota. J Saudi Chem Soc 18:200–207. https://doi.org/10.1016/j.jscs.2011.06.011
24. Sanchez-Prado L, Llompart M, Lores M, García-Jares C, Bayona JM, Cela R (2006) Monitoring the photochemical degradation of triclosan in wastewater by UV light and sunlight using solid-phase microextraction. Chemosphere 65(8):1338–1347. https://doi.org/10.1016/j.chemosphere.2006.04.025 [PMID: 16735047]
25. Srinivasan A, Viraraghavan T (2010) Decolorization of dye wastewaters by biosorbents: a review. J Environ Manag 91:1915
26. Alventosa-de Lara E, Barredo-Damas S, Alcaina-Miranda MI, Iborra-Clar MI (2012) Ultrafiltration technology with a ceramic membrane for reactive dye removal: optimization of membrane performance. J Hazard Mater 209, 492
27. Desai M, Mehta M (2014) Teritiary treatment for textile waste water: a review. Int J Eng Sci Res Tech 3:1579
28. Sanchez-Martin J, Beltran-Heredia J, Solera-Hernandez C (2010) Surface water and wastewater treatment using a new tannin-based coagulant. Pilot plant trials. J Environ Manag 91:2051
29. Khayet M, Zahrim AY, Hilal N (2011) Modelling and optimization of coagulation of highly concentrated grade leather dye by response surface methodology. Chem Eng J 77:77
30. Lopes A, Martins S, Morão A, Magrinho M, Goncalves I (2004) Degradation of textile dye CI direct Red 80 by electrochemical processes. Port Electrochim Acta 22:279
31. Barragan BE, Costa C, Carmen Marquez M (2007) Biodegradation of azo dyes by bacteria inoculated on solid media. Dyes Pigments 75:73–81
32. Bromley-Challenor KCA, Knapp JS, Zhang Z, Gray NCC, Hetheridge MJ, Evans MR (2000) Decolorization of an azo dye by unacclimated activated sludge under anaerobic conditions. Water Res 34:4410–4418
33. dos Santos AB, Cervantes FJ, van Lier JB (2007) Review paper on current technologies for decolourisation of textile wastewaters: perspectives for anaerobic biotechnology. Bioresour Technol 98:2369–2385
34. Frijters CTMJ, Vos RH, Scheffer G, Mulder R (2006) Decolorizing and detoxifying textile wastewater, containing both soluble and insoluble dyes, in a full scale combined anaerobic/aerobic system. Water Res 40:1249–1257
35. van der Zee FP, Villaverde S (2005) Combined anaerobic-aerobic treatment of azo dyes – a short review of bioreactor studies. Water Res 39:1425–1440
36. Zhang FM, Knapp JS, Tapley KN (1998) Decolourisation of cotton bleaching effluent in a continuous fluidized-bed bioreactor using wood rotting fungus. Biotechnol Lett 20:717–723
37. Rai HS, Bhattacharyya MS, Singh J, Bansal TK, Vats P, Banerjee UC (2005) Removal of dyes from the effluent of textile and dyestuff manufacturing industry: a review of emerging techniques with reference to biological treatment. Crit Rev Env Sci Technol 35:219–238
38. Husain Q (2006) Potential applications of the oxidoreductive enzymes in the decolorization and detoxification of textile and other synthetic dyes from polluted water: a review. Crit Rev Biotechnol 26:201–221
39. Kulla HG (1981) Aerobic bacterial degradation of azo dyes. FEMS Microbiol Lett 12:387–399

40. Sani R, Banerjee U (1999) Decolorization of acid green 20, a textile dye, by the white rot fungus Phanerochaete chrysosporium. Adv Environ Res 2:485–490
41. Delee W, O'Neill C, Hawkes FR, Pinheiro HM (1998) Anaerobic treatment of textile effluents: a review. J Chem Technol Biotechnol 73:323–335
42. Razo-Flores E, Luijten M, Donlon B, Lettinga G, Field J (1997) Biodegradation of selected azo dyes under methanogenic conditions. Water Sci Technol 36:65–72
43. van der Zee FP, Lettinga G, Field JA (2001) Azo dye decolourisation by anaerobic sludge. Chemosphere 44:1169–1176
44. Brown D, Hamburger B (1987) The degradation of dye stuffs. Part III. Investigations of Their Ultimate Degradability. Chemosphere 16:1539–1553
45. Brown D, Laboureur P (1983) The degradation of dyestuffs: Part I. Primary biodegradation under anaerobic conditions. Chemosphere 12:397–404
46. Stolz A (2001) Basic and applied aspects in the microbial degradation of azo dyes. Appl Microbiol Biotechnol 56:69–80
47. Gupta VK, Suhas (2009) Application of low-cost adsorbents for dye removal – a review. J Environ Manag 90:2313–2342
48. Babu SS, Mohandass C, Vijayaraj AS, Dhale MA (2015) Detoxification and color removal of Congo Red by a novel Dietzia sp. (DTS26)–a microcosm approach. Ecotoxicol Environ Saf 114:52–60. https://doi.org/10.1016/j.ecoenv.2015.01.002
49. Chequer FMD, Dorta DJ, de Oliveira DP (2011) Azo dyes and their metabolites: does the discharge of the azo dye into water bodies represent human and ecological risks in advances in treating textile effluent. InTech 27–49
50. Bumpus JA, Aust SD (1987) Biodegradation of DDT [1, 1, 1-trichloro-2, 2-bis (4-chlorophenyl) ethane] by the white rot fungus Phanerochaete chrysosporium. Appl Environ Microbiol 53:2001–2008
51. Lade H, Govindwar S, Paul D (2015) Low-cost biodegradation and detoxification of textile azo dye CI reactive blue 172 by *Providencia rettgeri* strain HSL1. J Chem
52. Placido J, Chanaga X, Ortiz-Monsalve S, Yepes M, Mora A (2016) Degradation and detoxification of synthetic dyes and textile industry effluents by newly isolated Leptosphaerulina sp. from Colombia. Bioresour Bioprocess 3:6. https://doi.org/10.1186/s40643-016-0084-x
53. Senthilkumar S, Perumalsamy M, Prabhu HJ (2014) Decolourization potential of white-rot fungus Phanerochaete chrysosporium on synthetic dye bath effluent containing Amido black 10B. J Saudi Chem Soc 18:845–853. https://doi.org/10.1016/j.jscs.2011.10.010
54. Singh L, Singh VP (2010) Microbial degradation and decolorization of dyes in semi-solid medium by the fungus–Trichoderma harzianum. Environ We Int J Sci Tech 5:147–153
55. Joutey NT, Bahafid W, Sayel H, El Ghachtouli N (2013) Biodegradation: involved microorganisms and genetically engineered microorganisms. In: Biodegradation-life of science. InTech
56. Verma AK, Raghukumar C, Parvatkar RR, Naik CG (2012) A rapid two-step bioremediation of the anthraquinone dye, Reactive Blue 4 by a marinederived fungus. Water Air Soil Pollu 223:3499–3509. https://doi.org/10.1007/s11270-012-1127-3
57. Corso CR, Almeida EJR, Santos GC, Morao LG, Fabris GSL, Mitter EK (2012) Bioremediation of direct dyes in simulated textile effluents by a paramorphogenic form of Aspergillus oryzae. Water Sci Tech 65:1490–1495. https://doi.org/10.2166/wst.2012.037
58. Kanagaraj J, Mandal AB (2012) Combined biodegradation and ozonation for removal of tannins and dyes for the reduction of pollution loads. Environ Sci Pollut Res 19:42–52. https://doi.org/10.1007/s11356-011-0534-0
59. Hruby CE, Soupir ML, Moorman TB, Shelley M, Kanwar RS (2016) Effects of tillage and poultry manure application rates on Salmonella and fecal indicator bacteria concentrations in tiles draining Des Moines Lobe soils. J Environ Manag 171:60–69. https://doi.org/10.1016/j.jenvman.2016.01.040
60. Singh S, Singh N, Kumar V, Datta S, Wani AB, Singh D, Singh J (2016) Toxicity, monitoring and biodegradation of the fungicide carbendazim. Environ Chem Lett 14:317–329. https://doi.org/10.1007/s10311-016-0566-2

61. Wang X, Cheng X, Sun D, Ren Y, Xu G (2014) Fate and transformation of naphthylaminesulfonic azo dye Reactive Black 5 during wastewater treatment process. Environ Sci Pollut Res 21:5713–5723. https://doi.org/10.1007/s11356-014-2502-y
62. Ali DM, Suresh A, Praveen Kumar R, Gunasekaran M, Thajuddin N (2011) Efficiency of textile dye decolorization by marine cyanobacterium, Oscillatoria formosa NTDM02. Afr J Basic Appl Sci 3:9–13
63. Elisangela F, Andrea Z, Fabio DG, de Menezes Cristiano R, Regina DL, Artur CP (2009) Biodegradation of textile azo dyes by a facultative Staphylococcus Arlette strain VN-11 using a sequential microaerophilic/aerobic process. Int Biodeterior Biodegrad 63:280–288. https://doi.org/10.1016/j.ibiod.2008.10.003
64. Verma P, Samanta SK (2018) Microwave-enhanced advanced oxidation processes for the degradation of water. Environ Chem Lett. https://doi.org/10.1007/s10311-018-0739-2
65. Sires I, Brillas E, Oturan MA, Rodrigo MA, Panizza M (2014) Electrochemical advanced oxidation processes: today and tomorrow. A review. Environ Sci Pollut Res. https://doi.org/10.1007/s11356-014-2783-1
66. Rahman N, Abedin Z, Hossain MA (2014) Rapid degradation of azo dyes using nano-scale zero valent iron. J Environ Sci 10:157
67. Adekunle AS, Ozoemena KI (2010) Comparative surface electrochemistry of Co and Co_3O_4 nanoparticles: nitrite as an analytical probe. Int J Electrochem Sci 5:1972
68. Chaudhary GR, Saharan P, Kumar A, Mehta SK, Mor S, Umar A (2013) Removal of water contaminants by iron oxide nanomaterials. J Nanosci Nanotechno 13:3240
69. Absalan G, Asadi M, Kamran S, Sheikhian L, Goltz DM (2011) Removal of reactive red-120 and 4-(2-pyridylazo) resorcinol from aqueous samples by Fe_3O_4 magnetic nanoparticles using ionic liquid as modifier. J Hazard Mater 192:476
70. Tsuji H, Yagi F, Hattori H, Kita H (1994) Self-condensation of n-butyraldehyde over solid base catalysts. J Catal 148:759
71. Jamil N, Mehmood M, Lateef A, Nazir R, Ahsan N (2015) Adv Mater Tech Connect Briefs 353
72. Dhal JP, Sethi M, Mishra BG, Hota G (2015) MgO nanomaterials with different morphologies and their sorption capacity for removal of toxic dyes. Mater Lett 141:267
73. Moazzam A, Jamil N, Nadeem F, Qadir A, Ahsan N, Zameer M (2017) Reactive dye removal by a novel Biochar/MgO nanocomposite. J Chem Soc Pak 39
74. Sun HM, Cao LY, Lu LH (2011) Magnetite/reduced graphene oxide nanocomposites: one step solvothermal synthesis and use as a novel platform for removal of dye pollutants. Nano Res 4:550
75. Shi XD, Ruan WQ, Hu JW, Fan MY, Cao RS, Wei XH (2017) Optimizing the removal of Rhodamine B in aqueous solution by reduced graphene oxide supported nanoscale zerovalent iron (nZVI/rGO) using an artificial neural network-genetic algorithm (ANN-GA). Nanomaterials 7:309. https://doi.org/10.3390/nano7060134
76. Ruan W, Jiwei Hu, Jimei Qi Yu, Hou CZ, Wei X (2019) Removal of dyes from wastewater by nanomaterials: a review. Adv Mater Lett 10(1):09–20
77. Kumar VS, Nagaraja BM, Shashikala V et al (2004) Highly efficient Ag/C catalyst prepared by electro-chemical deposition method in controlling microorganisms in water. J Mol Catal A Chem 223(1–2):313–319
78. Sondi I, Salopek-Sondi B (2004) Silver nanoparticles as antimicrobial agent: a case study on E. coli as a model for Gramnegative bacteria. J Colloid Interface Sci 275(1):177–182
79. Singh J, Dhaliwal AS (2021) Effective removal of methylene blue dye using silver nanoparticles containing grafted polymer of guar gum/acrylic acid as novel adsorbent. J Polym Environ 29:71–88
80. Ghaedi M, Khajehsharifi H, Yadkuri AH, Roosta M, Asghari A (2012) Oxidized multi-walled carbon nanotubes as efficient adsorbent for bromothymol blue. Toxicol Environ Chem 94(5):873–883
81. Seyyedi K, Jahromi MAF (2014) Decolorization of azo dye CI direct black 38 by photocatalytic method using TiO_2 and optimizing of process. APCBEE Proc 10:115

82. Marathe SD, Shrivastava VS (2015) Phototcatalytic removal of hazardous Ponceau S dye using nano structured Ni doped TiO_2 thin film prepared by chemical method. Appl Nanosci 5:229
83. Abualnaja KM, Alprol AE, Ashour M, Mansour AT (2021) Influencing multi-walled carbon nanotubes for the removal of ismate violet 2R dye from wastewater: isotherm, kinetics, and thermodynamic studies. Appl Sci 11:4786. https://doi.org/10.3390/app11114786
84. Madrakian T, Afkhami A, Ahmadi M, Bagheri H (2011) Removal of some cationic dyes from aqueous solutions using magnetic-modified multi-walled carbon nanotubes. J Hazard Mater 196:109–114. https://doi.org/10.1016/j.jhazmat.2011.08.078
85. Qu S, Huang F, Yu S, Chen G, Kong J (2008) Magnetic removal of dyes from aqueous solution using multi-walled carbon nanotubes filled with Fe_2O_3 particles. J Hazard Mater 160(2–3):643–647. https://doi.org/10.1016/j.jhazmat.2008.03.037
86. Mezohegyi G, van der Zee FP, Font J, Fortuny A, Fabregat A (2012) Towards advanced aqueous dye removal processes: a short review on the versatile role of activated carbon. J Environ Manag 102:148–164
87. Fathima NN, Aravindhan R, Rao JR, Nair BU (2008) Dye house wastewater treatment through advanced oxidation process using Cu-exchanged Y zeolite: a heterogeneous catalytic approach. Chemosphere 70:1146–1151. https://doi.org/10.1016/j.chemosphere.2007.07.033
88. Syafalni S, Abustan I, Wah IDCK, Umar G (2012) Treatment of dye wastewater using granular activated carbon and zeolite filter. Mod Appl Sci 6(2):2012
89. Selvanathan N, Subki NS, Sulaiman MA (2015) Dye adsorbent by activated carbon. J Trop Resour Sustain Sci 3:169–173
90. Sharma A, Bhattacharyya KG (2005) Utilization of a biosorbent based on Azadirachta indica Neem leaves for removal of water-soluble dyes. Indian J Chem Technol 12:285–295
91. Haluk Aydın A, Yavuz Ö (2004) Removal of acid red 183 from aqueous solution using clay and activated carbon. Indian J Chem Technol 11:89–94
92. Liang P, Qin PF, Lei M, Zeng QR, Song HJ, Yang J, Shao JH, Liao BH, Gua JD (2012) Modifying Fe_3O_4 nanoparticles with humic acid for removal of Rhodamine B in water. J Hazard Mater 209:193

Sustainable Development of Nanomaterials for Removal of Dyes from Water and Wastewater

Gaurav Yadav and Md. Ahmaruzzaman

Abstract Human beings will experience water scarcity in the future, which has become a prime threat to human life in recent years all over the world. For industry and our daily life, we need clean water. Due to the increase in the human population, urbanization, and industrialization, the water resources are contaminated. So, we need a reliable technique to purify wastewater. Due to this, dyes cause various harm to human beings as well as aquatic life. Recent studies show that dyes are carcinogenic to human beings. With the help of conventional methods, it is not possible to provide clean water to everyone, so we need an alternative, sustainable technique. Nanomaterials have special properties like large surface area and high surface reactivity, which help in the degradation of water contamination. In this chapter, the deletion of dyes from water effluents with the help of nanotechnology has been discussed thoroughly. There are various procedures through which we could remove harmful dyes from the waste, like photocatalysis, adsorption phenomena, etc.

Keywords Water scarcity · Dyes · Adsorption · Photocatalysis · Nanomaterials · Wastewater · Chemical reactions

1 Introduction

Water is the most generous natural asset all over the globe. However, due to the enormously growing population [3], enough water for human beings is not available. According to WHO, approximately half of the world's population will suffer from a water crisis by 2025. Polluted water consumption causes nearly 1.7 million deaths per year. Today, one of the major problems worldwide is the scarcity of fresh water. Various inorganic and organic materials are responsible for water contamination and these leads to health concerns of nearly 4 billion people as reported by Briggs et al. [12]. The traditional methods like chemical (flocculation, coagulation, electrochemical, and precipitation), physical (skimming, sedimentation, aeration, screening) as

G. Yadav · Md. Ahmaruzzaman (✉)
Department of Chemistry, National Institute of Technology, SilcharAssam 788010, India

S. S. Muthu and A. Khadir (eds.), *Advanced Oxidation Processes in Dye-Containing Wastewater*, Sustainable Textiles: Production, Processing, Manufacturing & Chemistry,
https://doi.org/10.1007/978-981-19-0987-0_8

well as biological procedures (enzyme treatment, aerobic and anaerobic processes) are not so efficient for fulfilling the needs of humans.

Nanotechnology is the most appropriate technology to provide an optimal solution for a different type of environmental pollution. Nanotechnology is the most advanced technique used to remove pollutants efficiently present in wastewater at a low cost [10, 117, 81]. Nanomaterials are helpful in the removal of aquatic pollutants. There are many unique characteristics of nanoparticles that assist in the adsorption of contaminants on the surface of metal nanoparticles. Many nanomaterials like activated carbon, graphene oxide, and carbon nanotubes eliminate pollutants by adsorption. Some sites are chemically active in adsorption, having high adsorption efficiency [51]. Some examples are composite materials [83], carbon-based materials, metal oxide-based materials, silica-based and graphene-based materials, etc. In addition, many photocatalysts like TiO_2, ZnO, and other metal oxides are used to degrade the pollutants.

Dyes are colored substances due to the absorbance of a certain kind of wavelength that is bound to the substrate. The majority of dyes are used in the textile industry to color clothes. In the twentieth century, the dyes mostly used were obtained from nature, which did not cause any contamination in water, but in the twenty-first century, we used synthetic dyes (artificial), which are harmful to human beings and aquatic life. Therefore, dyes used by the printing, textiles, and paper industries are the primary source of adulteration of lakes as well as rivers [13]. On an estimate, nearly 34 million tones are produced worldwide each year, and of these, about 11% go out as effluents. Based on the charge, chemical structure, applications, and color, we can classify dyes. Dyes are categorized into three parts based on charge: non-ionic, anionic, and cationic. The class of dyes can also be based on the source of materials. On the basis of source, dyes can be classified into two parts: synthetic dyes and natural dyes.

2 Adverse Effect of Dyes

Dyes are primarily used in the textile industry. Dyes are generally water-soluble organic compounds [60], and so, it is not easy to remove dyes from water. The color present in textile dyes is the cause of water body pollution. It inhibits the supply of sunlight in water [34], which causes low photosynthesis [40] and decreased DO (dissolved oxygen) in the water. The fabric stains also act as toxic, carcinogenic, and mutagenic agents [8, 47] in the environment as well as in the food chain through biomagnification [85]. The commonly used azo dyes used in the dyeing process are 15–50% that do not tie with fabric and are released into the water without any treatment.

Though the textile dyes are highly toxic, they cause environmental deterioration and cause various types of disease in humans and animals [46]. The recalcitrant nature of dyes leads to bioaccumulation in soil and sediment, which is further carried to the water systems [105]. Complex metal dyes are widely used in textiles, having a half-life of 2–13 years [19]. They are previously released into the water, lead to bioaccumulation in the tissue of fishes, and through the food chain, reaches the human body, generating several problems. Textile dyes cause dermatitis and affect the central nervous system [46], which results in enzyme inactivity [19]. Inhalation and oral ingestion of textile dyes cause acute toxicity and lead to irritations of the eyes and skin [18], allergic conjunctivitis, contact dermatitis, rhinitis, occupational asthma, and other allergic reactions [38].

3 Physiochemical Properties of Various Dyes

As we know, dyes are colored substances. The color of dyes is imparted due to the presence of chromophoric groups. Numerous physical properties of some dyes are given in Table 1.

4 Degradation/Removal of Dyes from Wastewater

In the last few years, many methods are developed for the deletion/degradation of dyes from water effluents with high efficiency based on various techniques. The adsorption and photocatalysis method are one of the clean methods used to eliminate stains from water. Some following processes give you the detailed information of elimination of dyes from water.

4.1 Adsorption

In the last few years, many methods have been developed for the deletion/degradation of dyes from water effluents with high efficiency based on various techniques. The adsorption and photocatalysis method are one of the clean methods used to eliminate stains from water. Some following processes give you detailed information on the elimination of dyes from water.

4.1.1 Activated Carbon

Activated carbon has high adsorption efficiency due to the high surface area for different kinds of contamination [21]. The sky-scraping adsorption of activated carbon is due to the inception of the material from natural materials like coal, lignite, and biomass. Different kinds of functional groups like epoxy, carboxylic, and hydroxyl are found on the exterior of activated carbon, which is helpful for adsorption. Due to adsorbed ions or the uncoupling of influential groups of carbon,

Table 1 Physical property of some dyes

Name	Cas No.	Molecular formula	Molar mass	Water solubility	Color	Uses
Methylene blue	61-73-4	$C_{16}H_{18}ClN_3S$	319.85	4.36 g/100 ml	Deep blue	As soft vegetable fibre
Methyl red	493-52-7	$C_{15}H_{15}N_3O_2$	269.3	800 g/L	Dark red	Paper printing and textile dyeing
Methyl Orange	547-58-0	$C_{14}H_{14}N_3NaO_3S$	327.33	0.5 g/100 ml	Red in acidic medium yellow in basic	Printing textiles
Congo Red	573-58-0	$C_{32}H_{22}N_6Na_2O_6S_2$	696.665	6.97 mg/ml	Below 3.0 pH blue and above 5 red	Textile industry
Napthol Green B	19381-50-1	$C_{30}H_{15}FeN_3Na_3O_{15}S_3$	878.46	160 g/l	Green	Stains wool, nylon, paper
Napthol Yellow S	846-70-8	$C_{10}H_6N_2NaO_8S$	337.22	0.5 g/10 ml	Yellow	Textiles industry
Orange G	1936-15-8	$C_{16}H_{10}N_2Na_2O_7S_2$	452.38	80 mg/ml	Orange	Staining formulations
Rose Bengal	4159-77-7	$C_{20}H_4Cl_4I_4O_5$	973.67	100 mg/ml	Red or rose color	Diagnose liver and eye cancer
Victoria blue B	2580-56-5	$C_{33}H_{32}ClN_3$	506.08	20 g/l	Reddish blue	Staining in microsocpy
Alcian blue	33864-99-2	$C_{56}H_{68}Cl_4CuN_{16}S_4$	1298.86	1 mg/ml	Solid phase-greenish black Aqueous phase-greenish blue	Staining
Azophloxine	3734-67-6	$C_{18}H_{13}N_3Na_2O_8S_2$	509.42	0.5 g/10 ml	Red-blue	Food additives
Alizarin	72-48-0	$C_{14}H_8O_4$	240.214	400 mg/l	Orange red	Dyeing textile fabrics
Crystal violet dye	548-62-9	$C_{25}H_{30}ClN_3$	407.99	50 mg/ml	Blue-voilet colour	Textile and paper

there is an electric charge on the surface of activated carbon that depends on conditions like pH and surface characteristics [21]. The capability of activated carbon is increased with the help of physical and chemical development. Activation of carbon with the use of the physical method is typically done with the help of inert gases, steam, and CO_2. This helps in the removal of non-carbonaceous atoms and opens new vents. Inorganic and organic chemicals are used to alter chemical activation to increase the efficiency of activated carbon. High temperature is generally required in physical activation, whereas, in chemical activation, low temperature is used. A study shows that activated carbon obtained from orange peel and modified with the help of H_3PO_4 is used to remove basic dyes, especially Methylene blue and Rhodamine B, from water [26].

Activated carbon is a versatile and economical material used for bleaching [68]. When activated carbon is used from agriculture waste and assisted chemically with the help of KOH, then the advanced materials are used to degrade acid yellow 17 dyes with a high rate of efficiency [76]. The highest disclosed monolayer adsorption efficiency is 215.05 mg/g. Activated carbon is prepared as the precursor of the date sphate and activated using phosphoric acid, then the material formed is used to degrade methyl orange from water [25]. In another study, activated carbon was prepared with the biomass of jacaranda and plum kernels,then, the activated carbon showed better removal capability of methylene blue and acid blue 25 [104].

4.1.2 Graphene-Based Materials

Graphene-based nanocomposites and nanostructures are being used for dye removal. Graphene oxides and other graphene materials show the greatest adsorption capacity. Graphene possesses various physical properties that are useful for the adsorption of particular harmful materials from water. Graphene oxide can be synthesized with the help of hummer methods [55]. Due to this process, carboxyl and hydroxyl groups are added, resulting in chemical stability, physical flexibility, mechanical rigidity, and a high exterior area of graphene oxide [97]. These belongings make graphene a suitable adsorbent.

Methylene blue and methyl violet degradation were observed by Liu et al. [56] with the help of 3D graphene oxide in an aqueous solution. Graphene oxide shows high degradation efficiency of up to 99.1% for methylene blue, and methyl violet is 98.8% within 2 min, having a primary concentration of 800 and 946 mg/l, respectively. A wide range of pH (6–10) is used to remove cationic dye arbitrated through the attraction between graphene oxide and dyes. Konicki et al. [49] observed the elimination of Direct Red 23 (DR23) and Acid Orange 8 (AO8) with the help of graphene oxide. GO shows that the maximum adsorption efficiency for DR23 is 15.3 mg/g and for AO8 is 29 mg/g. Based on graphene-related materials, chitosan and algitane gain exceptional attention. A novel nanocomposite based on graphene/algitane is used to degrade methylene blue [124]. The maximum adsorption capacity of the graphene/algitane double network is 7.2 mmol/g, and for graphene/algitane single networks, it shows an efficiency of 5.6 mmol/g.

4.1.3 Carbon Nanotubes (CNT)

CNT has a wide range of applications, including fluid filters, energy storage, and catalysis. CNTs exist in cylindrical graphite sheets having 2-dimensional and 3-dimensional structures. These are termed SWCNTs (single-walled CNT) and MWCNT (multi-walled CNT) based on sheets or layers. Because it has numerous active sites and large surface areas, SWCNTs play a crucial part in the clean-up approach. Solute particles get adsorbed on the adsorbent due to hydrophobic interaction. Therefore, these provide a promising solution for eliminating organic dyes from the water [79]. With the help of surface modification, we can enhance the adsorption capability of the CNTs. Surface medication of CNTs can be done with the help of absorption of some metals [101, 122], metal oxides like MnO_2, [54] Fe_2O_3, Fe_3O_4 [99], and Al_2O_3 [30] as well as by the help of acid treatment [39, 79].

MWCNT modified with alkali showed great adsorption efficiency for methylene blue (399 mg/g) and methyl orange (149 mg/g) [58]. In another experiment, MWCNT functionalized by glucose removes wide varieties of dyes [80]. The developed MWCNT shows higher efficiency than the parent one. Gupta and Suhas et al. [31], Madrakian et al. [59] show the adsorption of methylene blue (48.1 mg/g) and janus green (250 mg/g) on the surface of the CNT-nano-Fe_3O_4 composite.

4.1.4 Miscellaneous Nano-adsorbents

Apart from these nano-adsorbents, several other nano-adsorbents are also used for adsorption of harmful dyes from wastewater. Some of them are silica-based, metal oxides, chitosan derived, polymeric adsorbents, etc.

Like others, silica has unique properties such as high surface area, high selectivity, bio-degradability, bio-compatibility, and high absorption efficiency. Besides these, it is easy to prepare and has no toxicity. Many dyes are effectively adsorbed on the surface of silica with high efficiency.

Metal oxides like Fe_2O_3, Fe_3O_4, and their various forms like hematite and magnetite are used to adsorb the dyes via surface exchange. Iron oxide possesses a high surface area, so it has many active sites, which help them to adsorb the adsorbate. Neutral red dye can be effectively adsorbed (105 mg/g) by a hollow nanosphere of Fe_3O_4 within 60 min when the initial concentration is 200 mg/g [41]. In another experiment, when the initial concentration of acid red dye was 103 mg/g, it should be completely removed with the help of -Fe_2O_3 within 4 min (Nassar et al. 2010). Here we can also modify the surface of iron oxide to increase its capacity. Nano-sized MgO is also used to eliminate stains from contaminated water. Hu et al. [37] and Moassavi and Mahmoudi et al. [71] investigated the elimination of brilliant red X3B and Congo red. It shows adsorption efficiency up to 254.3 and 297 mg/g, respectively, greater than activated carbon. Literature shows that anionic dyes are also removed with the help of MgO. In the case of reactive red 198 and methylene blue 19, MgO was prepared by the sol–gel method, and it showed the highest adsorption efficiency, up to 123.25 and 166.7 mg/g [71].

Chitosan is a nature-based nanoparticle that is used against dyes and other pollutants. It has low toxicity and is easy to produce. It is one of the more optimistic adsorbents for the elimination of dyes. Some removal applications and the degradation efficiency of chitosan derivatives are given in Table 2. The nanosheet of MoS_2 (molybdenum disulfide) looks like a graphene structure which helps in the diffusion of dyes to the surface. Massey et al. [65] shows that the adsorption of methylene blue, rhodamine 6G, malachite green, congo red, and fuchsin acid are 297, 216, 204, 146, and 183 mg/g, respectively, in the presence of MoS_2. Molybdenum disulfide is also used as a photocatalyst.

4.2 *Photocatalysis*

The photocatalysis method is one excellent approach to eliminating dyes from the aqueous phase. In nanotechnology, photocatalysis is defined as "the use of light to remove or degrade pollutants." In the presence of light, the rate of reaction increases. The nanoparticles deliver diverse morphologies, large surface area, great surface states, which are helpful for photocatalysis [103]. Metal oxide nanomaterials such as TiO_2, CeO_2, MgO, Cu_2O, and ZnO have received significant attention in the last few years. Photocatalysis generally includes the following steps:

- Excitation of electrons by the help of light
- Excitation of electrons from VB (valence band) to CB (conduction band)
- VB has a hole that needs to be filled
- Hole migration to the catalysis surface
- The reaction of the hole with water forms a hydroxyl radical, which is very reactive
- Radicals react with pollutants and degrade them.

The primary causes of pollutant degradation are holes and electrons. Chemically, the detailed mechanism of photocatalyst is given below

$$TiO_2 + hv \rightarrow h^+_{VB} + e^-_{CB}$$

$$e^-_{CB} + O_2 \rightarrow {}^{\bullet}O_2$$

$$h^+_{VB} + H_2O \rightarrow {}^{\bullet}OH + H^+$$

The nanophotocatalyst has a size range between 1 and 100 nm. They produce various oxidizing species on the surface and increase the efficiency, therefore, beneficial for eliminating multiple pollutants in water [29].

Table 2 Elimination of various dyes with the help of nano-adsorbents

Dyes name	Nano-adsorbent	Quantity and dosage	pH	Removal efficiency	Time taken	References
Acid blue 129	CuO/Activated carbon	C_0-10 mg/l Dosage-0.9 g/l	2.0	65.4 mg/g	20–25 min	Nekouei et al. [75]
Bisphenol-A	SWCNTs	C_0-6 mg/l Dosage-0.5 g/l	7.0	71.0 mg/g	30 min	Dehghani et al. [20]
Congo red	O-MWCNTs	C_0-6 mg/l Dosage-0.022 g	1.0	357.1 mg/g	12 min	Sheibani et al. [89]
Reactive red 120	SWCNTs	C_0-50 mg/l Dosage-0.04 g/l	5.0	426.0 mg/g	180 min	Bazrafshan [11]
Methylene blue	MWCNTs	C_0-50 g/l Dosage-250 mg/l	7.0	400.0 mg/g	8 h	Szlachta and Wojto-wicz [95]
Reactive red 46	O-SWCNTs	C_0-150 mg/l Dosage-0.05 g	9.0	49.4 mg/g	100 min	Moradi [70]
Congo red	MWCNTs	C_0-40 mg/l Dosage-25 mg	5.0	165.2	60 min	Kamil et al. [43]
Congo red	ZnO with MWCNTs	C_0- Dosage-9 mg	7.0	249.5	50 min	Arabi et al. [9]
Methyl orange	Al doped MWCNT	C_0-50 mg/l Dosage-1.3 g/l	4.5	69.7	15 min	Kang et al. [44]
Methylene blue	MWCNTs	C_0-10 ppm Dosage-0.2 g/l	–	59.7	–	Wang et al. [107]
Reactive green 19	MWCNTs	C_0-250 ppm Dosage-0.4	–	152.0	–	Mishra et al. [69]

(continued)

Table 2 (continued)

Dyes name	Nano-adsorbent	Quantity and dosage	pH	Removal efficiency	Time taken	References
Congo red	PTSA with GO-CNTs	C_0-200 mg/l Dosage-20 mg	2.0	60.0	500 min	Ansari et al. [7]
Methylene blue	Magnetic graphene with CNTs	C_0-10 mg/l Dosage-10 mg	11.0	65.8	30 min	Wang et al. [108, 109]
Methyl blue	Polypyrrole/MWCNT/cobalt ferric magnetic nanocomposite	C_0-50 mg/l Dosage-1.0 g/l	3.0	137.0	30 min	Li et al. [53]
Congo red	Zinc oxide	C_0-250 mg/l Dosage-1.2 g/l	2.0–10	208.0	10 min	Chawla et al. [16]
Amaranth	MgO/Fe_3O_4	C_0- 0.2-24.28 mg/l Dosage- 0.1–0.4 g	9.0	37.9	60 min	Salem et al. [82]
Methylene blue	MgO/GO	C_0-20 mg/l Dosage-0.1–1 g	7.0	833.0	20 min	Heidarizad and sengor [35]
Methylene blue	CoO/SiO_2	C_0-29 mg/l Dosage-0.8 g/l	6-7	53.9	30 min	Abdel ghafar et al. [2]
Malachite green	ZnO	C_0-10 mg/l Dosage-80 mg/l	8.0	310.5	120 min	Kumar et al. [50]
Rhodamine B	Mesoporous silica	C_0-20 mg/l Dosage-1 g/l	5.8	234.6	30 min	Chen et al. [17]
Rhodamine 6G	Bifunctionalized mesoporous silica	C_0-0.05 mM Dosage-10 mg/10ml	–	9.9	6 h	Shinde et al. [91]
Basic blue 41	Nanoporous silica	C_0- 20-60 mg/l Dosage-0.01 g	7.0	345.0	60 min	Zarezadeh-mehrizi and Badiei [118]
Sulphur dyes	Iron oxide/silica gel	C_0-100 mg/l Dosage-0.2 g	4-5	11.1	20 min	Tavassolia et al. [100]
Xylenol orange	Cation modified silica	C_0-5 mg/l Dosage-20 mg	3.0	9.0	3 h	Zhang et al. [119, 120]

(continued)

Table 2 (continued)

Dyes name	Nano-adsorbent	Quantity and dosage	pH	Removal efficiency	Time taken	References
Acid orange	Silica	C_0-200 mg/l Dosage-1.0	5.0	230.0	4 h	Suzimara et al. [94]
Methylene blue	Engineered SiO_2	C_0-30 mg/l Dosage-0.08	6.0	9.5	20 min	Salimi et al. [84]
Crystal violet	Magnetic chitosan	C_0-77 mg/l Dosage-1 g	7.0	333.3	140 min	Massoudinejad et al. [66]
Reactive blue T	Chitosan	C_0-1000 ppm Dosage-2.0 g/l	4.6	357.1	130 min	Sarkheil et al. [86]
Methyl orange	Poly-HEMA chitosan-MWCNTs	C_0-35 mg/l Dosage-0.06 g	4.0	416.7	210 min	Mahmoodian et al. [62]
Acid red 88	Bio-silica/chitosan	C_0-50 mg/l Dosage-3 g/l	7.0	25.8	120 min	Soltani et al. [93]
Reactive brilliant red X-3B	Magnetic Fe_3O_4/chitosan	C_0-200 mg/l Dosage-0.6 g	2.0	76.8	5 h	Cao et al. [14]
Malachite green	Humic acid-Fe_3O_4	C_0- 20-60 mg/l Dosage-50 mg	8.0	79.3	50 min	Abate et al. [1]
Methylene blue	MoS_2	C_0-300 mg/l Dosage-1000 mg/l	–	303.0	8 h	Massey et al. [65]
Methylene blue	Calcined TNTs	C_0-50 mg/l Dosage-500 mg	–	93.46	60 min	Xiong et al. [110]
Methylene blue	Agar-based bimetallic NPs	C_0-5 mg/ Dosage-1000 mg/l	–	–	35 min	Patra et al. [77]
Methylene blue	$CeO_{2-\delta}$	C_0- 200 mg/ Dosage-2000 mg/l	–	106.3	240 min	Tomic et al. [102]
Mordant blue	$CeO_{2-\delta}$	C_0-200 mg/l Dosage-2000 mg/l	–	100	120 min	Tomic et al. [102]

(continued)

Table 2 (continued)

Dyes name	Nano-adsorbent	Quantity and dosage	pH	Removal efficiency	Time taken	References
Remazol red RB-133	Fe_2O_3/MgO	C_0-50 mg/l dosage-1000	–	28.35	60 min	Mahmoud et al. [63]
Indigo carmine	AgBr/CTAB	C_0- Dosage-200	–	144.93	40 min	Tang et al. [98]
Methylene blue	Cellulose-clay hydrogel	C_0-10 mg/l Dosage-50000	–	50.02	24 h	Peng et al. [78]

Table 3 Photocatalyst used for removal of dyes

Dye's name	Photocatalyst	Quantity and dosage	Light source	pH	Elimination	Time (min)	References
Reactive red And methylene blue	Nano-TiO_2	C_0-40 mg/l Dose-160 (RR) 500 mg/l (MB)	Solar	5.5 (RR) 7.4 (MB)	100%	95 (RR) 150 (MB)	Jeni and kanmani [42]
Rose Bengal	Co doped TiO_2	C_0-20 ppm Dose-0.06 g/100 ml	Visible	6.0	97.05%	150	Malini andAllen gnana raj [64]
Food black 1	TiO_2 and ZnO	C_0-50 mg/l Dose-0.8 (TiO_2) And 1.2 g/l (ZnO)	UV	5.0	96% (TiO_2) 95% (ZnO)	35	khezrianjoo et al. [48]
Methylene blue	Bi_2WO_6	C_0-5 mg/l	Visible	7.0	95%	30	Singh et al. [92]
Methyl orange and Methylene blue	Bismuth oxy- chloride	C_0-20 mg/l Dose-50 mg	UV-Vis	–	100%	30	Seddigi et al. [88]
Methyl orange	ZnO/SnO_2	C_0-20 ppm Dose-0.2 g	Visible	7.0	56%	100	Ali et al. [6]
Methylene blue	ZnS-ZnO/graphene	C_0-10 ppm Dose-10 mg	Visible	Neutral	99%	90	Lonkar et al. [57]
Acid red 88	CuO/ZnO	C_0-0.00005 M Dose-1.4 g/L	Visible	6.0	100%	45	Sathish kumar et al. [87]
Methyl orange	TiO_2/Fe_3O_4	C_0 Dose- 50 mg	UV	2.0	38%	60	Ahmadpour et al. [5]
Methylene blue And methyl orange	$TiO_2/ZnTiO_3/\alpha Fe_2O_3$	C_0-1000 mg/l Dose-10 g/L	UV	2.0	>95%	15	Mehrabi and Javanbakht [67]

(continued)

Table 3 (continued)

Dye's name	Photocatalyst	Quantity and dosage	Light source	pH	Elimination	Time (min)	References
Rhodamie B, Methyl orange, Methylene blue	Magnetic activated carbon	C0-4 mg/l (RhB) 45 (MB), 4 mg/l (MO) dose-0.44 g/l	UV-Vis, Solar	2.5	>90%	>90	Taghdiri et al. [96]
Napthol blue Black	TiO_2 NPs	C_0-5 mg/l Dose-2000 mg/l	Visible	–	92%	40	Nasr et al. [72]
Methylene blue	MoS_2/rGO	C_0-60 mg/L Dose-1000 mg/l	Visible	–	100%	60	Li et al. [52]
Rhodamie B	Cu^{2+}/Rgo	C_0-2.5 mg/l Dose-63 mg/l	Visible	–	91%	180	Xiong et al. [112]
Rhodamie B	Hierarchical $BiVO_4$	C_0-5 mg/l Dose-500 mg/l	Sunlight	–	97%	600	Dong et al. [22]
Basic fuchsing	C_3N_4/Ag_3VO_4	C_0-20 mg/l Dose-1000 mg/l	Visible	–	95%	150	Wang et al. [108, 109]
Methyl orang g	Ag modified C_3N_4	C_0-0.4 Dose-3000	Visible	–	90%	60	Yan et al. [113]
Methylene blue	ZnO NPs	C_0-100 mg/l Dose-100 mg/l	UV	–	90%	60	Shen et al. [90]
Methyl orange	Ag_2O/Ag_2CO_3	C_0-20 mg/l Dose-50 mg/l	Visible	–	100%	5	Yu et al. [115]

4.2.1 TiO_2

TiO_2 is extensively used as a photocatalyst due to its stability, low toxicity, low cost, and extreme reactive nature. TiO_2 exists in three forms, rutile, anatase, and brookite. However, due to the vast band gap energy of TiO_2 (3.2 eV), it is only applicable under UV radiation. Researchers use the doping of various metals, non-metals, and other transition metals to lower the bandgap energy. Non-metals like S, C, and N are used for doping to reduce the energy of the bandgap so that TiO_2 works under visible light [103]. C and S doped TiO_2 is effectively used to degrade reactive black 5, methylene blue, and methyl orange [15] in a short time. N-doped TiO_2 inhibits electron_hole recombination. Doping leads to an increase in the photocatalytic ability of TiO_2.

Nasr et al. [72] have observed the elimination of naphthol blue dye with the help of TiO_2. The efficiency of dye removal depends on the molecular structure of dyes and the surface property of titanium dioxide. With the help of $SrTiO_3$ (nano-strontium titanate), Karimi et al. [45] investigated the elimination of various azo dyes. Nano-strontium titanate is an advanced type of photocatalyst used to remove reactive orange 72 and direct green 6 dyes.

The effect of chloride ions was observed by Yuan et al. [116]. Although there is a favorable as well as a harmful effect of the chloride ion on dye removal, a high chloride ion concentration retards the elimination process of dyes. In contrast, low concentration magnifies the removal efficiency [116]. They observed the deletion of acid orange 7 dye in the presence of UV radiation.

4.2.2 ZnO

ZnO draws enormous attention as a photocatalyst because of its huge surface region, low cost, large capacity to absorb UV radiation [36], and large bandgap (3.37 eV). ZnO synthesized with the help of the sol–gel and co-deposition methods results in the high elimination of dyes under UV radiation. Deposition of ZnO on SiO_2 shows higher efficiency than ZnO synthesized by the sol–gel process [90] due to substantial surface area. Like TiO_2, the efficiency of ZnO can also be increased by inhibiting electron_hole recombination. Decolorization of organic dyes with the help of NiO-ZnO shows better efficiency than pristine ZnO and Degussa TiO_2 (P25) [32] in the presence of UV–visible radiation.

The association of ZnO and TiO_2 increases the physical properties and catalytic efficiency as compared to pure TiO_2 and ZnO. There is a migration of electrons from the CB of ZnO to the CB of TiO_2, which inhibits the reconnection process of electrons and holes and enhances the photolytic activity of organic dyes [4].

4.2.3 Other Photocatalysts

Along with promising adsorbents, graphene and graphene oxide would be good photocatalysts. Xiong et al. [112] observed the elimination of rhodamine B dye in the presence of visible light with the help of reduced graphene oxide. Graphene oxide efficiency would also be improved by modifying it with Cu ion [112] and Au [111], which stimulate the photolytic activity of graphene oxide and degrade rhodamine B effectively under visible light. MoS_2 incorporated with graphene results in better dye adsorption and helps with light absorption [52]. For example, the elimination of methylene blue with the help of MoS_2 incorporated by reduced graphene oxide (rGO) shows better photolytic results under visible light. MoS_2 quantum dots are very efficient photocatalysts for dye elimination due to their high reaction sites, high surface area, and increased stability.

Bismuth-based oxides like Bi_2O_3, $Bi_4Ti_3O_{12}$, $BiWO_6$, and $BiVO_4$ work as good photocatalysts for effective removal of dyes under visible light because of their small band gap energy. Bi-based catalysts like $BiVO_4$ are used efficiently to remove rhodamine B under natural sunlight [22, 24]. However, due to the high cost, bi-based photocatalyst has not been used widely.

g-C_3N_4 (graphite carbon nitride) has been used for dye removal under visible light due to its small bandgap energy. This catalyst has wide applications because of its low cost, high efficiency, low toxicity, and facile synthesis. The effective degradation of methyl orange is noticed by Yan et al. [113] in the presence of visible light. The natural g-C_3N_4 has low removal efficiency. But a variety of substances are coupled with g-C_3N_4 to increase its efficiency. For example, ZnO [106], Ag_3VO_4 [108–109], Ag [27], Co_3O_4 [33], modified g-C_3N_4 has dramatically enhanced their photolytic efficiency for removal of organic dyes. For example, Ag-loaded g-C_3N_4 removes methyl orange 11.5 times better than pristine g-C3N4 [27].

4.3 Chemical Reactions

Dyes can also be removed with the help of chemical reactions. For example, dye contamination can be effectively degraded by various oxidizing agents like $KMnO_4$, H_2O_2, peroxymonosulfate, etc., from industrial wastewater.

Ozonation is one of the best methods which is used for efficient degradation of dyes from wastewater effluents. The ozonation process is an effective technique due to the strong oxidation tendency of ozone. Free radicals like OH and O_2, formed due to ozone decomposition, react instantly with stains.

$$O_3 \rightarrow HO_4^{\bullet} + HO_3^{\bullet} + OH^{\bullet} + O_2^{-}$$

$$HO_3^{\bullet} + HO_4^{\bullet} + OH^{\bullet} + O_2^{-} + \text{Dyes} \rightarrow \text{Mineralized products}$$

The by-products formed by the ozonation process are quite toxic. So, the by-products that are created must be degraded by other methods. Degradation of by-products by ozonation is time-consuming and increases the cost of operation. To overcome this problem, numerous modifications are done. For example, ozonation of dyes with copper ferrite nanoparticles gives NO_3^- and SO_4^{2-} as a by-product [61].

H_2O_2 is a well-known oxidizing agent for the treatment of wastewater. Several catalysts are used with H_2O_2 for the elimination of dyes. The current study shows Fe alginate gel beads/H_2O_2/UV eliminate Acid Black 234 and Reactive Blue 222 [23], CuO/-Al_2O_3/H_2O_2 eliminates Amaranth dye [121] and TiO_2/UV-LED/H_2O_2 for 100% elimination of Rhodamine B [74].

Several other advanced types of oxidation are used for the effective treatment of water. The efficiency of elimination of Acid orange 7 with the help of H_2O_2, persulfate, and peroxymonosulfate is compared by Yang et al. [114]. Peroxides have a different degree of activation under UV radiation, heat, and anions. The ultrasound process is an advanced type of oxidation process. Geng and Thagard [28] observed the removal of Rhodamine B as a consequence of various parameters.

5 Conclusion

Water pollution is increasing continuously in the modern period. The various contaminations are mixed rapidly due to increased industrialization. So, the pure quality of water is not accessible for every human being as well as other living organisms. Nanotechnology is a better technology for water remediation than that of other conventional techniques. Various nano-adsorbents are found to possess excellent adsorption efficiency of multiple pollutants present in water. Dyes are generally soluble in water, and so, their removal is necessary. Activated carbon, graphene-based materials have good adsorption ability than any other materials. Several photocatalysts like TiO_2, ZnO, and MgO are also effective for degrading dyes in the water phase.

References

1. Abate GY, Alene AN, Habte AT, Addis YA (2021) J Polym Environ 29(3):967–984
2. Abdel Ghafar HH, Ali GA, Fouad OA, Makhlouf SA (2015) Enhancement of adsorption efficiency of methylene blue Rev Environ Sci Biotechnol on Co_3O_4/SiO_2 nanocomposite. Desalin Water Treat 53(11):2980–2989
3. Adeleye AS, Conway JR, Garner K, Huang Y, Su Y, Keller AA (2016) Engineered nanomaterials for water treatment and remediation: costs, benefits, and applicability. Chem Eng J 286:640–662
4. Agrawal M, Gupta S, Pich A, Zafeiropoulos NE, Stamm M (2009) A facile approach to fabrication of ZnO-TiO_2 hollow spheres. Chem Mater 21:5343–5348
5. Ahmadpour A, Zare M, Behjoomanesh M, Avazpour M (2015) Photocatalytic decolorization of methyl orange dye using nano-photocatalysts. Adv Environ Technol 3:121–127
6. Ali W, Ullah H, Zada A, Alamgir MK, Ahmad WMMJ, Nadhman A (2018) Effect of calcinations temperature on the photoactivities of ZnO/SnO_2 nanocomposites for the degradation of methyl orange, 2018. Mater Chem Phys 213:259–266

7. Ansari MO, Kumar R, Ansari SA, Ansari SP, Barakat MA, Alshahrie A, Cho MH (2017) Anion selective pTSA doped polyaniline@graphene oxide-multiwalled carbon nanotube composite for Cr(VI) and Congo red adsorption. J Colloid Interface Sci 496:407–415
8. Aquino JM, Rocha-Filho RC, Ruotolo LA, Bocchi N, Biag-gio SR (2014) Electrochemical degradation of a real textilewastewater using β-PbO_2 and DSA®anodes. Chem Engr-Ing J 251:138–145
9. Arabi SMS, Lalehloo RS, Olyai MRTB, Ali GA, Sadegh H (2019) Removal of congo red azo dye from aqueous solution by ZnO nanoparticles loaded on multiwall carbon nanotubes. Phys E Low Dimens Syst Nanostruct 106:150–155
10. Baruah S, Khan MN, Dutta J (2016) Perspectives and applications of nanotechnology in water treatment. Environ Chem Lett 14:1–14
11. Bazrafshan E, Mostafapour FK, Hosseini AR, Raksh Khorshid A, Mahvi AH (2013) Decolorisation of reactive red 120 dye by using single-walled carbon nanotubes in aqueous solutions. J Chem. https://doi.org/10.1155/2013/938374
12. Briggs AM, Cross MJ, Hoy DG, Blyth FH, Woolf AD, March L (2016) Musculoskeletal health conditions represent a global threat to healthy aging: a report for the 2015 world health organization world report on ageing and health. Gerontologist 56:243–255
13. Brindley L (2009) New solution for dye wastewater pollution. Chemistry World. Retrieved 2018-07-08
14. Cao C, Xiao L, Chen C, Shi X, Cao Q, Gao L (2014) In situ preparation of magnetic Fe_3O_4/chitosan nanoparticles via a novel reduction–precipitation method and their application in adsorption of reactive azo dye. Powder Technol 260:90–97
15. Chaudhary JP, Mahto A, Vadodariya N, Kholiya F, Maiti S, Nataraj SK, Meena R (2016) Fabrication of carbon and sulphur-doped nanocomposites with seaweed polymer carrageen an as an efficient catalyst for rapid degradation of dye pollutants using a solar concentrator. RSC Adv 6:61716–61724
16. Chawla S, Uppal H, Yadav M, Bahadur N, Singh N (2017) Zinc peroxide nanomaterial as an adsorbent for removal of Congo red dye from waste water. Ecotoxicol Environ Saf 135:68–74
17. Chen J, Sheng Y, Song Y, Chang M, Zhang X, Cui L, Zou H (2018) Multimorphology mesoporous silica nanoparticles for dye adsorption and multicolor luminescence applications. Sustain Chem Eng 6:3533–3545
18. Christie RM (2007) Environmental aspects of textile dyeing. Elsevier
19. Copaciu F, Opris¸ O, Coman V, Ristoiu D, Niinemets Ü, Copolovici L (2013) Diffuse water pollution by anthrax quinine and azo dyes in environment importantly alters foliage volatiles, carotenoids and physiology in wheat (*Triticum aestivum*). Water Air Soil Pollut 224(3):1478
20. Dehghani MH, Niasar ZS, Mehrnia MR, Shayeghi M, Al-Ghouti MA, Heibati B, Yetilmezsoy K (2017) Optimizing the removal of organophosphorus pesticide malathion from water using multi-walled carbon nanotubes. Chem Eng J 310:22–32
21. Dias JM, Alvim-Ferraz MCM, Almeida MF, Rivera-Utrilla J, Sánchez Polo M (2007) J Environ Manag 85:833–846. CrossRef CAS PubMed
22. Dong S, Feng J, Li Y, Hu L, Liu M, Wang Y, Pi Y, Sun J, Sun J (2014) Shape-controlled synthesis of BiVO4 hierarchical structures with unique natural-sunlight-driven photocatalytic activity. Appl Catal B-Environ 152–153:413–424
23. Dong Y, Dong W, Cao Y, Han Z, Ding Z (2011) Catal. Today 175:346–355
24. Dong S, Feng J, Fan M, Pi Y, Hu L, Han X, Liu M, Sun J, Sun J (2015) Recent developments in heterogeneous photocatalytic water treatment using visible light-responsive photocatalysts: a review. RSC Adv 5:14610–14630
25. Emami Z, Azizian S (2014) J Anal Appl Pyrolysis 108:176–184. CrossRef CAS PubMed
26. Fernandez ME, Nunell GV, Bonelli PR, Cukierman AL (2014) Ind Crops Prod 62:437–445. CrossRef CAS PubMed
27. Ge L, Han CC, Liu J, Li YF (2011) Enhanced visible light photocatalytic activity of novel polymeric g-C_3N_4 loaded with Ag nanoparticles. Appl Catal A 409:215–222
28. Geng M, Thagard SM (2013) Ultrason Sonochem 20:618–625

29. Gomez-Pastora J, Dominguez S, Bringas E, Rivero MJ, Ortiz I, Dionysiou DD (2017) Review and perspectives on the use of magnetic nanophotocatalysts (MNPCs) in water treatment. Chem Eng J 310:407–427
30. Gupta VK, Agarwal S, Saleh TA (2011) Synthesis and characterization of alumina-coated carbon nanotubes and their application for lead removal. J Hazard Mater 185:17–23
31. Gupta VK, Suhas (2009) Application of low-cost adsorbents for dye removal—a review. J Environ Manag 90:2313–2342
32. Hameed A, Montini T, Gombac V, Fornasiero P (2009) Photocatalytic decolourization of dyes on NiO-ZnO nano-composites. Photoch Photobio Sci 8:677–682
33. Han CC, Ge L, Chen CF, Li YJ, Xiao XL, Zhang YN, Guo LL (2014) Novel visible light induced Co3O4-g-C3N4 heterojunction photocatalysts for efficient degradation of methyl orange. Appl Catal B-Environ 147:546–553
34. Hassan MM, Carr CM (2018) A critical review on recent advancements of the removal of reactive dyes from dye house effluent by ion-exchange adsorbents. Chemosphere 209(1):201–219
35. Heidarizad M, Şngor SS (2016) Synthesis of graphene oxide/magnesium oxide nanocomposites with high-rate adsorption of methylene blue. J Mol Liq 224:607–617
36. Hong RY, Li JH, Chen LL, Liu DQ, Li HZ, Zheng Y, Ding J (2009) Synthesis, surface modification and photocatalytic property of ZnO nanoparticles. Powder Technol 189:426–432
37. Hu JC, Song Z, Chen LF, Yang HJ, Li JL, Richards R (2010) Adsorption properties of MgO(111) nanoplates for the dye pollutants from wastewater. J Chem Eng Data 55:3742–3748
38. Hunger K (2003) Industrial dyes: chemistry, properties and appli-cations. Willey-VCH, Weinheim
39. Ihsanullah Al-Khaldi FA, Abusharkh B, Khaled M, Atieh MA, Nasser MS, Saleh TA, Agarwal S, Tyagi I, Gupta VK (2015) Adsorptive removal of cadmium(II) ions from liquid phase using acid modified carbon-based adsorbents. J Mol Liq 204:255–263
40. Imran M, Crowley DE, Khalid A, Hussain S, Mumtaz MW, Arshad M (2015) Microbial biotechnology for decolorizationof textile wastewaters. Rev Environ Sci Biotechnol 14(1):73–92
41. Iram M, Guo C, Guan YP, Ishfaq A, Liu HZ (2010) Adsorption and magnetic removal of neutral red dye from aqueous solution using Fe_3O_4 hollow nanospheres. J Hazard Mater 181:1039–1050
42. Jeni J, Kanmani S (2011) Solar nanophotocatalytic decolorisation of reactive dyes using titanium dioxide Iran. J Environ Health Sci Eng 8(1):15–24
43. Kamil AM, Mohammed HT, Alkaim AF, Hussein FH (2016) Adsorption of Congo red on multiwall carbon nanotubes: effect of operational parameters. J Chem Pharm Sci 9:1128–1133
44. Kang D, Yu X, Ge M, Xiao F, Xu H (2017) Novel Al-doped carbon nanotubes with adsorption and coagulation promotion for organic pollutant removal. J Environ Sci 54:1–12
45. Karimi L, Zohoori S, Yazdanshenas ME (2014) J Saudi Chem Soc 18:581–588
46. Khan S, Malik A (2018) Toxicity evaluation of textile effluents and role of native soil bacterium in biodegradation of a textile dye. Environ Sci Pollut Res Interna-Tional 25(5):4446–4458
47. Khatri J, Nidheesh PV, Singh TA, Kumar MS (2018) Advanced oxidation processes based on zero-valent aluminium for treating textile wastewater. Chem Eng J 348:67–73
48. Khezrianjoo S, Lee J, Kim KH, Kumar V (2019) Eco-toxicological and kinetic evaluation of TiO_2 and ZnO nanophotocatalysts in degradation of organic dye. Catalysts 9:871. https://doi.org/10.3390/catal9100871
49. Konicki W, Aleksandrzak M, Moszyński D, Mijowska E (2017) Adsorption of anionic azo-dyes from aqueous solutions onto graphene oxide: equilibrium, kinetic and thermodynamic studies. J. Colloid Interf. Sci. 496:188–200
50. Kumar KY, Muralidhara HB, Nayaka YA, Balasubramanyam J, Hanumanthappa H (2013) Low-cost synthesis of metal oxide nanoparticles and their application in adsorption of commercial dye and heavy metal ion in aqueous solution. Powder Technol 246:125–136
51. Kyzas GZ, Matis KA (2015) Nanoadsorbents for pollutants removal: a review. J Mol Liq 203:159–168

52. Li J, Liu X, Pan L, QinW CT, Sun Z (2014) MoS_2-reduced graphene oxide composites synthesized via a microwave-assisted method for visible-light photocatalytic degradation of methylene blue. RSC Adv 4:9647–9651
53. Li X, Lu H, Zhang Y, He F (2017) Efficient removal of organic pollutants from aqueous media using newly synthesized polypyrrole/CNTs-$CoFe_2O_4$ magnetic nanocomposites. Chem Eng J 316:893–902
54. Liang J, Liu J, Yuan X, Dong H, Zeng G, Wu H, Wang H, Liu J, Hua S, Zhang S, Yu Z (2015) Facile synthesis of alumina decorated multi-walled carbon nanotubes for simultaneous adsorption of cadmium ion and trichloroethylene. Chem Eng J 273:101–110
55. Lingamdinne LP, Koduru JR, Roh H, Choi YL, Chang YY, Yang JK (2016) Adsorption removal of Co(II) from wastewater using graphene oxide. Hydrometallurgy 165:90–96
56. Liu F, Chung S, Oh G, Seo TS (2012) Three-dimensional graphene oxide nanostructure for fast and efficient water-soluble dye removal. Acs Appl Mater Inter 4:922–927
57. Lonkar SP, Pillai VV, Alhassan SM (2018) Facile and scalable production of heterostructured ZnS–ZnO/graphene nanophotocatalysts for environmental remediation. Sci Rep 8:13401. https://doi.org/10.1038/s41598-018-31539-7
58. Ma J, Yu F, Zhou L, Jin L, Yang MX, Luan JS, Tang YH, Fan HB, Yuan ZW, Chen JH (2012) Enhanced adsorptive removal of methyl orange and methylene blue from aqueous solution by alkali-activated multiwalled carbon nanotubes. Acs Appl Mater Inter 4:5749–5760
59. Madrakian T, Afkhami A, Ahmadi M, Bagheri H (2011) Removal of some cationic dyes from aqueous solutions using magnetic modified multi-walled carbon nanotubes. J Hazard Mater 196:109–114
60. Mahapatra NN (2016) Textile dyes. CRC Press, Boca Raton; Woodhead Publishing India Pvt., NewDelhi
61. Mahmoodi NM (2011) Desalination 279:332–337
62. Mahmoodian H, Moradi O, Shariatzadeha B, Salehf TA, Tyagi I, Maity A (2015) Enhanced removal of methyl orange from aqueous solutions by poly HEMA–chitosan-MWCNT nanocomposite. J Mol Liq 202:189–198
63. Mahmoud HR, El-Molla SA, Saif M (2013) Improvement of physicochemical properties of Fe_2O_3/MgO nanomaterials by hydrothermal treatment for dye removal from industrial wastewater. Powder Technol 249:225–233
64. Malini B, Allen Gnana Raj G (2018) Synthesis, characterization and photocatalytic activity of cobalt doped TiO_2 nanophotocatalysts for rose bengal dye degradation under day light illumination. Chem Sci Trans 7(4):687–695
65. Massey AT, Gusain R, Kumari S, Khatri OP (2016) Hierarchical microspheres of MoS_2 nanosheets: efficient and regenerative adsorbent for removal of water-soluble dyes. Ind Eng Chem Res 55:7124–7131
66. Massoudinejad M, Rasoulzadeh H, Ghaderpoori M (2019) Magnetic chitosan nanocomposite: fabrication, properties, and optimization for adsorptive removal of crystal violet from aqueous solutions. Carbohydr Polym 206:844–853
67. Mehrabi M, Javanbakht V (2018) Photocatalytic degradation of cationic and anionic dyes by a novel nanophotocatalyst of $TiO_2/ZnTiO_3/aFe_2O_3$ by ultraviolet light irradiation. J Mater Sci Mater Electron 29:9908–9919
68. Mezohegyi G, van der Zee FP, Font J, Fortuny A, Fabregat A (2012) J Environ Manag 102:148–164. CrossRef CAS PubMed
69. Mishra AK, Arockiadoss T, Ramaprabhu S (2010) Study of removal of azo dye by functionalized multi walled carbon nanotubes. Chem Eng J 162(3):1026–1034. https://doi.org/10.1016/j.cej.2010.07.014
70. Moradi O (2013) Adsorption behavior of basic red 46 by singlewalled carbon nanotubes surfaces. Fullerenes Nanotubes Carbon Nanostruct 21:286–301
71. Moussavi G, Mahmoudi M (2009) Removal of azo and anthraquinone reactive dyes from industrial wastewaters using MgO nanoparticles. J Hazard Mater 168:806–812
72. Nasr C, Vinodgopal K, Fisher L, Hotchandani S, Chattopadhyay AK, Kamat PV (1996) Environmental photochemistry on semiconductor surfaces. Visible light induced degradation

of a textile diazo dye, naphthol blue black, on TiO_2 nanoparticles. J Phys Chem-Us 100:8436–8442
73. Nassar NN (2010) Kinetics, mechanistic, equilibrium, and thermodynamic studies on the adsorption of acid red dye from wastewater by-Fe_2O_3 nanoadsorbents. Sep Sci Technol 45:1092–1103
74. Natarajan TS, Thomas M, Natarajan K, Bajaj HC, Tayade RJ (2011) Chem Eng J 169:126–134
75. Nekouei F, Nekouei S, Tyagi I, Gupta VK (2015) Kinetic, thermodynamic and isotherm studies for acid blue 129 removal from liquids using copper oxide nanoparticle modified activated carbon as a novel adsorbent. J Mol Liq 201:124–133
76. Njoku VO, Foo KY, Asif M, Hameed BH (2014) Chem Eng J 250:198–204. CrossRef CAS PubMed
77. Patra S, Roy E, Madhuri R, Sharma PK (2016) Agar based bimetallic nanoparticles as high performance renewable adsorbent for removal and degradation of cationic organic dyes. J Ind Eng Chem 33:226–238
78. Peng N, Hu D, Zeng J, Li Y, Liang L, Chang C (2016) Superabsorbent cellulose–clay nanocomposite hydrogels for highly efficient removal of dye in water. ACS Sustain Chem Eng 4:7217–7224
79. Ren X, Chena C, Nagatsu M, Wang X (2011) Carbon nanotubes as adsorbents in environmental pollution management: a review. J Chem Eng 170:395–410
80. Rong J, Rong XS, Qiu FX, Zhu XL, Pan JM, Zhang T, Yang DY (2016) Facile preparation of glucose functionalized multi-wall carbon nanotubes and its application for the removal of cationic pollutants. Mater Lett 183:9–13
81. Sadegh H, Shahryari-Ghoshekandi R, Kazemi M (2014) Study in synthesis and characterization of carbon nanotubes decorated by magnetic iron oxide nanoparticles. Int Nano Lett 4:129–135
82. Salem ANM, Ahmed MA, El-Shahat MF (2016) Selective adsorption of amaranth dye on Fe_3O_4/MgO nanoparticles. J Mol Liq 219:780–788
83. El Saliby IJ, Shon H, Kandasamy J, Vigneswaran S (2008) Nanotechnology for wastewater treatment: in brief. In: Encyclopedia of life support system (EOLSS)
84. Salimi F, Tahmasobi K, Karami C, Jahangiri A (2017) Preparation of modified nano-SiO_2 by bismuth and iron as a novel remover of methylene blue from water solution. J Mex Chem Soc 61:250–259
85. Sandhya S (2010) Biodegradation of azo dyes under anaerobic condition: role of azoreductase. In: Erkurt HA (ed) Biodegra-dation of azo dyes. The handbook of environmental chemistry, vol 9. Springer, Berlin, Heidelberg, pp 39–57
86. Sarkheil H, Noormohammadi F, Rezaei AR, Borujeni MK (2014) Dye pollution removal from mining and industrial wastewaters using chitson nanoparticles. In: International conference on agriculture, environment and biological sciences (ICFAE'14), Antalya, Turkey
87. Sathishkumar P, Sweena R, Wu JJ, Anandan S (2011) Synthesis of CuO–ZnO nanophotocatalyst for visible light assisted degradation of a textile dye in aqueous solution. Chem Eng J 171(1):136–140
88. Seddigi ZS, Gondal MA, Baig U, Ahmed SA, Abdulaziz MA, Danish EY et al (2017) Facile synthesis of light harvesting semiconductor bismuth oxychloride nano photo-catalysts for efficient removal of hazardous organic pollutants. PLoS ONE 12(2):e0172218. https://doi.org/10.1371/journal.pone.0172218
89. Sheibani M, Ghaedi M, Marahel F, Ansari A (2015) Congo red removal using oxidized multiwalled carbon nanotubes: kinetic and isotherm study. Desalin Water Treat 53:844–852
90. Shen WZ, Li ZJ, Wang H, Liu YH, Guo QJ, Zhang YL (2008) Photocatalytic degradation for methylene blue using zinc oxide prepared by codeposition and sol-gel methods. J Hazard Mater 152:172–175
91. Shinde P, Gupta SS, Singh B, Polshettiwar V, Prasad BL (2017) Amphi-functional mesoporous silica nanoparticles for dye separation. J Mater Chem A 5:14914–14921
92. Singh VP, Sharma M, Vaish R (2020) Enhanced dye adsorption and rapid photo catalysis in candle soot coated Bi2WO6 ceramics. Mater Chem Phys. https://doi.org/10.1016/j.matchemphys.2020.123311

93. Soltani RDC, Khataee AR, Safari M, Joo SW (2013) Preparation of bio-silica/chitosan nanocomposite for adsorption of a textile dye in aqueous solutions. Int Biodeterior Biodegrad 85:383–391
94. Suzimara R, Jonnatan JS, Paola C, Denise AF (2018) Highly pure silica nanoparticles with high adsorption capacity obtained from sugarcane waste ash. ACS Omega 3:2618–2627
95. Szlachta M, Wojtowicz P (2013) Adsorption of methylene blue and Congo red from aqueous solution by activated carbon and carbon nanotubes. Water Sci Technol 68:2240–2248
96. Taghdiri M (2017) Selective adsorption and photocatalytic degradation of dyes using polyoxometalate hybrid supported on magnetic activated carbon nanoparticles under sunlight. Visible, and UV Irradiation ID 8575096. 10.1155/ 2017/8575096
97. Taherian F, Marcon V, van der Vegt NF, Leroy F (2013) What is the contact angle of water on graphene? Langmuir 29:1457–1465
98. Tang L, Wang JJ, Wang L, Jia CT, Lv GX, Liu N, Wu MH (2016) Facile synthesis of silver bromide-based nanomaterials and their efficient and rapid selective adsorption toward anionic dyes. ACS Sustain Chem Eng 4:4617–4625
99. Tarigh GD, Shemirani F (2013) Magnetic multi-wall carbon nanotube nanocomposite as an adsorbent for preconcentration and determination of lead(II) and manganese(II) in various matrices. Talanta 115:744–750
100. Tavassolia N, Ansaria R, Mosayebzadeh Z (2017) Synthesis and application of iron oxide/silica gel nanocomposite for removal of sulfur dyes from aqueous solutions. Arch Hyg Sci 6:214–220
101. Tawabini BS, Khaldi SFA, Khaled MM, Atieh MA (2011) Removal of arsenic from water by iron oxide nanoparticles impregnated on carbon nanotubes. J Environ Sci Health 46:215–223
102. TomicNM D-M, PaunovicNM MDZ, Radic ND, Grbic BV, Askrabic SM, Babic BM, Bajuk-Bogdanovic DV (2014) Nanocrystalline CeO2-delta as effective adsorbent of azo dyes. Langmuir 30:11582–11590
103. Tong H, Ouyang SX, Bi YP, Umezawa N, Oshikiri M, Ye JH (2012) Nano-photocatalytic materials: possibilities and challenges. Adv Mater 24:229–251
104. Trevino-Cordero H, Juarez-Aguilar LG, Mendoza-Castillo DI, Hernandez-Montoya V, Bonilla-Petriciolet A, Montes-Moran MA (2013) Ind Crops Prod 42:315–323
105. Vikrant K, Giri BS, Raza N, Roy K, Kim KH, Rai BN et al (2018) Recent advancements in bioremediation of dye: Currentstatus and challenges. Biores Technol 253:355–367
106. Wang YJ, Shi R, Lin J, Zhu YF (2011) Enhancement of photocurrent and photocatalytic activity of ZnO hybridized with graphite-like C_3N_4. Energy Environ Sci 4:2922–2929
107. Wang S, Ng CW, Wang W, Li Q, Hao Z (2012) Synergistic and competitive adsorption of organic dyes on multiwalled carbon nanotubes. Chem Eng J 197:34–40. https://doi.org/10.1016/j.cej.2012.05.008
108. Wang SM, Li DL, Sun C, Yang SG, Guan Y, He H (2014) Synthesis and characterization of g C_3N_4/Ag_3VO_4 composites with significantly enhanced visible-light photocatalytic activity for triphenylmethane dye degradation. Appl Catal B-Environ 144:885–892
109. Wang P, Cao M, Wang C, Ao Y, Hou J, Qian J (2014) Kinetics and thermodynamics of adsorption of methylene blue by a magnetic graphene–carbon nanotube composite. Appl Surf Sci 290:116–124
110. Xiong L, Yang Y, Mai JX, Sun WL, Zhang CY, Wei DP, Chen Q, Ni JR (2010) Adsorption behavior of methylene blue onto titanate nanotubes. Chem Eng J 156:313–320
111. Xiong ZG, Zhang LL, Ma JZ, Zhao XS (2010) Photocatalytic degradation of dyes over graphene-gold nanocomposites under visible light irradiation. Chem Commun 46:6099–6101
112. Xiong Z, Zhang L, Zhao X (2011) Visible-light-induced dye degradation over copper-modified reduced graphene oxide. Chemistry 17:2428–2434
113. Yan SC, Li ZS, Zou ZG (2009) Photodegradation performance of g-C_3N_4 fabricated by directly heating melamine. Langmuir 25:10397–10401
114. Yang S, Wang P, Yang X, Shan L, Zhang W, Shao X, Niu R (2010) J Hazard Mater 179:552–558
115. Yu CL, Li G, Kumar S, Yang K, Jin RC (2014) Phase transformation synthesis of novel Ag_2O/Ag_2CO_3 heterostructures with high visible light efficiency in photocatalytic degradation of pollutants. Adv Mater 26:892–898

116. Yuan R, Ramjaun SN, Wang Z, Liu J (2012) Chem Eng J 192:171–178
117. Zare K, Najafi F, Sadegh H (2013) Studies of ab initio and Monte Carlo simulation on interaction of fluorouracil anticancer drug with carbon nanotube. J Nanostruct Chem 3:1–
118. Zarezadeh-Mehrizi M, Badiei A (2014) Highly efficient removal of basic blue 41 with nanoporous silica. Water Resour Ind 5:49–57
119. Zhang L, Zhang G, Wang S, Peng J, Cui W (2016) Cationfunctionalized silica nanoparticle as an adsorbent to selectively adsorb anionic dye from aqueous solutions. Environ Prog Sustain Energy 35:1070–1077
120. Zhang ZJ, Zhao AD, Wang FM, Ren JS, Qu XG (2016) Design of a plasmonic micromotor for enhanced photoremediation of polluted anaerobic stagnant waters. Chem Commun 52:5550–5553
121. Zhang G, Wang S, Zhao S, Fu L, Chen G, Yang F (2011) Appl Catal B 106:370–378
122. Zhang C, Sui J, Li J, Tang Y, Cai W (2012) Efficient removal of heavy metal ions by thiol functionalized superparamagnetic carbon nanotubes. Chem Eng J 210:45–52
123. Zhao LX (2011) Visible-light-induced dye degradation over copper-modified reduced graphene oxide. Chemistry 17:2428–2434
124. Zhuang Y, Yu F, Chen J, Ma J (2016) Batch and column adsorption of methylene blue by graphene/alginate nanocomposite: comparison of single-network and double-network hydrogels. J Env Chem Eng 4:147–156

Textile Wastewater Treatment Using Sustainable Technologies: Advanced Oxidation and Degradation Using Metal Ions and Polymeric Materials

Megha Bansal

Abstract Textile industries are considered as one of the major contributors for water pollution. Textile wastewater primarily contains organic dye molecules along with heavy metal ions and some polymeric waste material which causes adverse effects on aquatic system and human health. Insufficient and incomplete treatment of textile wastewater is a major concern especially when it is discharged directly into the water bodies. Removal of dye molecules from wastewater has been addressed by several researchers using various physical, chemical, and biological processes. This chapter highlights applications and challenges with conventional and advanced treatment processes. Advanced oxidation process using various combinations of oxidants and energy sources has emerged as a significant technique. Applications of nanomaterials and polymers have also been explored significantly for the degradation of pollutants present in textile wastewater. Integration of suitable technologies is proposed to achieve complete degradation of pollutants discharged in textile wastewater.

1 Introduction

India is one of the largest producers and exporters of textiles and garments in the world. In India, the textile industry plays a significant role towards the economic growth of the country and offers mutual exchange of technology between the countries. In addition to these benefits, the textile industry is also considered to be highly polluting industry and leading consumer of water [1]. The textile waste is inevitable outcome of industrialization. It uses more than 200 L of water/kg of textile product along with numerous other chemicals and discharge huge amount of wastewater to water bodies [2]. It is estimated that about 3×10^6 L of wastewater is reproduced after processing about 20,000 kg of textiles per day. India is not the only country to face challenges in addressing the issues of textile wastewater management. China produces approximately 170 million tons of municipal solid wastes annually where

M. Bansal (✉)
Department of Chemistry, Manav Rachna University, Faridabad, Haryana 121004, India
e-mail: megha@mru.edu.in

S. S. Muthu and A. Khadir (eds.), *Advanced Oxidation Processes in Dye-Containing Wastewater*, Sustainable Textiles: Production, Processing, Manufacturing & Chemistry,
https://doi.org/10.1007/978-981-19-0987-0_9

the contribution of textile waste is 10–15% [3]. According to US EPA, the contribution of textile waste is 9–10% in the total municipal solid waste generated in the United States [4].

In addition to textile industry, other industries such as pharmaceutical, leather, and plastic industry also discharge approximately 50,000 tons of organic dyes every year into water bodies [5]. Textile wastewater includes (a) suspended solid material, (b) mineral oils, (c) surfactants, (d) Phenols and resins, (e) halogenated organics produced, (f) dyes and polymeric substances, and (g) heavy metals such as lead, mercury, chromium, copper, and zinc. The heavy metals have tendency to accumulate in the biological tissues and are not easily degradable [6]. The disposal of these waste generates serious governance issues, thus there is a need to recycle them in sustainable manner [7].

It has been reported that processing of polyester and cotton requires approximately 100–350 kg of water per kg of fabric which contribute up to 80% total wastewater load [8]. Dyeing and finishing processes utilize large quantities of chemicals and release complex organic compounds and heavy metals to wastewater thus poses major threats [9]. In the wet processing of textile fibers, after dyeing the fibers are repeatedly rinsed, and treated with numerous surfactants, salts, recalcitrant and toxic chemicals which eventually discharged in water bodies [10].

In the dyeing process, the usage of reactive dyes is increasing exponentially due to high demand of bright colored fabric. These reactive dyes contain various reactive groups such as halotriazine, sulfone, pyrimidine, and dichloro-fluoropyrimidine [11]. These dyes are generally used to dye cotton fabric however their fixation rate is about 60–90% only. Thus, unfixed residual is discharged as the textile effluent along with other alkali, acids, organic dyes, and heavy metals [12]. The synthetic cationic and anionic dyes such as methylene blue, methyl orange, and rhodamine B are also not degraded easily thus adversely affect soft tissues and cause genetic changes [13]. Prolonged exposure to these toxicants leads to nervous system disorders, thyroid dysfunction, and may result as lethal as well [14]. In addition, due to the presence of various dyes and heavy metals, it is highly carcinogenic in nature and causes severe adverse effects [15].

Textile wastewater contains a wide range of organic pollutants with different chromophores and auxochromes (Fig. 1). Therefore, it is not possible to treat a wide range of pollutants ranging from nanoparticles to polymers, metallic to non-metallic, and organic to inorganic using single treatment methodology. In addition, textile wastewater has highly fluctuating pH, with high chemical oxygen demand, biological oxygen demand, and suspended solids.

Thus, it is imperative to design and adopt a sustainable and eco-friendly model of textile industry which reduces and overcomes all challenges of waste minimization and disposal [16]. It is needed to design effective techniques to understand the pathways of pollutant degradation and formation of intermediates or products. The methodologies should not only meet international standards, but also be able to protect the biodiversity and to ensure the availability of pure water for future generations. The Textile and Apparel industry in India contributes to 14% of total

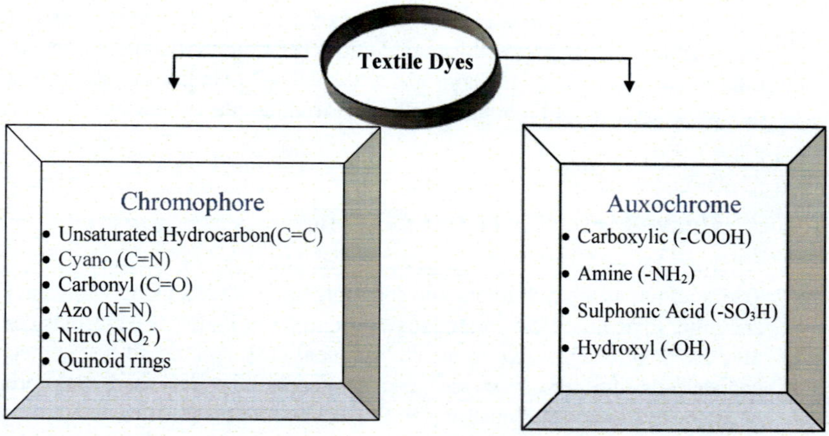

Fig. 1 List of few major chromophores and auxochromes present in textile dyes

industrial production, 4% of gross domestic product, and 15% of total export earnings (Textile Ministry, Make in India, TechSci Research, 2017). There are various treatments available for treating textile wastewater such as adsorption, photochemical oxidation, catalytic oxidation, flocculation, coagulation, and membrane filtration; however, these methods are compromised with certain limitation (Table 1). The major problem is sludge generation in large quantities which further needs to be treated thereby increasing the cost of treatment process [17].

Table 1 Methods available for the treatment of textile wastewater and their limitations

S. No	Method(s)	Major limitation
1	Flocculation, Fenton oxidation	Sludge production
2	Membrane filtration	Sludge disposal
3	Adsorption, phytoremediation	Disposal of concentrated pollutants
4	Bioremediation	Restricted bacterial growth
5	Photochemical oxidation	Generation of by-products
6	Ion exchange	Can be used for specific dyes only
7	Electrochemical oxidation	Expensive
8	Electrolytic precipitation	Need long contact time
9	Ozonation	Short life span

This chapter will focus upon discharge of heavy metals and polymeric fibers in textile effluent and their possible remediation. The chapter will also address modifications in the conventional methods and advanced technologies to address the removal of various organic and inorganic pollutants from textile wastewater.

2 Heavy Metals in Textile Effluent

Textile effluents are rich in organic and inorganic contents along with heavy metals and other toxicants. Heavy metal discharge to water bodies is of serious concern as they are nonbiodegradable in the environment [15]. Among all heavy metals lead, copper, cadmium, chromium, zinc, and manganese are widely used for the production of textile dyes and other materials in the dyeing process [18].

Several researchers have investigated the concentration of heavy metals in textile discharge and emphasized their adverse effects on aquatic system and living organisms [19]. Chromium pollution by textile industries is highly prevalent, as chromium is one of the main elements of metal-based dyes in textile industries [20]. Few essential metals such as iron, copper, zinc, and manganese are required in minute quantities to maintain normal cellular growth and functioning, however, excess quantity beyond permissible limit impose adverse effects on human health.

Direct and indirect discharge of heavy metals into the environment causes environmental pollution at various levels and poses threat to aquatic and human lives on prolonged exposure [21]. To propose possible solutions for the remediations of heavy metals it is needed to have complete understanding of their permissible limit, toxicological profile, biotransformation, and distribution in abiotic and biotic environment. Heavy metals are known to cause disturbances in neurological function, neurotoxicity, hepatotoxicity and nephrotoxicity in humans and animals, bone degeneration, damage to soft tissues, impaired heme synthesis, carcinoma to multiple organs eventually lead to cellular death [22, 23]. In addition, heavy metals also inhibit microbial growth, their activities and various biological processes such as nitrogen fixation, decomposition of organic matter in soil. Lead and chromium possess the ability to generate reactive radicals, which cause oxidative imbalance, resulting to cellular and DNA damage [24]. These metals also bioaccumulate in aquatic plants and animals and further enter to ecological cycle [25]. The heavy metals possess electron-sharing affinities and form covalent bond with sulfhydryl groups of proteins thereby inhibiting activity of antioxidant enzymes. Studies associated with the assessment of occupational exposure to textile wastewater have reported increased risk of hepatic, renal, and cardiovascular disorders [26]. Utilization of textile waste water for irrigation leads to accumulation of various heavy metals in crops which further enters to food chain undergoes biomagnification [27]. Thus, the toxins and pollutants discharged in water bodies by industries persist in the environment and decrease the quality of life. This necessitates the requirement of effective techniques for the removal of toxicants from industrial effluents prior to their discharge into water bodies.

3 Microplastic Pollution by Textile Effluent

Textile industry was earlier considered to release only dye containing wastewater along with some other organic and inorganic pollutants. However, in the recent past, it has been studied that it has also contributed significantly towards microplastic and polymeric pollution. During finishing process of textile, some urea-based resins such urea–formaldehyde, dimethylol ethylene urea are used for imparting anti-crease finish to final product which ultimately discharge formaldehyde to wastewater. Formaldehyde blocks respiratory tracts and causes serious respiratory ailments. Copolymers of acrylic acid such as acrylate and methacrylate containing free carboxyl groups are also released into water bodies which are resistant to biological attack. Plastic particles smaller than 5 mm in length are referred to as microplastics [28]. It has become a growing concern in recent years owing to its small size [29]. In order to propose effective solution for microplastics, it is imperative to evaluate possible sources of its pollution. These particles have derived from a wide range of industries such as polymer, personal care products, textile, and processing industries [30].

In 2016, textile washing processes generated approximately 5.4 million tons of synthetic fibers worldwide [31]. Among various types of microplastic, synthetic fibers are of most dominant type detected in water, sediments, and living organisms [32]. Over 90% of microplastics reaches global coastal environments via synthetic fibers [33]. Industrial and domestic textile laundry is considered as major sources of synthetic fibers, however, till date there are not many reports highlighting the impact of textile industries towards microplastic pollution. At industrial scale production of textile, fiber requires large amount of water which ultimately discharge into water bodies. The wastewater discharged by textile industries contains various toxic compounds, such as phenols, thiazoles, and phthalates, thus, considered as a primary source of water pollution source [34].

It has been estimated that a household washing machines release thousands of synthetic fibers after every wash cycle [35] and due to release of textile wastewater it adds approximately 160,000,000 microplastic particles per day to water bodies [36]. In wastewater treatment plants these particles undergo stress force, and mechanical mixing which further decreases their particle size and acts as carrier for various contaminates to food chain [37, 38].

Microplastics have been classified into five categories on the basis of their shape as films, fragments, fibers, foam, and pellets [39]. Among all the categories, the most harmful is microfiber due to its size specification. Usually, the length of this fiber ranges between 100 μm and 5 mm and width nearly 1.5 orders of magnitude shorter in comparison to length [40]. Various toxic chemicals such as heavy metals, organic dyes, poly aromatic hydrocarbons, and polychlorinated biphenyl are adsorbed over these fibers and easily enter food chain. Due to small size, they are easily ingested by aquatic animals and cause deleterious effects [41]. In Textile industries, a well-defined wastewater treatment plant removes various organic and inorganic impurities; however, the current available technologies retain major part of microplastics which are being continuously discharged to the water bodies. According to some reports, majorities of microfibers are removed in treatment stage before the aeration process only still a large amount enters the aquatic system because the quality of wastewater discharged to water bodies is very high [42]. Bioavailability of microfibers largely depends upon their shape and size therefore their accumulation is different in comparison to chemical pollutants.

In the last one-year, COVID-19 pandemic has created havoc worldwide and World Health Organization prescribed mandatory usage of face masks as one of the safety measures to prevent spread of in human beings. Thus, the usage of face masks and other accessories such as gloves, face shields, protective shoes, and suits increased thereby their manufacturing as well. All these materials are manufactured with various synthetic polymeric materials such as nylon, polyester, and copolymer of polyether–polyurea [43]. Thus, these types of textile industries have also contributed significantly to the polymeric pollution [44].

4 Advanced Oxidation Process (AOP) for Removal of Heavy Metals and Organic Compounds

Advanced Oxidation process is defined as a chemical treatment process for the removal of organic pollutants from wastewater by oxidation process using generation of highly reactive hydroxyl radical ($OH^{\cdot}$). If industrial wastewater contains nondegradable organic and inorganic compounds such as various dyes, polymeric substance, heavy metals, and their compounds, it is difficult to remove those pollutants by simple chemical or biological methods. Conventional oxidation techniques are not able to completely oxidize complex structure of organic compounds as they are resistant to oxidation. To efficiently remove such complex organic compounds from water bodies, more aggressive treatment methods are needed [45]. At high temperature the rate of oxidative degradation is limited owing to lesser availability of oxygen while at low temperature, though the availability of oxygen remains high however, degradation rate remains less due to limited activity of microorganisms. In addition, reducing chemicals discharged in wastewater also absorbs dissolved oxygen thereby retarding process of pollutant degradation. Thus, if there is sufficient supply of oxygen in terms of oxidizing radicals, it would help to enhance degradation of pollutants.

4.1 Fenton's Reagent

AOP is one of the most promising techniques to degrade organic pollutant from wastewater. In this method, highly reactive hydroxyl radical ($OH^{\cdot}$) are generated by using primary oxidants either O_2 or H_2O_2 (Fig. 2).

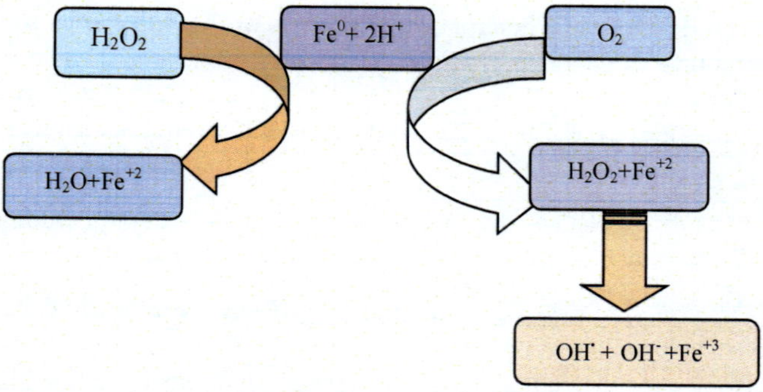

Fig. 2 Generation of hydroxyl free radical by advanced oxidation process using iron (Fe) and other primary oxidants. (Abbreviations used: H_2O_2—hydrogen peroxide; Fe^0—zero valent iron; $OH^{\cdot}$—hydroxyl radical; OH^-—hydroxyl ion)

In this process, all organic compounds are degraded to carbon dioxide (CO_2), H_2O, and inorganic salts [46]. This technique can remove persistent organic pollutants such as textile dyes, pharmaceuticals, polymeric substances, microplastics and serves as an effective treatment step in water purification or final stage before water discharge to water bodies [47]. Hydroxyl radical is highly reactive towards oxidative degradation, however, the major drawback is its non-selectivity. Along with the targeted molecules, it reacts with many other nontarget compounds present in the water matrix. Thus, demand of $OH^{\cdot}$ increases to complete the process of degradation.

On the basis of generation of $OH^{\cdot}$ Radical, the process can be defined as homogeneous and heterogeneous (Fig. 3). Homogeneous process requires chemical reagents

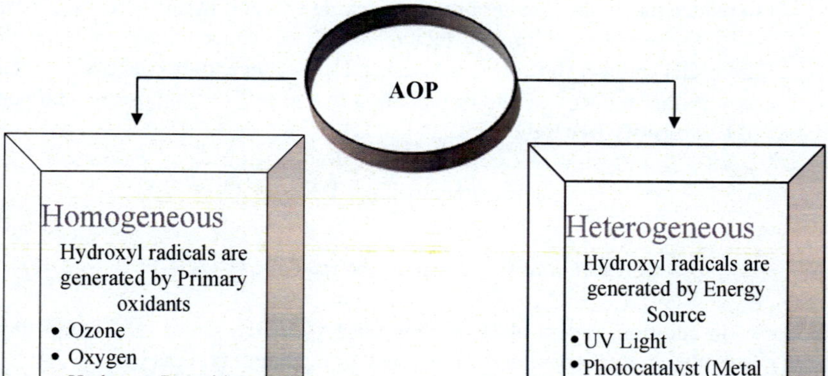

Fig. 3 Types of advanced oxidation process (AOP) on the basis of generation of hydroxyl free radicals

(O_2, O_3, H_2O_2) while heterogeneous process works with energy sources such as UV light source or photocatalyst.

$$Fe^{2+} + H_2O_2 \rightarrow Fe^{3+} + HO^{\cdot} + HO^{-} \quad \text{(Homogeneous Fenton)} \tag{1}$$

$$Fe^{3+} + H_2O_2 + hv \rightarrow Fe^{2+} + H^{+} + HO_2^{\cdot} \quad \text{(Photo Fenton reaction)} \tag{2}$$

$$H_2O_2 + hv \rightarrow 2HO^{\cdot} \quad \text{(Photo-peroxidation)} \tag{3}$$

Homogeneous Fenton and Photo-Fenton processes have few limitations as [48]:

- These reactions operate at pH less than 3 when Fe exist in soluble form
- Concentration of reagents need to be maintained optimum to avoid scavenger effects
- Sludge is produced as by-product thus disposal needs to be maintained
- As Fe salts are used for the supply of Fe thus, in the treated water Fe remains is present in traces which may adversely affect soft tissues
- Textile wastewater contains acidic radicals such as chloride (Cl^{-}), sulfate ($SO_4{}^{-2}$), phosphates ($PO_4{}^{-3}$), which may react with $OH^{\cdot}$ and decrease the efficiency of the oxidation process.

Fenton's reagents (H_2O_2 and Fe^{2+}) and ozone (O_3) under AOP have been used extensively to treat textile wastewater to achieve the best results. Fenton's oxidation was used to degrade various reactive dyes such as Remazol Yellow, Remazol Black 5, Remazol Blue and Remazol Red, and was found significantly effective in their decolorization (>99%) [49]. This process is also effective for the degradation of aromatic amines as well under acidic conditions. This oxidation process can be used to reduce the organic matter present in textile effluent. In this process, concentration of H_2O_2 serves as a limiting factor, i.e., higher concentration of H_2O_2 produces more $OH^{\cdot}$ radical; however, increasing H_2O_2 concentration above a certain level may further reduce the rate of oxidation due to self-decomposition and generation of less reactive free radical [50].

$$\text{Self-decomposition: } H_2O_2 \rightarrow O_2 + H_2O \tag{4}$$

$$\text{Generation of less reactive free radicals: } H_2O_2 + Fe^{3+} \rightarrow OH_2^{\cdot} \tag{5}$$

In addition, excess of H_2O_2 may react with $OH^{\cdot}$ radical and compete with the degradation and reducing the efficiency of treatment process [51].

4.2 Ozone (O_3)

Ozone is a potent oxidizing agent known for treating wastewater. Ozone gets easily dissolved in water and oxidize organic compounds and pollutants in two different ways:

- Direct oxidation as molecular ozone
- Indirect oxidation via formation of secondary oxidants like $OH^{\cdot}$.

Ozone due to its strong oxidizing property and high oxidizing potential (2.07) has been used for wastewater treatment since 1970s [52]. Germicidal efficacy of ozone is 100 times in comparison to hypo chloric acid disinfection property is 3,125 times faster. Ozone due to its highly unstable nature should always be generated on site. Oxidation potential of O_3 is −2.07 V which is greater than that of other potent oxidizing agents such as hypochlorite acid (−1.49 V) and chlorine (−1.26 V) (Table 2). With the help of ozone treatment, many nonbiodegradable products can also be decomposed. The challenges in using ozone treatment are the limited life span which is 20 min in water, and its reduction in the alkaline water [53]. Ozone function in two ways (a) a powerful disinfecting agent (b) a strong oxidizing agent and eliminate color, odor, and toxic organic compounds [54]. Ozone has been reported to decolorize all dyes, except insoluble vat and disperse dyes because these dyes have slow reaction rate which becomes a limiting factor in the reaction [55]. Decomposition of ozone produces free radical $OH_2^{\cdot}$ and $OH^{\cdot}$ which oxidizes various organic and inorganic impurities such as heavy metals and their salts, organic matter, and biological matter.

Ozone reacts via three specific reactions for the oxidation of organic molecules and polymeric substances: electrophilic addition of ozone to the Carbon–Carbon double bond: Ozone readily adds on to aliphatic unsaturated compounds, such as olefin and causes the oxidative cleaving of the alkene or alkyne.

Ozonolysis: Ozone gives addition reaction with alkenes to form ozonide. The reaction takes place in non-aqueous solvents, as presence of traces of water as well will hydrolyze ozone to other products before the reaction.

Table 2 Electrochemical oxidation potential of various oxidizing agents

S. No	Oxidizing agent	Electrochemical oxidation potential, V
1	Fluorine	−3.06
2	Hydroxyl radical	−2.80
3	Oxygen (atomic)	−2.42
4	Ozone	−2.07
5	Hydrogen peroxide	−1.78
6	Hypochlorite	−1.49
7	Chlorine dioxide	−1.26
8	Oxygen (molecular)	−1.23

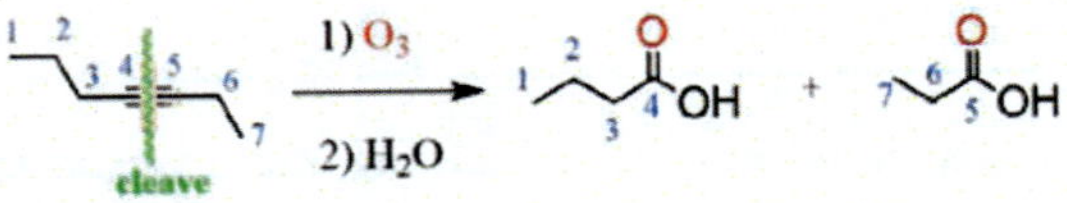

Substitution reaction: In this reaction, one atom or functional group is replaced with other groups. Ozone leads to cleavage of carbon–carbon bonds to generate fragmented organic compounds via inserting oxygen atom between the ring carbon and hydrogen atom [54].

The oxidation reaction using ozone is affected by various factors like pH, temperature, total organic carbon, and chemical oxygen demand. It has been reported that at acidic pH ozone reacts as intact O_3 molecule and the reaction is slow while at alkaline pH the reaction is rapid due to decomposition of ozone into hydroxyl free radicals. Thus, alkaline pH ranging from 8 to 10 is optimum for oxidation of organic molecules by ozone [54].

4.3 *Ozone (O_3)/Ultraviolet (UV)/Ultrasound*

Combination of two or more AOPs may provide synergistic effects for oxidative degradation by generating more hydroxyl radicals. The various combinations can be

- UV radiation + Ozone
- UV radiation + Hydrogen Peroxide
- Ozone + Ultrasound energy + Photochemical/photocatalytic oxidation.

The efficacy of combined system will depend upon the extent of synergism, number of generated hydroxyl radicals, and efficacy of contact between free radicals and pollutant molecules [56].

Treatment of organic compounds with ozone does not provide complete oxidation to CO_2 and H_2O in many cases. The intermediate remained in water after incomplete oxidation may be same or more toxic that initial toxicant. Thus, combination of O_3 with UV is more effective in comparison to simple ozonation. Photons of UV light activate ozone molecules, thus promote generation of hydroxyl radicals at a quicker rate thus increasing the efficacy of oxidation reaction [57]. For efficient ozone photolysis maximum radiation output of 254 nm from the UV lamp must be used. While proposing a combination of ultrasound with O_3, the operating frequency should not be more than 500 kHz which plays as one of the significant factors [56].

4.4 *Hydrogen Peroxide (H_2O_2)/Ultraviolet (UV) Radiation*

Hydrogen peroxide alone is not effective for oxidation of organic toxicants at both acidic and alkali pH [58], however, when H_2O_2 is irradiated with UV radiations it forms two hydroxyl radicals 2OH˙ that react with organic contaminants present in textile effluent [59]. The combination of H_2O_2 + UV is also highly effective to the scavenging effect of carbonate ions at alkaline pH. In the combination of H_2O_2 with UV radiation, the peroxide is activated by UV light. The extent of activation primarily depends upon concentration of H_2O_2, strength of UV radiation, pH of medium, and composition of dye. Acidic dyes are most easily decomposed by this system having accumulative number of azo groups; however, it is not effective for degradation of pigments [60]. This method has various advantages over other oxidation process and the most important is no sludge formation at any stage of process.

The process is effective at acidic pH; however, at alkaline pH H_2O_2 under UV radiation oxidizes alkalis to produce oxygen and water rather in place of hydroxyl radicals. Concentration of H_2O_2 also largely affects the reaction. At high concentration, it competes with the dye molecules for reaction with hydroxyl radicals and decreases the rate of oxidation. In addition, at high concentration, OH˙ radicals dimerize to form H_2O_2 [61]. Thus, it is essential to optimize concentration of H_2O_2 to have maximum efficacy of the reaction. The rate of dye degradation increases with the increase in intensity of UV light [62]. It has been found that increasing the power intensity of UV radiation from 18 to 54 W increased degradation efficiency from 90.69 to 100% due to excessive generation of OH˙ [63].

Advantages of H_2O_2/UV process are as follows:

- No Sludge formation
- Reaction takes place at ambient temperature
- Oxygen generated as by-product can be used for biological decay
- Rate of reaction depends upon pH, concentration of H_2O_2, and intensity of UV radiation.

4.5 *Ozone/Ultraviolet/Hydrogen Peroxide*

Among all AOPs, the combination of O_3/UV/H_2O_2 has been reported to be best for the purification of dye wastewater including degradation of polyester fiber, and degradation of dye molecules [64]. The addition of H_2O_2 to O_3/UV system enhances the rate of OH˙ generation by promoting decomposition of ozone and acting as a catalyst for the reaction [65]. In addition, at alkaline pH H_2O_2 reacts with O_3 to produce $OH_2^{\cdot}$ radical which dissociates O_3 more effectively than OH˙; however, at acidic pH the reaction is slow. This method is termed as peroxone method. At alkaline pH, H_2O_2 itself dissociates into OH_2^- ions which also initiate ozone decomposition more effectively than OH^- ion [66].

5 Advanced Techniques for the Degradation of Metallic and Polymeric Compounds

5.1 Photocatalysis

The principle of photocatalyst lies in the generation of electron as reducing agents and hole as oxidizing agents. The mode of action is described in Fig. 4. In photocatalysis, the design of photocatalytic reactor is also very important in order to have effective contact between the toxicants and photocatalyst and adequate absorption of photons. Thus, for an ideal photoreactor it must have high specific surface area, high mass transfer, and direct light irradiation of catalyst surface [67]. This process has various advantages as (a) It completely oxidize organic pollutants to CO_2, water and mineral acids, (b) moderate temperature and pressure is needed, (c) no sludge is generated thus free from sludge disposal problem.

Titanium dioxide (TiO_2) is the most commonly used photocatalyst owing to its high nontoxic nature, inertness towards biological and chemical reagents and high photocatalytic activity [68]. Photocatalytic degradation of various organic substances has been assessed using TiO_2 since the last six decades [69]. TiO_2 has been reported to degrade wide range of pollutants including dyes, phenols, plasticizers, polychlorinated biphenyls, surfactants, and dioxins [70]. The major challenge in using TiO_2 as a photocatalyst is its bandgap width (3.2 eV), which allows absorption of wavelengths lower than 380 nm only, i.e., in UV region. Various techniques such as dye

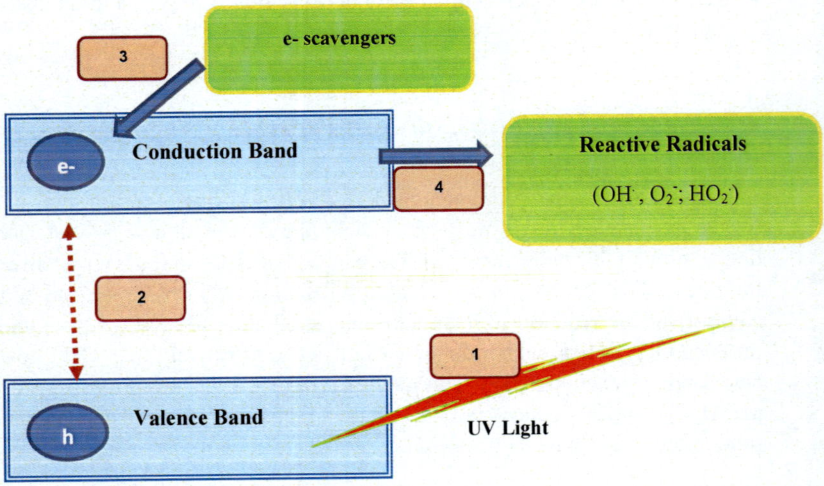

Fig. 4 Generation of free radicals in photocatalysis (1) With the help of UV light the electrons are excited, (2) electrons excited from valence band to conduction band generating behind positively charged holes (h^+) which act as oxidizing agent, (3) electrons are quenched by scavengers to prevent their combination with holes, and (4) electrons act as reducing agents and generate free radicals

sensitization, modification of surface, and doping have been used by researchers to enhance the activity of TiO_2 [71]. TiO_2 has been doped with copper, cobalt, molybdenum, and many other metals to increase its light absorption capacity to visible region [72]. TiO_2 doped with zinc showed better photocatalytic activity and reduced bandgap energy in visible light [73]. However using heavy metals for doping TiO_2 may produce toxic intermediates and create heavy metal pollution [74].

TiO_2 photocatalytic systems can be used as either photoreactors or immobilized on solid support system (Table 3). In photoreactors, the suspension system is maintained for the reaction. In suspension system, the catalyst has to be separated from treated wastewater and recycled before discharge which is a cost intensive and time-taking process. In addition, absorptions of UV light by catalyst particles limit its effective penetration to the desired depth thereby limiting complete oxidation [75]. This problem can be addressed by using immobilized photocatalyst over some suitable support system. However, immobilization of catalyst also decreases the rate of reaction due to availability of reduced surface area of catalyst for reaction. Researchers have studied degradation of Acid Blue 25 dye using immobilized titania nano photocatalysis and reported lower generation of hydroxyl radicals with increased concentration of dye [76]. At high concentration of dye, a smaller number of photons hit the surface of catalyst resulting in decreased generation of hydroxyl radicals [77]. Complete degradation of dye converts all carbon, nitrogen, and Sulfur atoms to CO_2, nitrate, and sulfate ions, respectively [78]. Immobilization of TiO_2 on polymeric sheets is one of the efficient methods for the photocatalytic implementations at large scale owing to easy availability and handling, flexibility, cost effective, and light weight of polymer [79]. The photocatalytic activity of TiO_2 is largely affected by various cations, anions, and organic compounds as they can be adsorbed over TiO_2 surface and decrease the activity [80, 81]. Photocatalytic degradation technique cannot be implemented for field applications in treating large-scale contaminated sites.

Table 3 Immobilization of TiO_2 onto various polymeric systems for the degradation of dyes

S. No	TiO_2/polymeric system	Degraded dye	Reference
1	TiO_2 immobilized on a glass slide/polyallylamine hydrochloride/polystyrene sulfonate system	Dimethoate	[82]
2	TiO_2/reduced graphene oxide/nylon-6 filter membrane	Methylene blue	[83]
3	TiO_2/cotton fabric	Rhodamine B	[84]
4	TiO_2/polydimethylsiloxane (PDMS)/	Methylene blue	[85]
5	TiO_2/cellulose acetate	Cytotoxin	[86]
6	TiO_2/sodium alginate/carboxymethyl cellulose/	Congo red	[87]

5.2 Metallic Nanoparticles

Metals play a very important role in removal of dyes from textile wastewater. One of the areas is by using magnetic nanoparticle as adsorbents treatment [88]. In the recent past, many researchers have used various magnetic materials for the removal of dyes from wastewater [89]. Usually, applications of nanoparticles are limited for the removal of anionic dyes due to similar charges both on dye and surface of nanoparticle there for several modifications were done to change surface properties of these particles (Table 4).

Jabbar et al. [93] synthesized cobalt aluminate ($CoAl_2O_4$) nanoparticles by pyrolysis for the dissociation of Novacron Deep Night S-R dye in combination with UV radiation. They reported that in the presence of H_2O_2 the dye degraded 67% within 50 min. Several metal oxides have also been reported for the photocatalytic degradation of organic dyes and pollutants [94]. Spinel oxides are stable semiconductors which have sufficient band gap energies suitable for photocatalytic degradation of organic dyes [95]. The catalytic efficiency of such metal oxides depends on their specific surface area, morphology, crystallinity, particle size, and chemical composition [96]. Several routes of metal oxide preparation and their benefits are presented in Fig. 5.

Oxidation process by using Zero Valent iron has added advantages as it involves various pathways comprising of reduction and oxidation reactions which are specific for the nature of pollutant and reaction conditions [97]. The generation of free radicals occurs in three stages:

$$Fe^0 + O_2 + 2H^+ \rightarrow Fe^{2+} + H_2O_2 \tag{6}$$

$$Fe^0 + H_2O_2 + 2H^+ \rightarrow Fe^{2+} + 2H_2O \tag{7}$$

$$Fe^{2+} + H_2O_2 \rightarrow HO^{\cdot} + OH^- + Fe^{3+} \tag{8}$$

Nanotechnology is playing an imperative role in environmental remediation of organic and inorganic pollutants since the last decade [98]. Several researchers have reported efficacy of nano zero-valent iron towards degradation of organic textile

Table 4 Modified nanoparticles for the oxidative degradation of dyes

S. No	Modified nanoparticles	Dye	Reference
1	Ammonium-functionalized silica nanoparticle	Methyl orange	[90]
2	Modified Fe_3O_4 magnetic nanoparticles	Reactive red-120 and resorcinol	[91]
3	Surface modified zinc ferrite nanoparticle	Direct red 23, Direct red 31 and Direct green 6	[92]

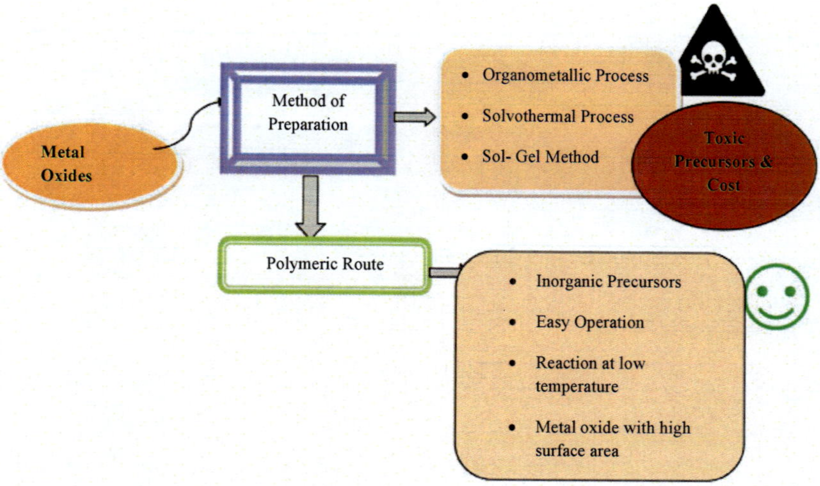

Fig. 5 Various methods for the preparation of metal oxide along with their benefits and challenges

pollutants [99]. It has been observed that specific area of nano zero-valent iron is much larger (29 m^2/g) in comparison to bulk particles (0.2 m^2/g) which improves the rate of oxidation and degradation of organic pollutants [100]. The degradation of various organic and inorganic pollutants by using nano zero-valent iron is summarized in Fig. 6. In this process, pH of the reaction media, dose of nanoparticle, and initial concentration of dye play a very important role in degradation process of organic pollutants. If pH is alkaline, Fe^{2+} ions and OH^- ions may precipitate as Fe $(OH)_2$ on the surface of zero-valent iron and occupying the reactive sites [101]. At alkaline pH, the reducing pathway is prevalent which is less effective in comparison to oxidation process. Aggregation of these particles is a major challenge in this process as due to high surface area these particles are highly reactive and aggregate at the faster rate. These particles get oxidized and aggregated in the presence of atmospheric air thus are highly unstable [102].

Various solid supports can be used to prevent aggregation of nano zero-valent iron and increase their efficiency for dye degradation (Fig. 7). The solid support system maintains nanoparticles in dispersed phase, protects them from air oxidation, and increases contact area, however, reduced their activity towards degradation of textile dyes. Few supports such as biochar, kaolin and rectorite, rather than degrading the pollutants simply adsorb them and transfer them to the surrounding environment. Hence, degradation or oxidation of pollutants is recommended instead of their adsorption.

Incorporation of heavy metals as a support system for the synthesis of nanoparticles may further introduce these toxic substances to the environment. Thus, researchers proposed the synthesis of nano zero-valent iron using various green support materials such as extracts of green tea, eucalyptus leaf, grape leaf, and other

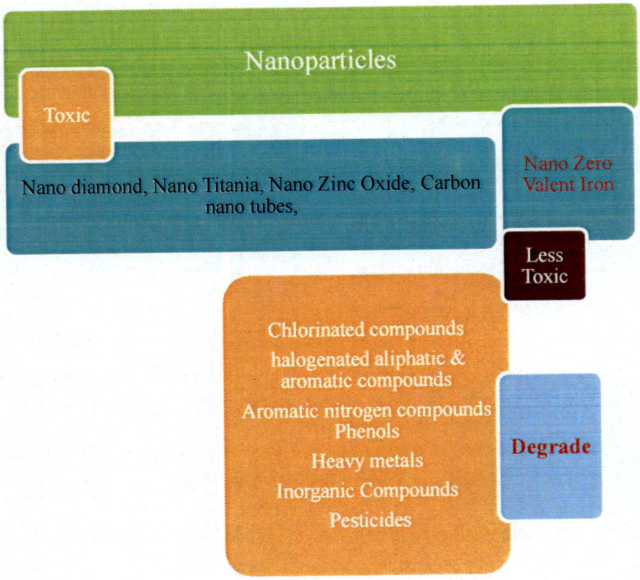

Fig. 6 Nanoparticles synthesized for the degradation of textile dyes. Toxicity of nano zero-valent iron is less in comparison with other particles

Fig. 7 Solid support system for the synthesis of nano zero-valent iron particles

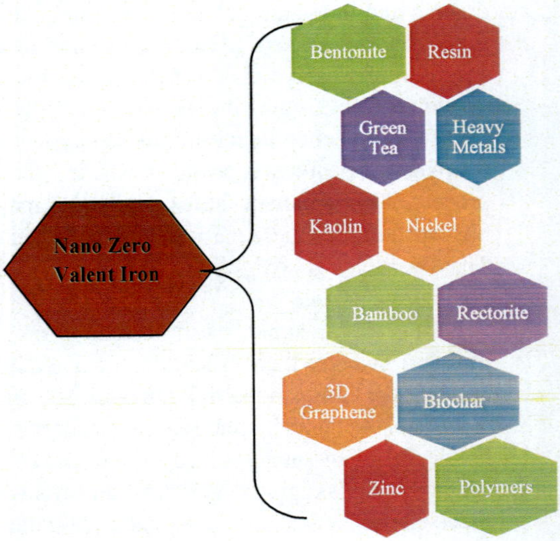

tree leaf extracts. These extracts provide stabilization to nanoparticles and enhance their efficacy as well for the degradation of textile dyes and other organic pollutants [103]. Zhang et al. [104] reported removal of Pb(II) by using nano zero-valent iron supported over kaolin while Ali et al. [105] synthesized nano zero-valent iron using porous cation exchange resin as a solid support reported the efficacy for the reduction of Cr(VI) to Cr(III) in aqueous solution.

The process of pollutant degradation is based on reducing nature of these particles. Nano zero-valent iron particles are potential reducing agent (electron donors) while organic dye molecules are brilliant oxidizing agent (electron acceptors). Therefore, they reduce Fe^{2+} ions, and generate hydroxyl ions which further dissociate chromophore bond and auxochrome bond thereby decolorize the dye molecule. The intermediates on complete mineralization produce CO_2, H_2O, and inorganic ions. The efficiency and reactivity of nanoparticles can be improved, by integrating with other treatments such as irradiation with microwave, UV, or ultrasonic irradiation and using H_2O_2 oxidation [106]. Partial removal (up to 55%) of organic pollutants using nano zero-valent iron has been reported, however, by using additional oxidizing agent H_2O_2 and integration of UV technique the removal increases up to 90.0% [107]. Mao et al. [106] reported decolorization of reactive yellow and solvent blue dyes up to 60.0% and 94.0%, respectively, in only 5 min using nano iron along with microwave radiations. Studies are being conducted using scrap zero-valent iron as well for the treatment of wastewater as a method to propose reutilization of by-product from machinery industries [108]. These methods are one of the promising and cost-effective way for treating textile wastewater.

5.3 *Natural and Modified Polymers*

Natural polymers plant or animal-based serves as an efficient method for treating textile wastewater and dye removal. Removal of organic and inorganic pollutants using natural adsorbents is considered to be economic, and sustainable alternative over other physical and chemical processes [109]. Biosorption is a passive process in which metal ions present in the textile wastewater interact with the functional groups on biological material's surface. This process ensures various advantages such as cost effectiveness, reduced sludge generation, easy regeneration of adsorbent, and higher removal capacity [110]. The specific chemical structure and composition of the polymer play a significant role towards effective removal of dyes from textile wastewater. Natural polymers are synthesized from various plant sources which then further subjected to chemical and physical treatment to give modified natural polymer with enhanced efficacy for the removal of textile dyes (Fig. 8).

The natural polymers contain various anionic groups such as hydroxyl (OH^-), phosphate (PO_4^-), and carboxyl (COO^-). These groups show affinity towards cationic dye molecules present in wastewater by using electrostatic force of attraction [111]. Usually, these polymers are effective in acidic environment, however, removal of methylene blue dye is reported in alkaline conditions (pH-8) as well

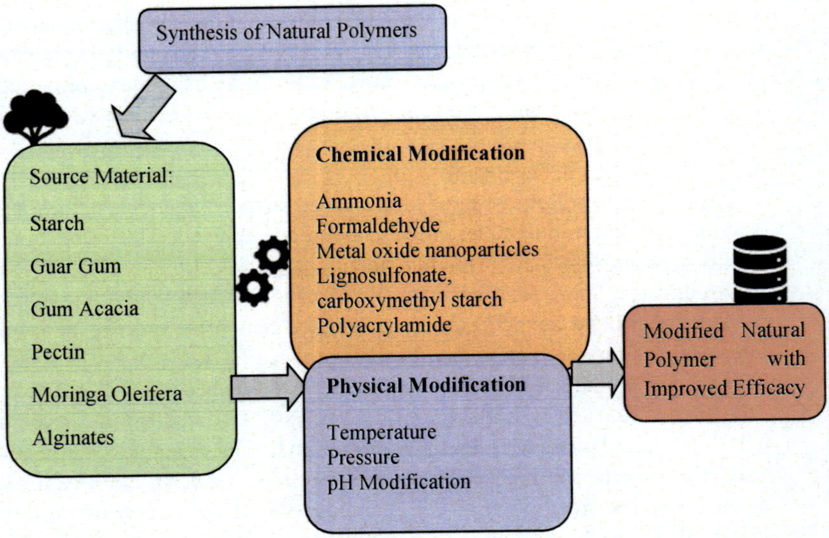

Fig. 8 Synthesis of modified natural polymers using various physical and chemical modification techniques

[112]. Seeds of nirmali show excellent binding properties for dye molecules due to the presence of OH^- groups along with galactomannan and the galactan molecular chains. Polymers differ in molecular weights based on their molecular structure. High molecular weight polymers carryout removal of organic dye molecules more effectively due to charge neutralization, bridging, and electrostatic attraction. Chemical modifications have been done in cellulose with sulfur-bearing functional groups such as thiols, dithiocarbamates, dithiophosphates, and xanthates for metal bonding as these groups have higher affinities for heavy metals ions and least affinities for other light metals [113]. Xanthates offer high stability with heavy metal ions, in addition they are easily prepared and cost effective as well. The sludge formed using xanthate metal complexes is easy to settle down thereby can be disposed of with minimal efforts. On the other hand, the precipitates as metal hydroxide may release metals ions into the environment thus, they need to be treated before disposal and can be pretreated using solidification or stabilization [114].

In recent years researchers have focused on the extraction of microbial polymers owing to their low cost and easy processing. The microbes which have been used are Pseudomonas pseudoalcaligenes, Pseudomonas plecoglossicida, and Staphylococcus aureus. These polymers have been effectively used for those dyes which are resistant towards decolorization due to their acidic nature such as fawn dyes, mediblue, mixed dyes, and whale dyes [115].

5.4 *Nanocomposites*

Metal oxide nanoparticles and nanocomposites show excellent adsorption properties for heavy metal ions present in textile effluent [116]. Magnesium-based nanoparticles with different morphologies have been used for the removal of various heavy metal ions and organic pollutants. The mechanism of adsorption using Mg-based nanoparticles includes electrostatic attraction and surface complexation between the dye molecule and hydroxyl groups which are present at the surface of the adsorbent. These nanoparticles carry out oxidative degradation of organic molecules by breaking down P–S or P–O bond of organophosphates and other organic dye molecules [117]. Magnesium oxide (MgO) embedded fiber-based substrate have been used successfully as an adsorbent for the removal of toxic dyes from textile water [118]. MgO nanoparticles in combination with carbon nanofibers have been used as a significant adsorbent for the removal of heavy metal ions from textile effluent [118]. Adsorption of cadmium (Cd^{2+}) ions have been studied using polyacrylonitrile-based carbon nanofibers using MgO as adsorbent [119]. Reinforcement of MgO nanoparticles increases the adsorption capacity of carbon nanofibers. Hybrid nanofibers of MgO with polypropylene glycol have been also prepared for the removal of heavy metals [120]. These nanofibers had specific surface area of 185 m^2/g and found to be effective for the removal of Pb, Cu, and Cd at pH of 7.5. In addition, regeneration experiments showed that efficacy remained high even after seven cycles.

5.5 *Graphene and Its Composites*

Graphene is a covalently bonded two-dimensional lattice having specific properties. Graphene is oxidized to form graphene oxide which can be further chemically modified to have several functional groups such as epoxies, carboxyl, and hydroxyl [121]. Presence of such functional groups increases negative charge on graphene oxide surface and helps to interact with dye molecules through H-bonding. Graphene oxide due to its conjugated structure has been studied as a potent option for the adsorption of dye molecules from textile effluent [122].

Recently, many researchers have given much consideration to the polymeric materials to be used as nano-adsorbents. The polymers such as polyaniline, polystyrene, and polypyrrole can be coated over the surface of nanoparticles to enhance their photoelectrical properties environmental suitability. Polymeric nanocomposites of graphene have been reported to be effective for the degradation of inorganic pollutants and organic dye molecules [123]. Noreen et al. [124] synthesized a range of nanocomposites of graphene oxide with polyaniline, polypyrrole, and polystyrene to achieve a higher removal rate of Actacid orange-RL dye. Composite of graphene oxide with polyaniline has been reported as potential adsorbent for the removal of Hg^{2+} ion and dye removal owing to its loose porous structure [125]. However, the applications of these nanocomposites are limited for the removal of heavy metals due to their expensive and complex synthesis and troublesome recycling [126]. In addition, graphene oxide due to its high clustering tendency is less effective in the adsorption process [127]. Thus, a sustainable adsorbent should be cost effective, show high adsorption efficacy, ease of separation and high reusability.

Natural polysaccharide, Chitosan has also been used to improve mechanical strength and adsorption ability of nanocomposites. The free $-NH_2$ group in chitosan reacts with other metal/nonmetal oxides and provides chemical stability [128]. The –COOH group in graphene oxide easily interact with highly reactive $-NH_2$ group of chitosan to give a stable nanocomposite with improved adsorption capacity [129]. Combination of biopolymer chitosan with graphene oxide enhances thermal and mechanical stability, improves hydrophilic property of composite, and reduces entangling tendency during separation [130].

Poly vinyl alcohol (PVA) is a biodegradable polymer, with hydrophilic nature and high fiber-forming ability. It can easily cross-link with graphene oxide/chitosan nanocomposite to improve its adsorption efficacy for dye removal [131, 132]. The separation of graphene oxide/chitosan–PVA biopolymer from treated water is very simple and can be completed with limited technology. Das et al. [133] studied effective removal of Congo red dye using this nano-polymer composite. Under acidic condition, the amino group ($-NH_2$) in chitosan get protonated and then ionized further to ammonium ion thereby increasing swelling properties of GO/Chitosan–PVA polymer resulting in increased weight of polymer [133]. However, at high pH swelling decreased due to deprotonation. At high concentration of dye, removal efficiency decreases due to competitive adsorption on active sites of adsorbent [134]. At low pH due to increased ionization, electrostatic attraction between negatively charged dye molecules and positively charged polymer molecules increases which accelerates adsorption of dye molecules [135].

5.6 Waste Textile Fiber Copolymer

The prime component of textile industry is fiber, either natural or synthetic which can be converted to branched copolymer by free radical or ionic polymerization of chemical groups. The cellulose molecules in textile fiber show strong interaction with adjacent molecules due to the presence of –OH and develop hydrophilicity, degradability, and absorption ability [136]. Conversion of waste textile into adsorbent material is a novel approach which can be achieved via two ways (a) incorporation of metal-binding groups to the fiber backbone (b) Graft copolymerization (Fig. 9).

Incorporation of metal-binding functionalities into textile fiber produces heavy metal adsorbents [137]. In order to form branched copolymer, the monomers are grafted to fiber backbone [138]. Heating at specific temperature enhances the rate of graft copolymerization thus, among available heating processes the microwave heating is considered as effective and reliable method. Zhou et al. [139] used both microwave and UV radiation and converted waste textiles fiber to an adsorbent by grafting acrylic acid. Cerium(IV) ammonium nitrate was selected as free radical

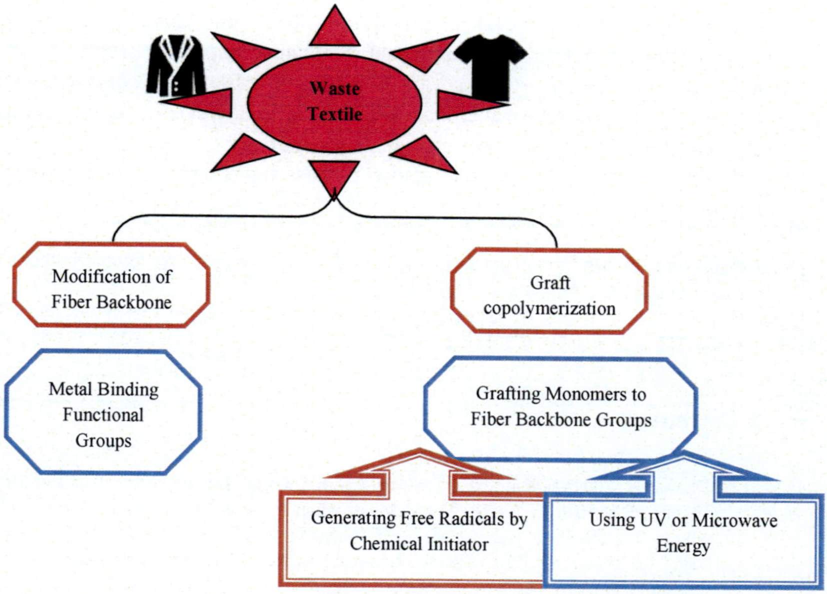

Fig. 9 Various methods for the conversion of waste textile material to copolymer

initiator and adsorption was carried out at high pH. Under acidic conditions, H^+ ions compete with metal ions for the adsorption and decrease the efficiency of process [140]. However, at alkaline pH deprotonation takes place and the surface of waste textile fiber becomes negative, which further enhances adsorption of cationic species. [141], prepared graft copolymer of textiles waste by the reaction with poly-acrylic acid by the process of free radical polymerization. The chelating properties were developed by the reaction polymer with diamine solution. They studied the removal of Pb(II) and Cr(VI) ions and reported removal as 11.81 mg/g and 2.19 mg/g, respectively. The advantage of such polymers is they can be easily regenerated and adsorbed metals can be recovered by elution.

6 Conclusion

Enormous number of toxicants have been generated and released into the environment due to insufficient treatment techniques adopted for the treatment of textile wastewater. The challenge is to design a process which is cost effective, easy to operate, less time consuming and able to degrade a wide range of pollutants ranging from nano to polymer and inorganic to organic level. Advanced oxidation process using potent oxidants and high energy radiations either alone or in combination is one of the promising methods. In addition, due to the advancement of nanotechnology, the incorporation of nanoparticle and nanocomposites have provided significant solutions to this problem. Various modified polymers of natural and synthetic origin provide high removal efficacy of pollutants. However, it can be proposed that the best design for textile wastewater treatment can be a hybrid technology which utilizes principles of advanced oxidation process and materials ranging from metallic nanoparticles to polymers.

Acknowledgements Author thanks Prof. (Dr.) I. K. Bhat Vice chancellor Manav Rachna University for providing all kinds of support for writing this chapter.

References

1. Chavan R (2001) Indian textile industry-environmental. Indian J Fibre Text Res 26:11–21
2. Vajnhandl S, Valh JV (2014) The status of water reuse in European textile sector. J Environ Manag 141:29–35. https://doi.org/10.1016/j.jenvman.2014.03.014
3. Sala M, Carmen M (2012) Electrochemical techniques in textile processes and wastewater treatment. Environ Photocatal 2012:1–12. https://doi.org/10.1155/2012/629103
4. U.S. EPA (2013) Report on the 2013 U.S. Environmental Protection Agency (EPA) international decontamination research and development conference. Research Triangle Park, NC, November 05–07, 2013. U.S. Environmental Protection Agency, Washington, DC, EPA/600/R-14/210, 2014
5. Johann F, Osma José L, Toca-Herrera S-C (2018) Uses of laccases in the food industry. Enzym Res 2010:1–8. https://doi.org/10.4061/2010/918761
6. Xiang Y, Xiang Y, Wang L, Li X (2018) Effects of sewage sludge modified by coal gasification slag and electron beam irradiation on the growth of Alhagi sparsifolia Shap and transfer of heavy metals. Environ Sci Pollut Res 25:11636–11645
7. Haule LV, Carr CM, Rigout M (2016) Preparation and physical properties of regenerated cellulose fibres from cotton waste garments. J Clean Prod 112:4445–4451
8. Yao L (2014) An evaluation of the carbon footprint and carbon reduction measures of textile raw material stage. Tianjin Polytech Univ J 33:71–76
9. Mustafa I, Delia TS (2006) Biological treatment of acid dyeing wastewater using a sequential anaerobic/aerobic reactor system. Enzym Microb Technol 38:887–892. https://doi.org/10.1016/j.enzmictec.2005.05.018
10. Hasanbeigi A, Price L (2015) A technical review of emerging technologies for energy and water efficiency and pollution reduction in the textile industry. J Clean Prod 95:34–48. https://doi.org/10.1016/j.jclepro.2015.02.079
11. Tehrani-Bagha AR, Amini FL (2010) Decolorization of a reactive dye by UV-enhanced ozonation. Prog Color Color Coat 3:1–8
12. Körbahti BK, Aktaş N, Tanyolaç A (2007) Optimization of electrochemical treatment of industrial paint wastewater with response surface methodology. J Hazard Mater 5:83–90. https://doi.org/10.1016/j.jhazmat.2007.02.005
13. Ullah A, Farooq M, Nadeem F, Rehman A, Hussain M, Nawaz A, Naveed M (2020) Zinc application in combination with zinc solubilizing Enterobacter sp. MN17 improved productivity, profitability, zinc efficiency, and quality of desi chickpea. J Soil Sci Plant Nutr 20:2133–2144
14. Korniłłowicz-Kowalska T, Rybczyńska K (2014) Anthraquinone dyes decolorization capacity of anamorphic Bjerkandera adusta CCBAS 930 strain and its HRP-like negative mutants. World J Microbiol Biotechnol 30:1725–1736

15. Flora SJS, Mittal M, Mehta A (2008) Heavy metal induced oxidative stress and its possible reversal by chelation therapy. Indian J Med Res 128:501–530
16. Robinson BH, Brooks R, Howes AW, Kirkman JH, Gregg PEH (1997) The potential of the high-biomass nickel hyperaccumulator Berkheya coddii for phytoremediation and phytomining. J Geochem Explor 60:115–126
17. Domínguez A, Couto SR, Sanromán MÁ (2005) Dye decolorization by *Trametes hirsuta* immobilized into alginate beads. World J Microbiol Biotechnol 21:405–409. https://doi.org/10.1007/s11274-004-1763-x
18. Correia VM, Stephenson T, Judd SJ (1994) Characterisation of textile wastewaters—a review. Environ Technol 15:917–929
19. Nahar K, Chowdhury MAK, Chowdhury MAH (2018) Heavy metals in handloom-dyeing effluents and their biosorption by agricultural byproducts. Environ Sci Pollut Res 25:7954–7967. https://doi.org/10.1007/s11356-017-1166-9
20. Banat IM, Nigam P, Singh D, Marchant R (1996) Microbial decolorization of textile-dye containing effluents: a review. Bioresour Technol 58:217–227
21. Sharifuzzaman SM, Rahman H, Ashekuzzaman SM, Islam MM, Chowdhury SR, Hossain MS (2016) Heavy metals accumulation in coastal sediments. In: Environmental remediation technologies for metal-contaminated soils. Springer, Tokyo, pp 21–42
22. Gautam RK, Mudhoo A, Lofrano G, Chattopadhyaya MC (2014) Biomass-derived biosorbents for metal ions sequestration: adsorbent modification and activation methods and adsorbent regeneration. J Environ Chem Eng 2:239–259
23. Mittal M, Flora SJS (2006) Effects of individual and combined exposure to sodium arsenite and sodium fluoride on tissue oxidative stress, arsenic and fluoride levels in male mice. Chem Biol Interact 162:128–139
24. Mittal M, Chatterjee S, Flora SJS (2018) Combination therapy with vitamin C and DMSA for arsenic–fluoride co-exposure in rats. Metallomics 10:1291–1306
25. Mathur N (2005) Mutagenicity assessment of effluents from textile/dye industries of Sanganer, Jaipur (India): a case study. Ecotoxicol Environ Saf 61:105–113. https://doi.org/10.1016/j.ecoenv.2004.08.003
26. Akhtar MF, Ashraf M, Ahmad AA, Javeed A, Sharif A, Saleem A, Akhtar B (2016) Textile industrial effluent induces mutagenicity and oxidative DNA damage and exploits oxidative stress biomarkers in rats. Environ Toxicol Pharmacol 41:180–186. https://doi.org/10.1016/j.etap.2015.11.022
27. Ahmed S, Rasul MG, Martens WN (2011) Advances in heterogeneous photocatalytic degradation of phenols and dyes in wastewater: a review. Water Air Soil Pollut 215:3–29. https://doi.org/10.1007/s11270-010-0456-3
28. Thompson RC, Olsen Y, Mitchell RP, Davis A, Rowland SJ, Mcgonigle D, McConigle D, Russell AE (2004) Lost at sea: where is all the plastic? Science 304:838–838
29. Galafassi S, Nizzetto L, Volta P (2019) Plastic sources: a survey across scientific and grey literature for their inventory and relative contribution to microplastics pollution in natural environments, with an emphasis on surface water. Sci Total Environ 693:133499. https://doi.org/10.1016/j.scitotenv.2019.07.305
30. Mahon AM, O'Connell B, Healy MG, O'Connor I, Officer R, Nash R, Morrison L (2017) Microplastics in sewage sludge: effects of treatment. Environ Sci Technol 51:810–818. https://doi.org/10.1021/acs.est.6b04048
31. Carr SA (2017) Sources and dispersive modes of micro-fibers in the environment. Integr Environ Assess Manag 13:466–469
32. Abbasi S, Soltani N, Keshavarzi B, Moore F, Turner A, Hassanaghaei M (2018) Microplastics in different tissues of fish and prawn from the Musa Estuary, Persian Gulf. Chemosphere 205:80–87
33. Barrows APW, Cathey SE, Petersen CW (2018) Marine environment microfiber contamination: global patterns and the diversity of microparticle origins. Environ Pollut 237:275–284
34. Avagyan R, Luongo G, Thorsén G, Östman C (2015) Benzothiazole, benzotriazole, and their derivates in clothing textiles—a potential source of environmental pollutants and human exposure. Environ Sci Pollut Res 22:5842–5849

35. Napper IE, Thompson RC (2016) Release of synthetic microplastic plastic fibres from domestic washing machines: effects of fabric type and washing conditions. Mar Pollut Bull 112:39–45. https://doi.org/10.1016/j.marpolbul.2016.09.025
36. Magni S, Binelli A, Pittura L, Avio CG, Della Torre C, Parenti CC, Gorbi S, Regoli F (2019) The fate of microplastics in an Italian wastewater treatment plant. Sci Total Environ 652:602–610. https://doi.org/10.1016/j.scitotenv.2018.10.269
37. Enfrin M, Lee J, Gibert Y, Basheer F, Kong L, Dumée LF (2020) Release of hazardous nanoplastic contaminants due to microplastics fragmentation under shear stress forces. J Hazard Mater 384:121393
38. Li J, Zhang K, Zhang H (2018) Adsorption of antibiotics on microplastics. Environ Pollut 237:460–467. https://doi.org/10.1016/j.envpol.2018.02.050
39. Miller RZ, Ajr W, Winslow BO, Galloway TS, Apw B (2017) Mountains to the sea: river study of plastic and nonplastic microfiber pollution in the northeast USA. Mar Pollut Bull 124:245–251
40. Fischer EK, Paglialonga L, Czech E, Tamminga M (2016) Microplastic pollution in lakes and lake shoreline sediments—a case study on Lake Bolsena and Lake Chiusi (central Italy). Environ Pollut 213:648–657
41. Dris R, Gasperi J, Rocher V, Tassin B (2018) Synthetic and non-synthetic anthropogenic fibers in a river under the impact of Paris megacity: sampling methodological aspects and flux estimations. Sci Total Environ 618:157–164
42. Lares M, Ncibi MC, Sillanpää M, Sillanpää M (2018) Occurrence, identification and removal of microplastic particles and fibers in conventional activated sludge process and advanced MBR technology. Water Res 133:236–246
43. Shruti VC, Pérez-Guevara F, Elizalde-Martínez I, Kutralam-Muniasamy G (2020) First study of its kind on the microplastic contamination of soft drinks, cold tea and energy drinks-future research and environmental considerations. Sci Total Environ 726:138580
44. Kutralam-Muniasamy G, Pérez-Guevara F, Elizalde-Martínez I, Shruti VC (2020) Review of current trends, advances and analytical challenges for microplastics contamination in Latin America. Environ Pollut 115463
45. El-Shahawi MS (2010) An overview on the accumulation, distribution, transformations, toxicity and analytical methods for the monitoring of persistent organic pollutants. Talanta 80:1587–1597. https://doi.org/10.1016/j.talanta.2009.09.055
46. Deng Y, Zhao R (2015) Advanced oxidation processes (AOPs) in wastewater treatment. Curr Pollut Rep 1:167–176
47. Zangeneh H, Zinatizadeh AAL, Feizy M (2014) A comparative study on the performance of different advanced oxidation processes ($UV/O_3/H_2O_2$) treating linear alkyl benzene (LAB) production plant's wastewater. J Ind Eng Chem 20:1453–1461
48. Lucas MS, Peres JA (2007) Degradation of reactive black 5 by Fenton/UV-C and ferrioxalate/H_2O_2/solar light processes. Dye Pigment 74:622–629
49. Meriç S, Selçuk H, Belgiorno V (2005) Acute toxicity removal in textile finishing wastewater by Fenton's oxidation, ozone and coagulation–flocculation processes. Water Res 39:1147–1153
50. GilPavas E, Dobrosz-Gómez I, Gómez-García MÁ (2019) Optimization and toxicity assessment of a combined electrocoagulation, H_2O_2/Fe^{2+}/UV and activated carbon adsorption for textile wastewater treatment. Sci Total Environ 651:551–560
51. Ghanbari F, Moradi M (2015) A comparative study of electrocoagulation, electrochemical Fenton, electro-Fenton and peroxi-coagulation for decolorization of real textile wastewater: electrical energy consumption and biodegradability improvement. J Environ Chem Eng 3:499–506
52. Koch M, Yediler A, Lienert D, Insel G, Kettrup A (2002) Ozonation of hydrolyzed azo dye reactive yellow 84 (CI). Chemosphere 46:109–113
53. Arslan I, Balcioglu IA (2001) Advanced oxidation of raw and biotreated textile industry wastewater with O_3, H_2O_2/UV-C and their sequential application. J Chem Technol Biotechnol Int Res Process Environ Clean Technol 76:53–60

54. Langlais B, Reckhow DA, Brink DR (1991) Ozone in water treatment—application and engineering. Lewis Publishers, Michigan
55. Rajeswari R, Kanmani S (2009) Degradation of pesticide by photocatalytic ozonation process and study of synergistic effect by comparison with photocatalysis and UV/ozonation processes. J Adv Oxid Technol 12(2):208–214
56. Saharan VK, Pinjari DV, Gogate PR, Pandit AB (2014) Advanced oxidation technologies for wastewater treatment: an overview. Elsevier, Butterworth, Heinemann, UK, pp 141–191
57. Al-Kdasi A, Idris A, Saed K, Guan CT (2004) Treatment of textile wastewater by advanced oxidation processes—a review. Glob NEST Int J 6:222–230
58. Ebrahiem EE, Al-Maghrabi MN, Mobarki AR (2017) Removal of organic pollutants from industrial wastewater by applying photo-Fenton oxidation technology. Arab J Chem 10:S1674–S1679
59. Crittenden JC, Hu S, Hand DW, Green SA (1999) A kinetic model for H_2O_2/UV process in a completely mixed batch reactor. Water Res 33(10):2315–2328
60. Andreozzi R, Caprio V, Marotta R (2001) Oxidation of benzothiazole, 2-mercaptobenzothiazole and 2-hydroxybenzothiazole in aqueous solution by means of H_2O_2/UV or photo assisted Fenton systems. J Chem Technol Biotechnol Int Res Process Environ Clean Technol 76:196–202
61. Aleboyeh A, Moussa Y, Aleboyeh H (2005) The effect of operational parameters on UV/H_2O_2 decolorization of acid blue 74. Dyes Pigm 66:129–134
62. Shen YS, Wang DK (2002) Development of photoreactor design equation for the treatment of dye wastewater by UV/H_2O_2 process. J Hazard Mater 89:267–277
63. Yang Y, Wyatt II, Travis D, Bahorsky M (1998) Decolorization of dyes using UV/H_2O_2 photochemical oxidation. Text Chem Color 30:27–35
64. Azbar NURI, Yonar T, Kestioglu K (2004) Comparison of various advanced oxidation processes and chemical treatment methods for COD and color removal from a polyester and acetate fiber dyeing effluent. Chemosphere 55:35–43
65. Singh R, Verma RS (2010) Advance oxidation processes for textile waste water treatment-At a glance. Curr World Environ 5:317
66. Glaze WH, Kang JW (1989) Advanced oxidation processes. Description of a kinetic model for the oxidation of hazardous materials in aqueous media with ozone and hydrogen peroxide in a semibatch reactor. Ind Eng Chem Res 28:1573–1580
67. Oliveira DF, Batista PS, Muller PS Jr, Velani V, França MD, De Souza DR, Machado AE (2012) Evaluating the effectiveness of photocatalysts based on titanium dioxide in the degradation of the dye Ponceau 4R. Dye Pigment 92:563–572
68. Konstantinou IK, Albanis TA (2004) TiO_2-assisted photocatalytic degradation of azo dyes in aqueous solution: kinetic and mechanistic investigations: a review. Appl Catal B Environ 49:1–14
69. Horikoshi S, Serpone N (2020) Can the photocatalyst TiO_2 be incorporated into a wastewater treatment method? Background and prospects. Catal Today 340:334–346
70. Singla P, Pandey OP, Singh K (2016) Study of photocatalytic degradation of environmentally harmful phthalate esters using Ni-doped TiO_2 nanoparticles. Int J Environ Sci Technol 13:849–856
71. Šuligoj A, Štangar UL, Ristić A, Mazaj M, Verhovšek D, Tušar NN (2016) TiO_2–SiO_2 films from organic-free colloidal TiO_2 anatase nanoparticles as photocatalyst for removal of volatile organic compounds from indoor air. Appl Catal B Environ 184:119–131
72. Janczarek M, Wei Z, Endo M, Ohtani B, Kowalska E (2016) Silver-and copper-modified decahedral anatase titania particles as visible light-responsive plasmonic photocatalyst. J Photonics Energy 7:012008
73. Singla P, Sharma M, Pandey OP, Singh K (2014) Photocatalytic degradation of azo dyes using Zn-doped and undoped TiO_2 nanoparticles. Appl Phys A 116:371–378
74. Manivel A, Naveenraj S, Kumar S, Selvam P, Anandan S (2010) CuO-TiO_2 nanocatalyst for photodegradation of acid red 88 in aqueous solution. Sci Adv Mater 2(1):51–57

75. Ray AK, Beenackers AA (1998) Development of a new photocatalytic reactor for water purification. Catal Today 40:73–83
76. Mahmoodi NM, Arami M (2009) Degradation and toxicity reduction of textile wastewater using immobilized titania nanophotocatalysis. J Photochem Photobiol B Biol 94:20–24
77. Mahmoodi NM, Arami M, Limaee NY, Tabrizi NS (2006) Kinetics of heterogeneous photocatalytic degradation of reactive dyes in an immobilized TiO_2 photocatalytic reactor. J Colloid Interface Sci 295:159–164
78. Mahmoodi NM, Borhany S, Arami M, Nourmohammadian F (2008) Decolorization of colored wastewater containing azo acid dye using photo-Fenton process: operational parameters and a comparative study. J Color Sci Technol 2:31–40
79. Essawy AA, Aleem SAE (2014) Physico-mechanical properties, potent adsorptive and photocatalytic efficacies of sulfate resisting cement blends containing micro silica and nano-TiO_2. Constr Build Mater 52:1–8
80. Abdullah M, Low GK, Matthews RW (1990) Effects of common inorganic anions on rates of photocatalytic oxidation of organic carbon over illuminated titanium dioxide. J Phys Chem 94:6820–6825
81. Parent Y, Blake D, Magrini-Bair K, Lyons C, Turchi C, Watt A, Prairie M (1996) Solar photocatalytic processes for the purification of water: state of development and barriers to commercialization. Sol Energy 56:429–437
82. Thanekar P, Lakshmi NJ, Shah M, Gogate PR, Znak Z, Sukhatskiy Y, Mnykh R (2020) Degradation of dimethoate using combined approaches based on hydrodynamic cavitation and advanced oxidation processes. Process Saf Environ Prot 143:222–230
83. Pant HR, Pant B, Pokharel P, Kim HJ, Tijing LD, Park CH, Kim CS (2013) Photocatalytic TiO_2–RGO/nylon-6 spider-wave-like nano-nets via electrospinning and hydrothermal treatment. J Membr Sci 429:225–234
84. Wang JC, Lou HH, Xu ZH, Cui CX, Li ZJ, Jiang K, Shi W (2018) Natural sunlight driven highly efficient photocatalysis for simultaneous degradation of rhodamine B and methyl orange using I/C codoped TiO_2 photocatalyst. J Hazard Mater 360:356–363
85. EL-Mekkawi DM, Abdelwahab NA, Mohamed WA, Taha NA, Abdel-Mottaleb MSA (2020) Solar photocatalytic treatment of industrial wastewater utilizing recycled polymeric disposals as TiO_2 supports. J Clean Prod 249:119430
86. Pinho LX, Azevedo J, Brito A, Santos A, Tamagnini P, Vilar VJ, Boaventura RA (2015) Effect of TiO_2 photocatalysis on the destruction of *Microcystis aeruginosa* cells and degradation of cyanotoxins microcystin-LR and cylindrospermopsin. Chem Eng J 268:144–152
87. Thomas M, Naikoo GA, Sheikh MUD, Bano M, Khan F (2016) Effective photocatalytic degradation of Congo red dye using alginate/carboxymethyl cellulose/TiO_2 nanocomposite hydrogel under direct sunlight irradiation. J Photochem Photobiol A Chem 327:33–43
88. Ambashta RD, Sillanpää M (2010) Water purification using magnetic assistance: a review. J Hazard Mater 180:38–49
89. Qadri S, Ganoe A, Haik Y (2009) Removal and recovery of acridine orange from solutions by use of magnetic nanoparticles. J Hazard Mater 169:318–323
90. Peng X, Yan Z, Cheng X, Li Y, Wang A, Chen L (2019) Quaternary ammonium-functionalized rice straw hydrochar as efficient adsorbents for methyl orange removal from aqueous solution. Clean Technol Environ Policy 21:1269–1279
91. Kamran S, Tavallali H, Azad A (2014) Fast removal of reactive red 141 and reactive yellow 81 from aqueous solution by Fe_3O_4 magnetic nanoparticles modified with ionic liquid 1-octyl-3-methylimidazolium bromide. Biquarterly Iran J Anal Chem 1:78–86
92. Mahmoodi NM, Abdi J, Bastani D (2014) Direct dyes removal using modified magnetic ferrite nanoparticle. J Environ Health Sci Eng 12(1):1–10
93. Jabbar EI, ElHafdi M, Benchikhi M, El Ouatib R, Er-Rakho L, Essadki A (2019) Photocatalytic degradation of navy blue textile dye by nanoscale cobalt aluminate prepared by polymeric precursor method. Environ Nanotechnol Monit Manag 12:100259
94. Rahimi-Nasrabadi M, Pourmohamadian V, Karimi MS, Naderi HR, Karimi MA, Didehban K, Ganjali MR (2017) Assessment of supercapacitive performance of europium tungstate nanoparticles prepared via hydrothermal method. J Mater Sci Mater Electron 28:12391–12398

95. Tatarchuk T, Al-Najar B, Bououdina M, Ahmed MA (2019) Catalytic and photocatalytic properties of oxide spinels. In: Handbook of ecomaterials. Springer International Publishing, New York, pp 1701–1750
96. Abdel-Khalek EK, Rayan DA, Askar AA, Maksoud MA, El-Bahnasawy HH (2021) Synthesis and characterization of $SrFeO_{3-\delta}$ nanoparticles as antimicrobial agent. J Sol Gel Sci Technol 97:27–38
97. Katsoyiannis IA, Ruettimann T, Hug SJ (2008) pH dependence of Fenton reagent generation and As(III) oxidation and removal by corrosion of zero valent iron in aerated water. Environ Sci Technol 42:7424–7430
98. Kamat PV, Meisel D (2003) Nanoscience opportunities in environmental remediation. Comptes Rendus Chim 6:999–1007
99. Tan C, Dong Y, Fu D, Gao N, Ma J, Liu X (2018) Chloramphenicol removal by zero valent iron activated peroxymonosulfate system: kinetics and mechanism of radical generation. Chem Eng J 334:1006–1015
100. Shih YH, Tso CP, Tung LY (2010) Rapid degradation of methyl orange with nanoscale zerovalent iron particles. Nanotechnology 7:7
101. Chen JL, Al-Abed SR, Ryan JA, Li Z (2001) Effects of pH on dechlorination of trichloroethylene by zero-valent iron. J Hazard Mater 83:243–254
102. Tiraferri A, Chen KL, Sethi R, Elimelech M (2008) Reduced aggregation and sedimentation of zero-valent iron nanoparticles in the presence of guar gum. J Colloid Interface Sci 324:71–79
103. Machado S, Pacheco JG, Nouws HPA, Albergaria JT, Delerue-Matos C (2015) Characterization of green zero-valent iron nanoparticles produced with tree leaf extracts. Sci Total Environ 533:76–81
104. Zhang X, Lin S, Chen Z, Megharaj M, Naidu R (2011) Kaolinite-supported nanoscale zero-valent iron for removal of Pb^{2+} from aqueous solution: reactivity, characterization and mechanism. Water Res 45(11):3481–3488
105. Ali SW, Mirza ML, Bhatti TM (2015) Removal of Cr(VI) using iron nanoparticles supported on porous cation-exchange resin. Hydrometallurgy 157:82–89
106. Mao H, Qiu Z, Shen Z, Huang W (2015) Hydrophobic associated polymer based silica nanoparticles composite with core–shell structure as a filtrate reducer for drilling fluid at ultra-high temperature. J Pet Sci Eng 129:1–14
107. Shu HY, Chang MC, Chang CC (2009) Integration of nanosized zero-valent iron particles addition with UV/H_2O_2 process for purification of azo dye acid black 24 solution. J Hazard Mater 167:1178–1184
108. Wang D, Ma W, Han H, Li K, Hao X (2017) Enhanced treatment of Fischer-Tropsch (FT) wastewater by novel anaerobic biofilm system with scrap zero valent iron (SZVI) assisted. Biochem Eng J 117:66–76
109. Ishak SA, Murshed MF, Md Akil H, Ismail N, Md Rasib SZ, Al-Gheethi AAS (2020) The application of modified natural polymers in toxicant dye compounds wastewater: a review. Water 12:2032
110. Vilar VJ, Botelho CM, Boaventura RA (2007) Methylene blue adsorption by algal biomass based materials: biosorbents characterization and process behaviour. J Hazard Mater 147:120–132
111. Gunatilake SK (2015) Methods of removing heavy metals from industrial wastewater. Methods 1:14
112. Aisyah SI, Norfariha S, Azlan M, Norli I (2014) Comparison of synthetic and natural organic polymers as flocculant for textile wastewater treatment. Iran J Energy Environ 5:436–445
113. Janaki V, Vijayaraghavan K, Oh BT, Ramasamy AK, Kamala-Kannan S (2013) Synthesis, characterization and application of cellulose/polyaniline nanocomposite for the treatment of simulated textile effluent. Cellulose 20:1153–1166. https://doi.org/10.1007/s10570-013-9910-x
114. Othmani A, Kesraoui A, Boada R, Seffen M, Valiente M (2019) Textile wastewater purification using an elaborated biosorbent hybrid material (luffa–cylindrica–zinc oxide) assisted by alternating current. Water 11:1326

115. Buthelezi SP, Olaniran AO, Pillay B (2012) Textile dye removal from wastewater effluents using bioflocculants produced by indigenous bacterial isolates. Molecules 17:14260–14274
116. Gao SL, Mäder E, Plonka R (2008) Nanocomposite coatings for healing surface defects of glass fibers and improving interfacial adhesion. Compos Sci Technol 68:2892–2901
117. Lange LE, Obendorf SK (2012) Effect of plasma etching on destructive adsorption properties of polypropylene fibers containing magnesium oxide nanoparticles. Arch Environ Contam Toxicol 62:185–194. https://doi.org/10.1007/s00244-011-9702-y
118. Woo DJ, Obendorf SK (2014) MgO-embedded fibre-based substrate as an effective sorbent for toxic organophosphates. RSC Adv 4:15727–15735
119. Othman FEC, Yusof N, Jaafar J, Ismail AF, Abdullah N, Hasbullah H (2016) Adsorption of cadmium(II) ions by polyacrylonitrile-based activated carbon nanofibers/magnesium oxide as its adsorbents. Malays J Anal Sci 20:1467–1473
120. Almasian A, Giahi M, Fard GC, Dehdast SA, Maleknia L (2018) Removal of heavy metal ions by modified PAN/PANI-nylon core-shell nanofibers membrane: filtration performance, antifouling and regeneration behavior. Chem Eng J 351:1166–1178
121. Banerjee P, Mukhopadhyay A, Das P (2019) Graphene oxide–nanobentonite composite sieves for enhanced desalination and dye removal. Desalination 451:231–240
122. Guo H, Jiao T, Zhang Q, Guo W, Peng Q, Yan X (2015) Preparation of graphene oxide-based hydrogels as efficient dye adsorbents for wastewater treatment. Nanoscale Res Lett 10:1–10
123. Jiao T, Liu Y, Wu Y, Zhang Q, Yan X, Gao F, Li B (2015) Facile and scalable preparation of graphene oxide-based magnetic hybrids for fast and highly efficient removal of organic dyes. Sci Rep 5:1–10
124. Noreen S, Tahira M, Ghamkhar M, Hafiz I, Bhatti HN, Nadeem R, Younas F (2021) Treatment of textile wastewater containing acid dye using novel polymeric graphene oxide nanocomposites (GO/PAN, GO/PPy, GO/PSty). J Mater Res Technol 14:25–35
125. Guo R, Jiao T, Li R, Chen Y, Guo W, Zhang L, Peng Q (2018) Sandwiched Fe_3O_4/carboxylate graphene oxide nanostructures constructed by layer-by-layer assembly for highly efficient and magnetically recyclable dye removal. ACS Sustain Chem Eng 6:1279–1288
126. Zhang H, Hines D, Akins DL (2014) Synthesis of a nanocomposite composed of reduced graphene oxide and gold nanoparticles. Dalton Trans 43:2670–2675
127. Petosa AR, Jaisi DP, Quevedo IR, Elimelech M, Tufenkji N (2010) Aggregation and deposition of engineered nanomaterials in aquatic environments: role of physicochemical interactions. Environ Sci Technol. https://doi.org/10.1021/es100598h
128. Fan H, Wang L, Zhao K, Li N, Shi Z, Ge Z, Jin Z (2010) Fabrication, mechanical properties, and biocompatibility of graphene-reinforced chitosan composites. Biomacromolecules 11:2345–2351. https://doi.org/10.1021/bm100470q.
129. Chang X, Wang Z, Quan S, Xu Y, Jiang Z, Shao L (2014) Exploring the synergetic effects of graphene oxide (GO) and polyvinylpyrrodione (PVP) on poly(vinylylidenefluoride) (PVDF) ultrafiltration membrane performance. Appl Surf Sci 117:43–56. https://doi.org/10.1016/j.apsusc.2014.07.202
130. Banerjee P, Barman SR, Mukhopadhayay A, Das P (2017) Ultrasound assisted mixed azo dye adsorption by chitosan–graphene oxide nanocomposite. Chem Eng Res Des. https://doi.org/10.1016/j.cherd.2016.10.009
131. Sharma R, Singh N, Gupta A, Tiwari S, Tiwari SK, Dhakate SR (2014) Electrospun chitosan-polyvinyl alcohol composite nanofibers loaded with cerium for efficient removal of arsenic from contaminated water. J Mater Chem A. https://doi.org/10.1039/c4ta02363c
132. Xu X, Tian M, Qu L, Zhu S (2016) Graphene oxide/chitosan/polyvinyl-alcohol composite sponge as effective adsorbent for dyes. Water Environ Res. https://doi.org/10.2175/106143016x14609975746127
133. Das L, Das P, Bhowal A, Bhattachariee C (2020) Synthesis of hybrid hydrogel nano-polymer composite using graphene oxide, chitosan and PVA and its application in waste water treatment. Environ Technol Innov 18:100664. https://doi.org/10.1016/j.eti.2020.100664
134. Santhi T, Manonmani S, Smitha T (2010) Removal of methyl red from aqueous solution by activated carbon prepared from the *Annona squamosa* seed by adsorption. Chem Eng Res Bull 14:11–18

135. Mahmoodi NM, Abdi J, Taghizadeh M, Taghizadeh A, Hayati B, Shekarchi AA, Vossoughi M (2019) Activated carbon/metal-organic framework nanocomposite: preparation and photocatalytic dye degradation mathematical modeling from wastewater by least squares support vector machine. J Environ Manag 233:660–672
136. Wang S, Lu A, Zhang L (2016) Recent advances in regenerated cellulose materials. Prog Polym Sci 53:169–206
137. Singh V, Tiwari A, Tripathi DN, Sanghi R (2006) Microwave enhanced synthesis of chitosan-graft-polyacrylamide. Polymer 47:254–260
138. Littunen K, Hippi U, Saarinen T, Seppala J (2013) Network 611 formation of nanofibrillated cellulose in solution blended 612 poly(methyl methacrylate) composites. Carbohydr Polym 91:183–190
139. Zhou G, Byun JH, Oh Y, Jung BM, Cha HJ, Seong DG, Chou TW (2017) Highly sensitive wearable textile-based humidity sensor made of high-strength, single-walled carbon nanotube/poly(vinyl alcohol) filaments. ACS Appl Mater Interface 9:4788–4797
140. Zhu Y, Hu J, Wang J (2014) Removal of Co^{2+} from radioactive wastewater by polyvinyl alcohol (PVA)/chitosan magnetic composite. Prog Nucl Energy 71:172–178
141. Racho P, Waiwong W (2020) Modified textile waste for heavy metals removal. Energy Rep 6:927–932

Multiphase Reactors in Photocatalytic Treatment of Dye Wastewaters: Design and Scale-Up Considerations

Suman Das and Hari Mahalingam

Abstract Treatment of dye wastewaters is important to recover and reuse the water in order to mitigate the impending freshwater crisis precipitated by a growing population, industrialization and declining freshwater reserves. Photocatalysis is very effective in complete mineralization of the different pollutants present, but the complex design, construction and scale-up of photocatalytic reactors for industrial-scale applications is still an open problem. Among the different configurations of reactors studied, the work on multiphase photocatalytic reactors is comparatively less. In this chapter, after a brief look at the basic fundamentals, a comprehensive review of the different multiphase photocatalytic reactors is presented with the aim of showing why this type could be a better option for the degradation of toxic dyes. The important operational parameters are discussed followed by an overview of the issues encountered in scale-up. Finally, the future aspects concerning the use of multiphase reactors for photocatalytic treatment of dye wastewaters are given.

Keywords Dye wastewater · Multiphase · Reactor · Photocatalyst · Design · Development · Scaleup · Gas · Liquid · Solid

1 Introduction

Water is an essential component of life on earth, and access to clean drinking water is a basic human right [1]. Due to the rapid industrial growth, climate change, and demographic expansion, depletion in potable water sources is increasingly a global

S. Das
Department of Desalination and Water Treatment, The Zuckerberg Institute for Water Research, Ben-Gurion University of the Negev, Beersheba, Israel
e-mail: sumand@post.bgu.ac.il

H. Mahalingam (✉)
Department of Chemical Engineering, National Institute of Technology Karnataka, Surathkal, Mangalore, India
e-mail: mhari@nitk.edu.in

S. S. Muthu and A. Khadir (eds.), *Advanced Oxidation Processes in Dye-Containing Wastewater*, Sustainable Textiles: Production, Processing, Manufacturing & Chemistry,
https://doi.org/10.1007/978-981-19-0987-0_10

issue [2]. Moreover, many existing and even new anthropogenic harmful chemical contaminants are added to the fresh water bodies every year thus reducing the quality of drinking water [3]. Consequently, almost a third of the world's population suffers from water scarcity and over one billion people lack access to potable water [4]. The presence of toxic contaminants in drinking water resources leads to water-borne diseases, and a huge amount of the population is affected [5]. According to the World Health Organization (WHO), 3.2 million people (including children) die every year worldwide due to polluted water consumption [6]. The colored or color-containing wastewater, contributing majorly to this contaminated water, comes from several sources such as the textile industry, dyes and pigments industry, pulp and paper industry, tannery industry, etc., which is not only harmful to human, but aquatic animals as well. The details of colored wastewater, the eco-friendly modern photocatalytic remediation of wastewater, and multiphase reactors for photocatalytic wastewater treatment are discussed in the next few subsections.

1.1 Colored Wastewater

Organic dye colors are one of the main components of toxic compounds discharged into wastewater from industries such as textile, dye and pigment, pulp and paper, and tannery industries [6–9]. About 10–15% of the dyes used during the manufacturing processes in the world are lost as waste and discharged as effluents into the environment [10, 11]. Globally, over 0.7 million tons of organic dyes are produced each year mainly to be used in the textile, leather goods, printing, food, and cosmetics industries. The main contaminants in dye wastewaters are suspended solids, COD, color, acidity, and other dissolvable substances [12].

The textile industry utilizes around 21–377 m^3 of water for each ton of textile produced and consequently produces very substantial amounts of toxic, colored wastewater [13]. This wastewater is considered to be the dirtiest among the wastewaters produced from various industries. The textile industry uses around 10,000 distinct dyes and pigments, over half of which are azo dyes (~50 to 70%) [14].

The textile wastewater treatment is an intense issue because of a few reasons which are described as follows:

- Presence of toxic metals like Cr, As, Cu, etc.(reason for strong colors in synthetic dyes) [15, 16]
- The presence of large amounts of dissolved solids in the effluent [17]
- Dyes are recalcitrant molecules meaning that the molecules are complex in structure and therefore very difficult or resistant to break down easily [18]
- The presence of dissolved silica and free chlorine [19].

Most dyes utilized in the textile industries are not affected by light and are not naturally degradable since dyes usually have a synthetic origin and complex aromatic molecular structures which make them highly stable and more difficult to biodegrade. Pagga and Brown [20] reported that out of 87 dyes tested, only 62% are biodegradable.

The same authors estimated that about 12–15% of these dyes are discharged as effluents during the manufacturing process.

Wastewater treatment and reuse is a crucial part; several conventional economical processes such as physical, chemical, and biological treatments [21] are in use for a long time. The major ecological problem is the mineralization of color containing wastewater from textile and dyestuff effluents. Traditional treatment processes such as coagulation, anaerobic process, membrane separation have been utilized for the treatment of textile effluents [22]. However, these treatment processes are not effective as the dyes are recalcitrant in nature and the dye wastewaters carry a high salt content. Chlorination and ozonation are very effective processes in this regard, but their high operating cost makes them unpopular [23, 24]. The traditional physical techniques, for example, adsorption on activated carbon, reverse osmosis, ultrafiltration, coagulation with chemical agents, ion exchange, and so on (Fig. 1), have been in use for the removal of dyes from wastewater [25–28]. These methods are effective to some extent to transfer the organic pollutants from the liquid phase to the solid phase, thereby creating secondary contamination requiring additional treatment of solid wastes and recovery or regeneration of the solid phase making the process costlier. The following fundamental factors must be considered in the dye wastewater treatment processes such as (a) Treatment capability, (b) Complete mineralization of parent and intermediate contaminants, (c) The total productivity of the wastewater treatment process, (d) Recycling capacity and potential utilization of treated water, (e) Cost Adequacy and environmental safety.

For these reasons, advanced oxidation processes (AOP) are the most appropriate, modern, and popular method to completely mineralize the organic pollutants present in wastewater. Based on the catalyst and pollutant phase, AOPs can be classified as shown in Fig. 2 [29]. AOPs basically use three different oxidizing reagents (oxygen, ozone, and hydrogen peroxide) in different combinations or applied with UV irradiation and/or with various types of homogeneous or heterogeneous catalyst mixtures [30–32]. The generation of $\dot{O}H$ radicals or super oxidant radicals ($\dot{O}^{2-}$) leads to

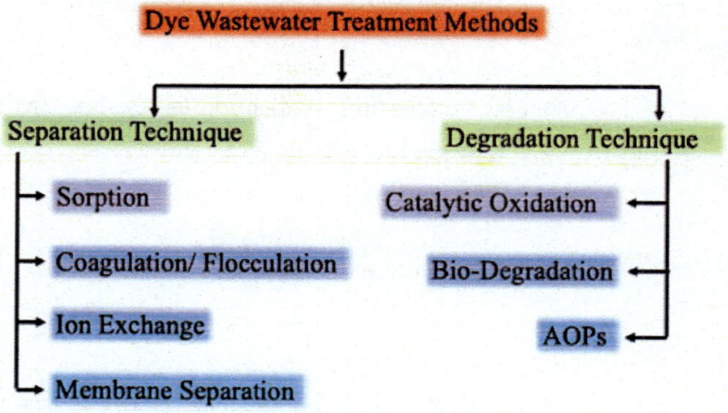

Fig. 1 Dye wastewater treatment techniques

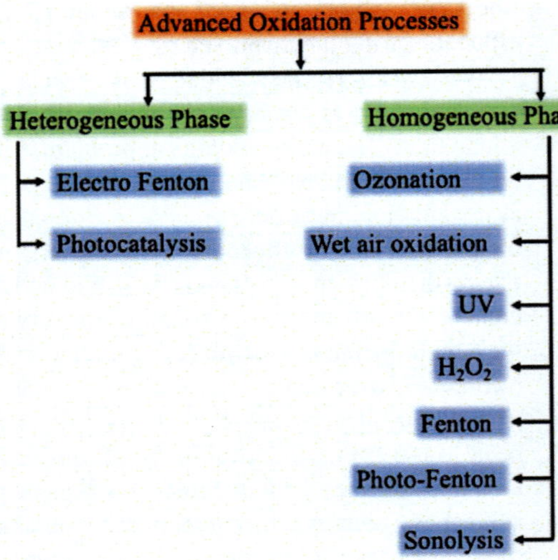

Fig. 2 Classification of advanced oxidation processes

higher oxidation rates. Among the several AOPs, photocatalysis is one of the widely used wastewater treatment processes for its various beneficial aspects and inherent advantages [33, 34].

1.2 Photocatalysis

The photocatalysis process can be described by four simple steps, namely, (a) absorption of light to generate electron-hole pairs, (b) separation of excited charges, (c) transfer of electrons and holes to the surface of photocatalysts, (d) the utilization of charges on the surface for oxidation reactions [35, 36]. In general, the photocatalytic reaction for wastewater treatment can be described by the following equations (Eqs. 1–5):

$$\text{Semiconductor (hv)} \rightarrow \text{electron } (e^-) + \text{hole } (h^+) \rightarrow \text{reformation of semiconductor} + \text{heat} \tag{1}$$

$$e^- + O_2 \rightarrow \dot{O}_2^- \tag{2}$$

$$h^+ + OH^- \rightarrow \dot{O}H \tag{3}$$

$$\dot{O}H + \text{pollutant} \rightarrow \text{Intermediate} \rightarrow CO_2 + H_2O \tag{4}$$

$$\dot{O}_2^+ + \text{pollutant} \rightarrow \text{Intermediate} \rightarrow CO_2 + H_2O \quad (5)$$

There are two types of photocatalysis, i.e., homogeneous and heterogeneous photocatalysis. In homogeneous photocatalysis [37, 38], the catalyst, as well as the reactant, are in the same phase. In the case of heterogeneous photocatalysis [39, 40], the catalyst and reactant exist in different phases. Catalyst usually exists in a solid phase while reactant is in the liquid phase. Semiconductor metal oxides, such as ZnO, TiO_2, WO_3, are generally used as a heterogeneous photocatalyst.

There are several advantages of the heterogeneous photocatalytic process [41]

- Green technology: this does not produce any sludge or harmful product after the waste treatment.
- Versatility: it can be used for multiple applications in one operation; H_2 production as well as pollutant degradation, water splitting as well as electricity production, etc. Organic or inorganic contaminants can be removed simultaneously from wastewater by oxidation or reduction reaction [42].
- Energy efficiency: since multiple applications are possible in the same operation, it saves time as well as energy. In a typical photocatalytic wastewater treatment process, lesser energy is required compared to other processes such as reverse osmosis.
- Cost-effectiveness: since this process needs only light energy to activate the catalyst for carrying out the degradation process, it is cost effective.
- Higher resistance to toxic pollutants: the semiconductor photocatalysts are capable of mineralizing the toxic pollutants without affecting itself.
- The mineralization of low concentrations of contaminants: According to recent research, photocatalysis can be used to treat very small or trace amounts (in ppm or even ppb levels) of pollutants present in wastewater. [43, 44].
- Simple, light reactor apparatus: the photocatalytic process is very simple, and reactors consist of only a few simple parts. It can be easily fabricated at low cost.

An ideal photocatalyst should have the following properties [35], (a) highly stable, (b) economical, (c) non-toxic (to environment or humans), (d) high turnover, (e) can be supported on various substrates easily, (f) complete destruction of organic pollutants into harmless compound, (g) high catalytic activity, (h) strong oxidizing power, (i) stable against photo-corrosion, and (j) chemically and biologically inert.

TiO_2 is one of the most widely used photocatalysts as it has almost all the characteristics of an ideal photocatalyst. There are other photocatalysts that work as good as TiO_2, namely, graphene oxide, g-C_3N_4, ZnO, SnO_2, ZrO_2, WO_3, Si, CdS, ZnS, $SrTiO_2$, Fe_2O_3, etc. [45, 46]. Some applications of photocatalysts are shown in Fig. 3.

Even though the photocatalytic process has several advantages, it has a few drawbacks listed below with strategies to overcome these difficulties such as

- Catalyst recovery: The separation of powdered photocatalysts after the operation in the slurry reactor is a big challenge as of now. It is also time-consuming, expensive, and difficult. To overcome this disadvantage, photocatalysts can be immobilized into different substrate materials [47, 48].

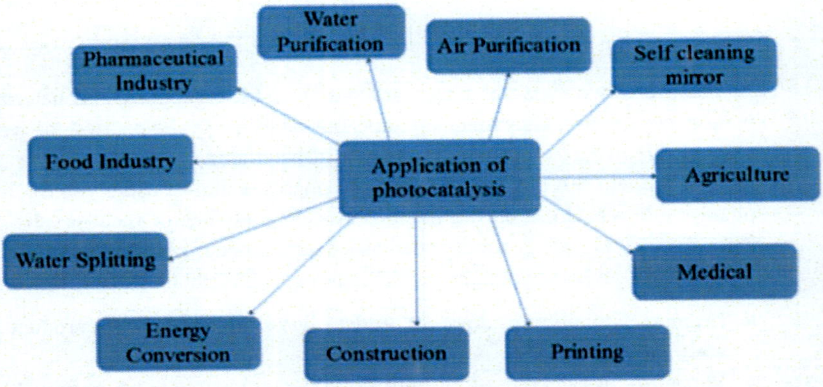

Fig. 3 Applications of photocatalytic process

- Reusability: The adsorption and photocatalytic activity both are taking place on the catalyst surface; there is a great possibility of reduction in catalyst activity. To overcome this problem, photocatalyst can be reactivated by catalyst surface cleaning [49, 50].
- Activation energy: Most of the photocatalysts work under UV light, which means high operating cost. To defeat this problem photocatalysts can be doped with some other metal or non-metallic substance [51].
- Low photon quantum yield: The quantum yield is the ratio of photons emitted to photon absorbed. In case of high bandgap photocatalysts, the quantum yield is very low since high activation energy is required. To solve this problem, photocatalysts with low bandgap or composite or doped or co-doped materials can be used instead [52].

In the next section, multiphase reactors for wastewater treatment are described.

1.3 Basic Fundamentals of Multiphase Photocatalytic Reactors

Multiphase reactors usually refer to a reactor having more than one phase such as liquid–solid, liquid–gas, solid–gas, and in some cases, all the phases (liquid–solid-gas). Based on the water quality (temperature, pH, dissolved oxygen, conductivity, turbidity, etc.), and contaminants (organic, inorganic, biological), the treatment procedure as well as the reactor type is decided. Other parameters that need to be considered before choosing the reactor are photocatalyst, photocatalyst attachment mode, light source, etc. Based on these parameters, requirement of reactors can be chosen as depicted in Fig. 4 [53, 54].

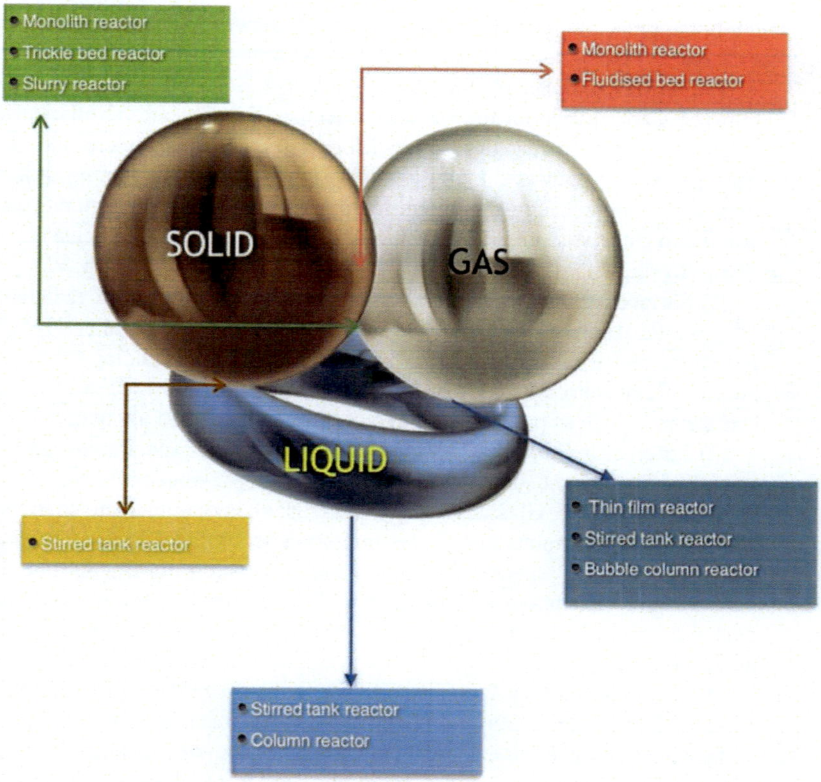

Fig. 4 Conventional multiphase reactor selection criteria

The multiphase reactors usually have a few significant properties such as high mass transfer rate, facile processing of solids, high efficiency, low maintenance, simple design, which makes it a valuable and worthy choice [55]. There are several multiphase conventional reactors, which can be used as a photocatalytic reactor for wastewater treatment, but it might need slight modification since light irradiation is needed to carry out the photocatalytic process. Those conventional reactors include bubble column reactor, fluidized bed reactor, trickle bed reactor, slurry photocatalytic reactor, thin film reactor, and airlift reactor. The photocatalytic wastewater treatment process is not limited to those conventional reactors; various innovative photocatalytic reactors have been developed according to the need and purpose such as annular falling film reactor, optical fiber reactor, photo capillary reactor, submerged membrane reactor, recirculating batch reactor, and compound parabolic reactor [56]. In these examples, the reactor contains at least one interface.

The basic requirement of a photocatalytic reactor is a light source, which should irradiate the catalyst surface to initiate and sustain the reaction. Some photocatalytic multiphase reactors recently reported for dye wastewater decontamination are listed in Table 1.

The above-mentioned photocatalytic reactors, mostly studied at laboratory scale or pilot scale, are still waiting to cross the barrier to industrial scale. An efficient scale-up of the photocatalytic reactor is still a difficult task and needs to consider important parameters [72] such as the kind of light source needed, maximum utilization of light, type of photocatalyst: immobilized or suspended, and optimum ratio of reactor volume to catalyst surface area required to achieve maximum efficiency, etc. [56]. The following common difficulties [73] are encountered in scale-up:

(1) Surface area to volume ratio: To produce the high ratio of irradiated surface area to total volume is difficult especially in case of immobilized catalyst [74, 75]. For slurry operation, the nano-size particles can be a good choice for high surface area to volume ratio, whereas for immobilized operation, substrate material with high surface area, such as pumice stone and perlite can be a good alternative [76, 77].

(2) External light source and transparent material: A specific design is needed according to the light source used. If an external light source is used, a transparent outer shell is needed. Glass, especially quartz glass, is a popular choice in such cases but with some constraints such as size limitation, sealing problem, and breakage risk. Instead of glass, any other transparent material, which is not fragile (like perspex), can be used for the photocatalytic reactor [78]. The need for transparent material of construction can be eliminated by placing the light source inside the reactor [72, 79]; however, the internal elements need to be carefully designed.

(3) Light distribution: It is very difficult to distribute the light uniformly on the surface of the catalyst into the photocatalytic reactor. Nanoparticles have a large surface area; however, in a vertical column photocatalytic reactor, only a portion of suspended nanoparticles gets to absorb the light from an external light source. In case of immobilized nanoparticles, a very thin film of catalyst is coated in a carrier surface; thus, the active part of the catalyst which comes in contact with light is very less in spite of a high catalyst surface area of catalyst [80, 81]. The light source can be maintained as close as possible to the catalyst surface and in case the light source is outside the reactor vessel, the transparent vessel should not absorb the irradiated light, which is meant for the catalyst activation. Use of a reflector can also help in this regard.

(4) Catalyst surface area: large catalyst surface area is needed for photocatalytic reactors driven by solar light. As it has already been discussed that nanoparticles have large surface area but only a small portion of it will be active; hence, nanoparticles with larger surface area will be more preferable [82, 83].

(5) Efficient catalyst: Solar light focused photocatalytic reactors need efficient catalysts that show appreciable photocatalytic activity under solar light. Photocatalyst with a large surface area and reusable for several times, will be an innovative idea [84, 85].

(6) Pollutant concentration: Based on pollutant concentration, the required amount of catalyst should be present in the reactor. This problem may restrict the capacity of the reactor and the necessary time required to achieve high conversion [86, 87].

Table 1 Studies done using different types of photocatalytic reactors for dye wastewater treatment

Sl. no.	Photocatalytic reactor type	Photocatalyst (slurry/immobilized)	Pollutant	Light source	Findings	References
1	Microchannel photocatalytic continuous reactor	100 nm thickness of TiO_2 immobilized	MB (1 µM, 18 mL, flowrate: 1.2 to 5.8 cm3 min-1)	UV	Almost complete decolorization achieved in 90 min	[57]
2	Airlift reactor	TiO_2/rGO/g-C_3N_4 nanocomposite-polystyrene film	Remazol turquoise blue (10 ppm, 1 L)	Solar	~55% degradation occurred in 90 min	[58]
3	Nanofiber membrane reactor	Ag/TiO_2	MB (10 ppm, 50 mL)	Solar	>80% MB removal within 30 min of irradiation, whereas complete degradation was observed after 80 min	[59]
4	Compound parabolic reactor	TiO_2 immobilized in glass spheres	15 different emerging contaminants (0.1 ppm, 10 L)	Sunlight	85% of the pollutants were degraded within 2 h	[60]
5	Tubular reactor	TiO_2 (0.3–1.2 g/L)	MB (15–60 ppm)	UV	98% dye removal in <1 h	[61]
6	Pebble bed reactor	TiO_2 coated silica rich white pebbles	Mixture of six different reactive dyes (83 ppm TOC, 10 L)	Sunlight	15% TOC reduction was achieved in 5 h	[62]

(continued)

Table 1 (continued)

Sl. no.	Photocatalytic reactor type	Photocatalyst (slurry/immobilized)	Pollutant	Light source	Findings	References
7	Flat plate reactor	ZnO	MB (5 ppm, 5 L)	UV	>80% degradation achieved in 6.5 h	[63]
8	Solar photoreactor	TiO_2 coated in plastic granules	MB (25 ppm, 10 L)	Solar light	Almost complete decolourization occurred within 20 h	[64]
9	Packed bed reactor	N-doped TiO_2	MB (7 ppm, liquid flow rate: 50 mL/min)	Visible light	Almost total decolourization of the dye within 2 h	[65]
10	Rotating disk reactor	TiO_2 immobilized in ceramic balls	MO (10 ppm, 0.045–5 L)	UV	>97.5% decolourization and >75% TOC removal	[66]
11	Capillary array reactor	P-25 TiO_2 powder and TiO_2 film immobilized onto a quartz tube (0.8 g/L, flowrate: 1 mL/min, pH 3)	MO (5 ppm, 50 mL)	UV	Almost complete degradation occur after 2 h of irradiation	[67]
12	Continuous fixed bed reactor	N-TiO_2 immobilized in glass spheres (0.34 wt%, flowrate: 1.45 L/h)	MB (10 ppm)	UV	>70% MB conversion after 4.5 h	[68]
13	Flat plate reactor	N-TiO_2 (372 g, flowrate: 150 mL/min)	MB (7 ppm, 300 mL)	Visible light	45% TOC removal after 3 h of irradiation	[69]
14	Packed bed reactor	N-TiO_2 (325 g, flowrate: 6.3–74.15 mL/min)	MB (10 ppm)	Visible light	Almost total decolorization after 30 min of irradiation	[65]
15	Translucent packed bed reactor	TiO_2 (1.9 g/L)	MB (10 ppm, 50 mL)	LED light	Photocatalytic space–time yield 0.657 $m^{3\ day-1\ m-3}$ reactor kW^{-1}	[70]
16	Ultrafiltration membrane reactor	TiO_2 (1.5 mg/L)	Acid red B (50 ppm, 1.2 L)	UV	>80% decolorization and 23.5% TOC removal occurred after 2 h	[71]

(7) Residence time: Longer residence time for the catalyst is beneficial for the pollutant degradation, but the use of baffles or any such strategy can reduce the light irradiation on catalyst surface and hence bring down the photocatalytic efficiency [88, 89]. Highly efficient photocatalyst, which shows better productivity at less residence times, can be a good solution in this regard.

(8) Weather dependency: Solar light-driven photocatalytic reactors face another problem, i.e., dependence on weather. Hence, the reactor should be designed in such a way that the simulated solar light can also be used for the same purpose [90, 91].

(9) Reuse: The reuse of a photocatalyst beyond four times is usually not worthwhile. In case, the photocatalyst is coated on the reactor inner surface, the catalyst has to be replenished after three or four recycles [92, 93].

(10) Temperature: Maintaining the equilibrium temperature in an industrial-scale reactor is difficult, especially in case of transparent reactor outer surface, because the cooling jacket can block the light irradiation. Cool air can be circulated into the reactor, but that can again increase the operating cost [94, 95].

(11) Catalyst immobilization: The immobilization of the catalyst can be done using various substrate materials (brick, rock, polymer, glass, metal, ceramic, perlite, cork, etc.) [96]. The shape, size, and physical properties of the substrate material also play an important role in this regard. Also, the way of using the substrate material can also be a vital factor for the efficient outcome. Based on the material packing (irregular shape, regular shape, film, floatability, etc.), the reactor performance can change [97].

Most of the photocatalytic reactors are capable of complete degradation or removal of contaminants but from the above discussion, it is obvious that still there are some parameters that can be improved for an ideal photocatalytic reactor. Those parameters are (a) reduction of residence time, (b) reuse of photocatalyst for several times without reducing efficiency, c) high-performance catalyst, (d) effective design of photocatalytic reactor, (f) effective use of solar light, (g) reduce the fabrication cost of a photocatalytic reactor, (h) reduce the reactor surface area, (i) high mass transfer between gas and liquid, (j) selection of suitable substrate material for the immobilization of the catalyst, (k) reduce the fabrication and operating cost of the reactor.

In the next section, different multiphase photocatalytic reactors for dye wastewater treatment in the last few decades will be discussed along with their design and scale-up consideration in detail.

2 Multiphase Reactors for Dye Wastewater Treatment

As indicated earlier, photocatalytic multiphase reactors can be mainly divided into two kinds: slurry and immobilized. The advantages and disadvantages of the slurry and immobilized process are listed in Tables 2 and 3.

From the slurry/immobilized multiphase reactors tabulated in Table 1, it is clear that the outcome from each reactor is different, because of the design as well as

Table 2 Advantages of slurry and immobilized photocatalytic system

Advantages of slurry system	Advantages of immobilized system
High surface area of catalyst	Easy separation process
High mass transfer rate between catalyst and pollutant	Reuse of catalyst is also possible
High photocatalytic activity	Extended photocatalyst lifetime
High rate of electron–hole pair generation	Economically viable
High throughput	Proper utilization of the irradiated light
Better mixing	

Table 3 Disadvantages of slurry and immobilized photocatalytic system

Disadvantages of slurry system	Disadvantages of immobilized system
Catalyst recovery is tedious and expensive	Low surface area
Catalyst agglomeration takes place	Mass transfer rate is relatively low
Opacity of the wastewater solution increases, can block the light irradiation on the catalyst surface	Low photocatalytic yield (can compensate by increasing the irradiation time)
Continuous agitation at a particular rate is necessary	Low photon accessibility
Large scale application is quite difficult	Waste of small portion of the catalyst

other operating parameters. A few popular reactors used commonly for wastewater treatment are shown below (Fig. 5).

The modified conventional reactors can be used efficiently as a photocatalytic reactor with a basic modification by adding a light source placed inside or outside (with a transparent reactor body). These modified conventional reactors were used very often for dye wastewater treatment, but recently, there are many novel reactors which are in use for the same purpose. The newly developed multiphase reactors for dye wastewater treatment are discussed below.

2.1 *Novel and Modified Multiphase Reactors for Dye Wastewater Treatment*

2.1.1 Bubble Column Reactor

It is a vertical cylinder filled with liquid, into which air is passed from the bottom (Fig. 5a). Mixing takes place primarily due to the gas-induced bulk liquid phase circulation. The bubble column reactor has several advantages: efficient contact between the phases including the solid catalyst which is generally in slurry form, high liquid

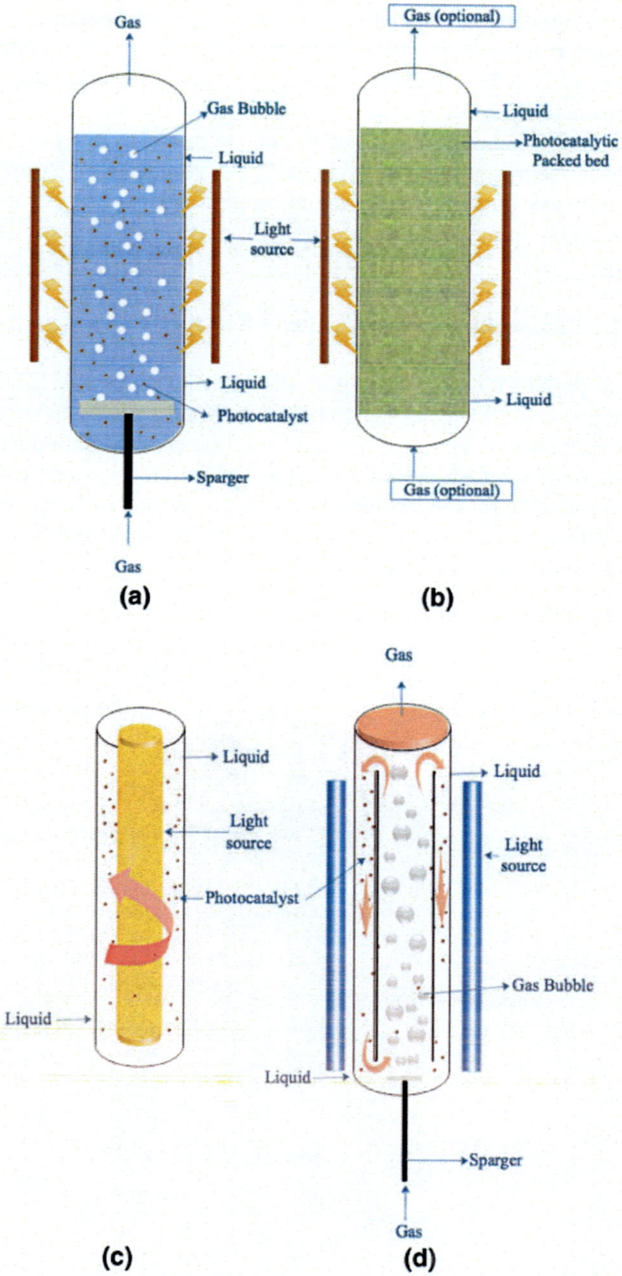

Fig. 5 Conventional multiphase reactors for dye wastewater treatment: **a** bubble column photocatalytic reactor, **b** packed bed photocatalytic reactor, **c** slurry photocatalytic reactor (with inside light source), **d** airlift photocatalytic reactor

hold ups, moderate mass transfer rates, easy temperature control, low cost of construction, and maintenance. Few limitations are also there in the slurry bubble column reactor such as high pressure drop with respect to packed bed reactors, and less residence time of gas.

A modified photocatalytic bubble column reactor is designed [98] for the degradation of methylene blue dye using Fe^{3+}/TiO_2 photocatalyst in slurry form under UV light. The use of microbubbles in their work improved the photocatalytic efficiency by increasing the mass transfer.

2.1.2 Photoelectrochemical Column Reactor

Boron-doped TiO_2 nanotubes [99] were used to remove Acid Yellow 1 dye from wastewater. A 500 mL cylindrical glass photoelectrochemical reactor at 25 °C was used as shown in Fig. 6. and the photocatalytic-anode (boron-doped TiO_2) was irradiated by a Hg high pressure lamp to initiate the reaction. This type of reactor can be suitable for the small-scale operation, but the scale-up consideration needs detailed assessment of efficient and stable photocatalytic-anode, catalyst reusability, and operating cost analysis.

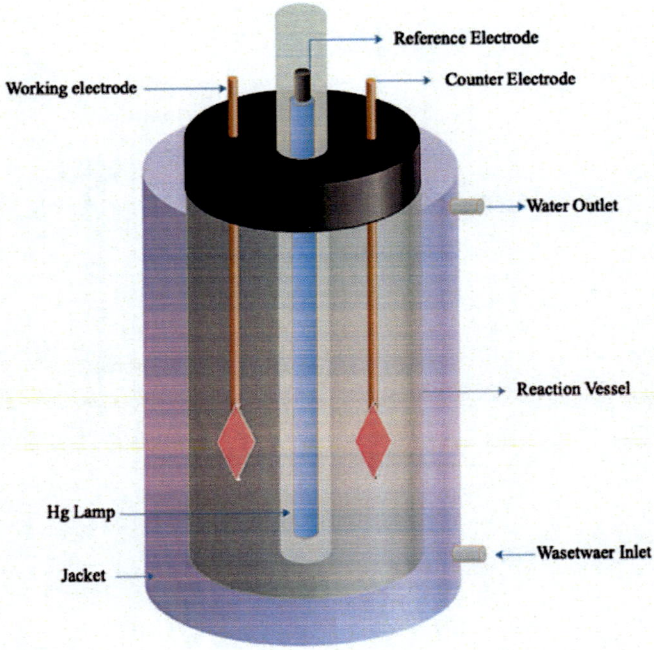

Fig. 6 Schematic of photoelectrochemical reactor

2.1.3 Fluidized Bed Reactor

In this type of reactor, gas passes through the solid or liquid phase or both from the bottom of the cylindrical reactor with a high velocity so that fluidization occurs. Advantages of this reactor are uniform particle mixing, uniform temperature gradient, high surface area between solid and fluid per unit bed volume, and can be used in continuous mode. Important parameters that affect the fluidization are minimum fluidization velocity, particle terminal velocity, void fraction, minimum bubbling velocity, etc.

The photocatalytic degradation of reactive brilliant red dye by TiO_2/activated carbon in a fluidized bed photocatalytic reactor was performed [100]. The optimum parameters (pH, initial dye concentration, effect of Na_2SO_4 concentration, etc.) for the degradation of brilliant red dye were also identified.

2.1.4 Airlift Reactor

A cylindrical column contains a draft tube which helps the liquid to circulate into the reactor. Mixing of liquid is done by gas passing from the bottom. Solid catalysts can also be used in suspended form. Main advantages of this reactor are simple design, low construction cost, low maintenance cost, no agitation required, easy scale-up, high mass transfer, uniform shear stress, etc. Disadvantages include no bubble breaker, foaming, and lower oxygen transfer rates than bubble columns.

Based on the configuration or structure, an airlift reactor (ALTR) can be divided into two types (i) internal-loop ALTR and (ii) external loop ALTR. In an internal ALTR, baffles are placed strategically, which separates the riser and downcomer medium from each other. with a uniform average superficial liquid velocity throughout the reactor. An external ALTR consists of a bubble column as a riser, and downcomer section is provided at the outside of the bubble column. The density difference between the two sections is the reason for continuous recirculation and mixing.

The gas–liquid mass transfer rate is the most important factor to estimate the efficiency of air-lift reactors. An ALTR has maximum mass transfer rate at minimum power consumption with efficient mixing. There are certainly other parameters, smaller bubble diameter, higher gas holdup, and higher gas–liquid interface renewal frequency which is very effective to increase the mass transfer rate; hence, it is profitable to search for efficient ways to decrease the bubble size and enhance the gas holdup to increase mass transfer and mixing processes.

Wastewater treatment using ALTR has received a lot of attention nowadays, as it is one of the very efficient ways to affect degradation of toxic chemicals in wastewater.

A recent study [101] using ALTR (Fig. 7) on the degradation of remazol turquoise blue dye shows that the recyclable TiO_2/rGO/g-C_3N_4 nanocomposite-polystyrene film used in an airlift reactor under solar light is quite efficient and effective. Their work also includes the optimization of several photocatalytic parameters and confirmation of the degradation by Total Organic Carbon analysis, and HPLC–MS (High

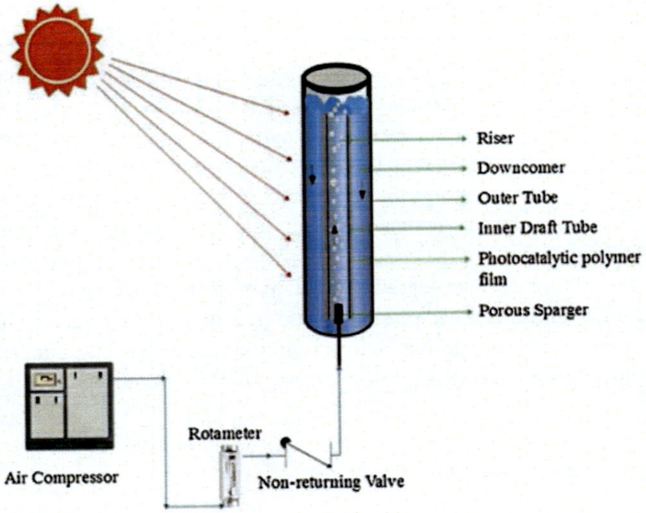

Fig. 7 Multiphase airlift photocatalytic reactor. (Reprinted from J. Environmental Chemical Engineering, 7, Das S; Mahalingam H, Dye degradation studies using immobilized pristine and waste polystyrene-TiO_2/rGO/g-C_3N_4 nanocomposite photocatalytic film in a novel airlift reactor under solar light, 103,289, Copyright (2019), with permission from Elsevier)

Performance Liquid chromatography–mass spectrometry) analysis. This work also gives insights of the degraded by-products and the future possibility of scale-up of the reactor.

Other similar studies using TiO_2/rGO/g-C_3N_4 were used under UV irradiation for the degradation of dye [58, 102].

2.1.5 Fixed Bed/ Packed Bed Reactor

A simple, horizontally placed, packed bed reactor in batch mode as well as continuous mode was studied [65] as shown in Fig. 8. The packed bed reactor consists of

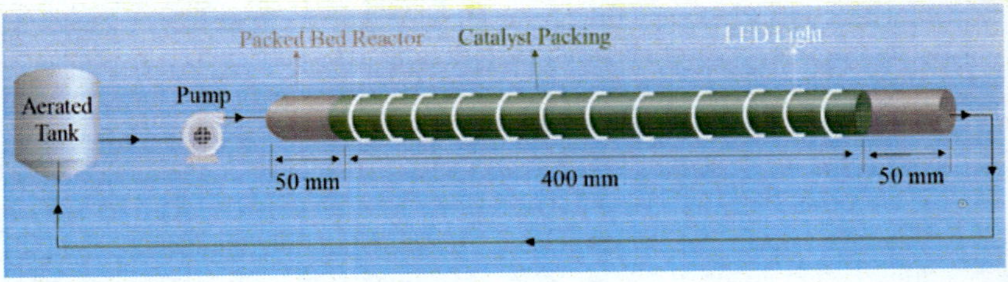

Fig. 8 Multiphase packed bed reactor

a cylindrical pyrex glass tube to facilitate the maximum incident irradiation intensity (length: 40 cm, external diameter: 5 cm, and total working volume: 500 mL) and N-TiO_2-polystyrene spheres as a photocatalyst (325 g) packing material. The photocatalytic experiments were performed by varying the flow rate in the range of 6.3–74.15 mL/min. This work also demonstrates the degradation of methylene blue dye (and few other pollutants) under LED light (400–700 nm) which was surrounded by the outer shell of the reactor. This type of photocatalytic reactor is encouraging for scale-up considerations while keeping other parameters (dead zone, residence time, optimum outer shell diameter and thickness, etc.) in mind.

2.1.6 Membrane Reactor

Membrane technology for photocatalytic wastewater treatment can be used in two different ways, (a) immobilization of the photocatalyst on the membrane surface [37], and (b) separation of slurry photocatalyst after the treatment [103]. [37] used a ZnO/polyester and Fe^{3+}-doped-ZnO/polyester membrane for the degradation of RB5 dye under simulated visible light. In their study, the used reactor of 100 mL working volume showed the color removal efficiency of more than 80% for the first time and the used photocatalyst also exhibited excellent reusability even after the fifteenth run.

There are other reported studies on dye degradation using membrane reactors, which shows promising outcomes but most of the membrane reactors possess high operating cost which is the main limitation of this process. The simplified version of two different types of membrane reactors is shown below (Figs. 9 and 10).

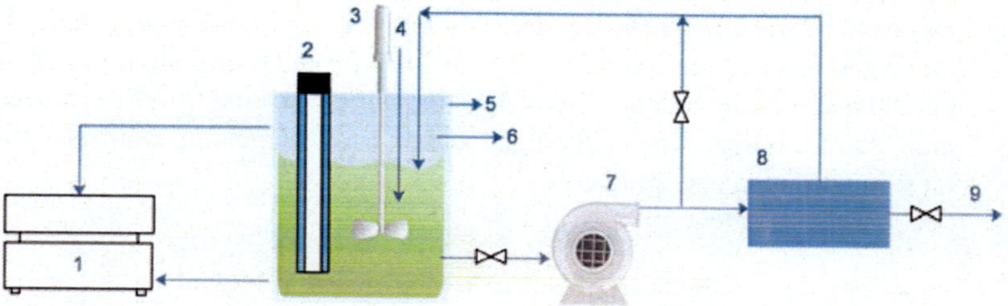

1. Water Chiller, 2. UV light, 3. Stirrer, 4. Feed, 5. Cooling Jacket, 6. Photocatalytic Reactor, 7. Pump, 8. Membrane Filtartion Unit, 9. Treatted Water Outlet.

Fig. 9 Membrane photocatalytic slurry reactor (membrane was used for the separation of the photocatalyst after operation)

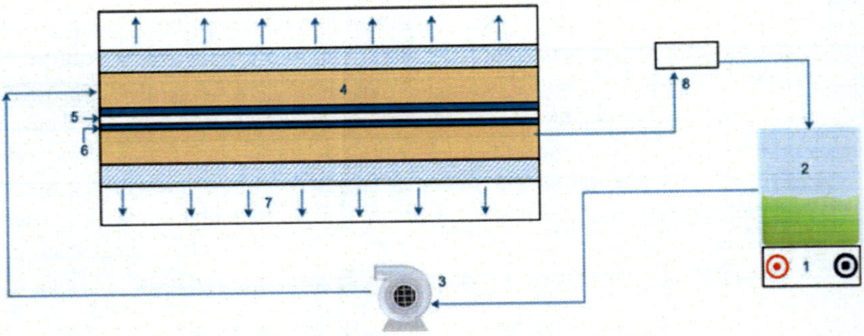

Fig. 10 Photocatalyst embedded membrane photocatalytic reactor

2.1.7 Optical Fiber Photocatalytic Reactor

An optical fiber photoelectrochemical reactor has been designed [104] for the degradation of azo orange dye II. TiO_2-based photoanode is used as an optic fiber support with a novel arrangement of TiO_2 layer, positioned on top of an optical fiber substrate. This arrangement was used in an internally illuminated setup under UV irradiation. The amount of photo-generated H_2O_2 helps to decolourize the dye completely with more than 56% Total Organic Carbon removal efficiency. The complicated design of this type of reactor makes it difficult to scale-up for industrial-scale application.

A flexible and efficient support material, plastic optical fiber fabrics [105] were used to support TiO_2 by dip-coating. The TiO_2 coated plastic optical fiber fabric was used for the photocatalytic degradation of methylene blue dye under UV light in a batch reactor. This process can be helpful for the remediation of contaminated groundwater in almost any kind of reactor. But the scaling up of the process needs more careful observation of the light source, coating method, catalyst activity, and most importantly, operating cost.

2.1.8 Photo-Capillary Reactor

A complicated photo-capillary reactor [67] was developed (Fig. 11). The reactor consists of two tetrafluoro-ethylene tubes, fixed on two brackets up and down, and the TiO_2 catalyst loaded capillaries were inserted into the tetrafluoroethylene tubes. There were a total of 18 capillary columns (each 15 cm length and 0.012 cm^3 volumes) and an UV lamp to provide irradiation. The reactor performed better for methyl orange degradation compared to a conventional batch reactor.

Fig. 11 Photo capillary multiphase reactor

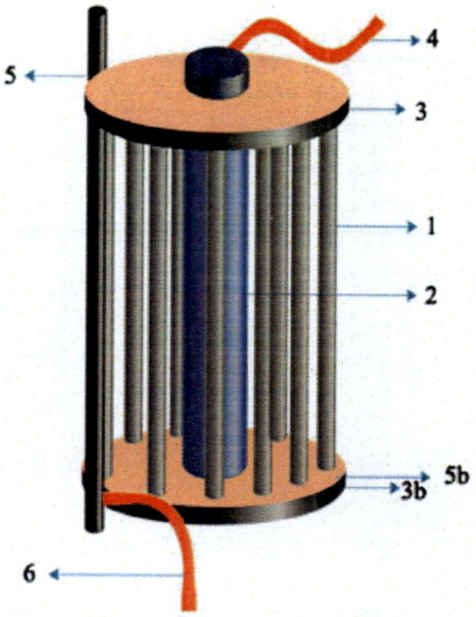

1: capillary with TiO_2 film; 2: UV lamp; 3a and b: up and down brackets; 4: outlet; 5a and b: up and down tetrafluoroethylene tubes; 6: inlet.

2.1.9 Recirculating Batch Reactor

Activated carbon doped TiO_2 was immobilized in glass plates using Polyethylene glycol as binder with heat treatment [106]. The recirculating batch photocatalytic reactor used in this study has a 254-nm UV lamp and is equipped with a quartz or a pyrex glass sleeve, with an effective volume of 1 L. The temperature of the wastewater solution is maintained at ~32 °C by cooling water flow inside the UV lamp sleeve. Mixing was achieved by using air circulation (100 ml/s). The distance between the vertically placed UV lamp and the surface of the immobilized catalyst is approximately 2.5 cm. At optimum conditions, 90% color removal in synthetic wastewater and 86.4% color removal in actual textile wastewater has been observed.

Another work on a recirculation study [107] with 550 mL reactor volume and 2350 mL solution in the reservoir is conducted under UV light at different flow rate, as shown in Fig. 12. The below figure also shows the remazol turquoise blue dye wastewater before and after treatment. Almost 75% decolourization was observed in 6 h.

2.1.10 Microchannel Reactor

Nanofibrous TiO_2 photocatalyst in a novel photocatalytic microreactor [108] was used for the degradation of Methylene blue dye under UV irradiation. The TiO_2 film/nanofiber was immobilized on the glass surface and covered with a microchannel polydimethyl siloxane film. The fast diffusion rate into the microfluidic channel, and large surface area of the nanofibrous photocatalyst, demonstrates

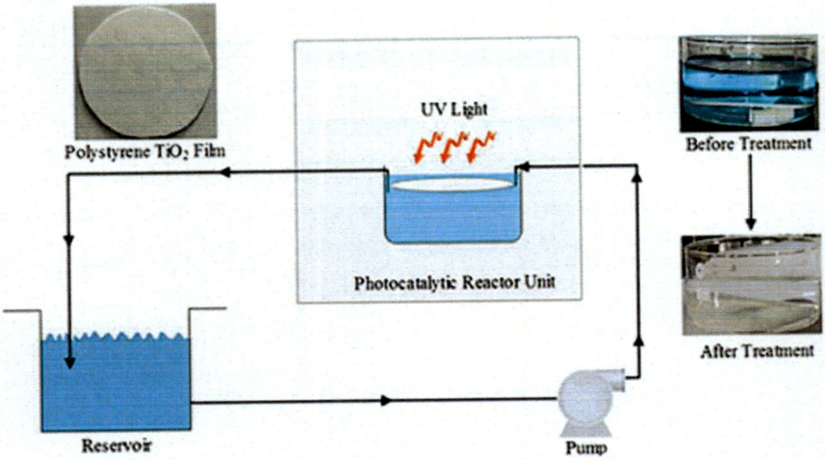

Fig. 12 Recirculating batch photocatalytic reactor. (Reprinted from J. Chemical Technology and Biotechnology, 94(8), Das S; Mahalingam H, Reusable floating polymer nanocomposite photocatalyst for the efficient treatment of dye wastewaters under scaled-up conditions in batch and recirculation modes, 2597–2608, Copyright (2019), with permission from John Wiley and Sons)

higher photocatalytic activity. This type of reactor is suitable for micro/small-scale applications.

There are other reported studies on microchannel reactors for the degradation of methylene blue under visible light irradiation using different types of microreactors [109, 110].

2.1.11 Tubular Reactor

A cylindrical narrow diameter (1.5 cm) tubular reactor (46 cm height) used for the degradation of Methylene blue dye under UVC irradiation [63]. The reactor column was placed in between two UVC lights, and was connected to a reservoir-aeration tank of 1 L as shown in Fig. 13. There was a 98% removal efficiency for this batch recirculating type tubular reactor. The scale-up consideration of this type of reactor can be approachable after considering various factors, such as continuous flow of wastewater, and low operating cost.

2.1.12 Rotating Disc Reactor

A rectangular tank plug flow rotating disc reactor (operating volume ~5 L) with TiO_2 nanowire arrays deposited on a metal Ti disc (two nos.) was developed for the removal of methyl orange [66]. The fixed rotating shaft (1 cm above the water surface) in the middle of the reactor is connected to a motor (1–40 rpm speed). The UV lamps (three

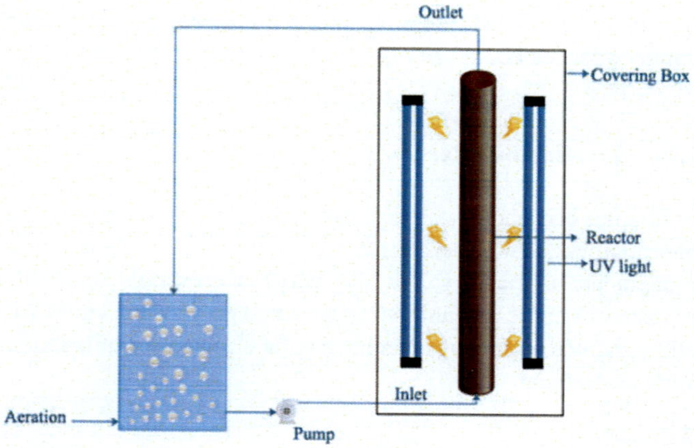

Fig. 13 Multiphase tubular photocatalytic reactor

Fig. 14 Rotating disc photocatalytic reactor

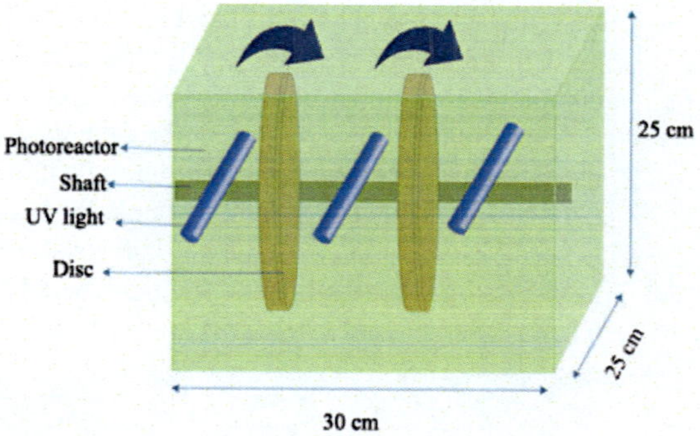

nos.) were attached on the wall of the reactor (as can be seen in Fig. 14). The distance between the photocatalytic disc and lamp was fixed at 5 cm. This reactor showed >97% decolorization efficiency for methyl orange but the degradation percentage was less than half of the decolourization efficiency. The scale-up consideration of this kind of reactor is possible for continuous processes as well, while an efficient catalyst is involved for the complete degradation of the pollutant.

2.1.13 Pebble Bed Reactor

Silica rich white pebbles coated with TiO_2 particles fixed on a plane surface are used as a photocatalytic bed [62] to remove dye contaminated water. This low-cost pebble bed photocatalytic reactor was used under sunlight for the decolourization of several types of textile dye. The results show that the total organic carbon removal of maximum 35%, indicates the requirement of improvement in this regard. This type of reactor can be scaled up easily, but it needs more surface area compared to bubble column or packed bed reactors.

2.1.14 Floating Bed Reactor

The floating bed photocatalytic reactor (Fig. 15) [111] is very similar to the principle of bubble column reactor. In their work, this multiphase reactor was used for the degradation of acid yellow dye which successfully removed >99% dye from the wastewater. The polyaniline-TiO_2 coated polystyrene cubes with optimum TiO_2 loading showed promising outcome but the diffusional limitations encountered need further assessment.

2.1.15 Continuous Submerged Solid Small-Scale Photoreactor

A Continuous Submerged Solid Small-Scale Laboratory Photoreactor (Fig. 16) was designed and used [112] for the degradation of rhodamine B dye. The photocatalyst used in this work was TiO_2/NO_3^- on ceramic substrate under UV light. Almost complete breakdown of the dye was achieved but there were several byproducts formed (toxic/non-toxic) during the operation. This reactor efficiently decolourizes the dye solution but the complete mineralization of the dye can be a better and suitable option to attain environmental sustainability. Also, the scale-up of this type of reactor is possible but for a cost-efficient industrial-scale operation, replacing the UV light to visible light source or solar light can be a better option.

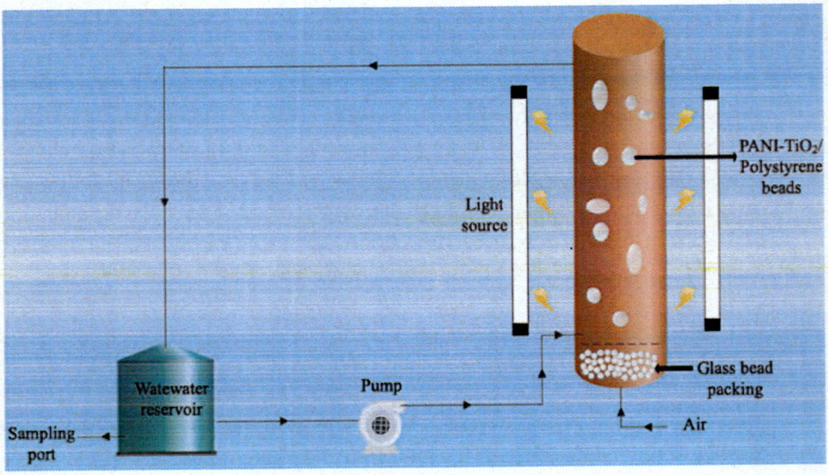

Fig. 15 Floating bed photocatalytic reactor

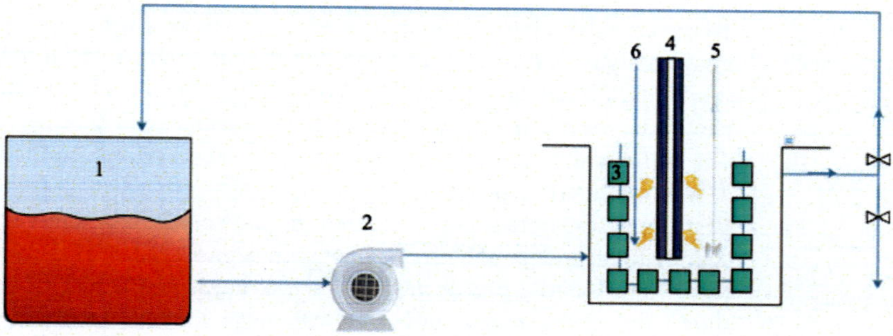

1. Sample tank; 2. Pump; 3. Locus of TiO_2/NO_3-Ceramic plates; 4. UV-Lamp; 5. Stirrer; 6. O_2.

Fig. 16 Continuous submerged solid small-scale photoreactor

There are several studies reported [76, 113–118] on photocatalytic decolourization/degradation of synthetic or real dye wastewater using multiphase reactors but there is very little work focused on scale-up studies. The implementation of lab scale reactors for the real world, large-scale applications is very much needed as photocatalysts are capable of breaking down the dye molecules in a completely eco-friendly way.

3 Difficulties Encountered in Large Scale Photocatalytic Reactors

An efficient design and scale-up of the photocatalytic reactor is still a difficult task. To design a photocatalytic reactor, it is very essential to consider important parameters, such as the kind of light source needed, maximum utilization of light, type of photocatalyst (pure, doped, co-doped, composite, etc.), mode of catalyst utilization (immobilized or suspended), catalyst-pollutant contact time, mixing, reaction kinetics, mass transfer, temperature, pH, and minimum volume of reactor required to achieve maximum efficiency. From the detailed literature analysis presented above, the following common difficulties that a photocatalytic reactor designer faces are listed below.

I. To preserve the same ratio of irradiated surface to total volume [71].
II. Photocatalytic reactor needs to be specifically designed according to the light source used. The choice of a light source very often decides what kind of photocatalytic reactor could be used or vice-versa.
III. The uniform distribution of the light on the catalyst surface is difficult when a large volume is involved.
IV. Higher catalyst surface area is needed for photocatalytic reactors driven by solar light regardless of slurry or immobilized mode.
V. Solar light focused photocatalytic reactors need an efficient catalyst that shows appreciable photocatalytic activity under solar light.
VI. Solar light driven photocatalytic reactors face another problem, i.e., dependence on weather. Hence, interest in research using simulated visible or solar light is growing.
VII. In multiphase reactors, uniform mixing and/or oxidation due to gas or air in the reactor volume is also difficult.
VIII. For a photocatalytic reactor, having a light source outside, a transparent outer shell is needed. Also, the transparent material with low light absorbance property is essential, as most of the photon energy should be transmitted to the catalyst surface instead of being absorbed by the reactor shell. The opacity of the shell, depth of irradiation penetration, and light scattering are key parameters here.
IX. A longer residence time is needed for the recirculating batch process to ensure complete mineralization of the pollutant.
X. In a large-scale slurry reactor, separation of the catalyst after use is challenging (almost impossible), since photocatalysts are mostly nanoparticles. In case, the photocatalyst is coated on the inner surface of the reactor, the catalyst lining has to be replenished after three or four recycles in most of the cases as it loses its effectiveness after continuous use/reuse. Due to this reason, the development of highly efficient photocatalysts is also recommended.
XI. Maintaining the equilibrium temperature throughout the photoreactor.
XII. The strong coupling of physicochemical phenomena can be achieved by transport processes, light absorption, and reaction kinetics in a photocatalytic reaction medium leads to a highly efficient treatment process but it is the major obstacle faced in the development of a photocatalytic reactor.

The photocatalytic reactor scale-up has been severely restricted by the fact that the reactor alignments have not been capable of addressing the crucial variables such as the light distribution in the liquid phase that passes to the catalyst surface, and providing high catalyst coating surface areas per unit reactor volume. The newly developed reactor design concepts should provide a high ratio of immobilized catalyst to irradiated surface and must include optimum amount of active catalyst in contact with the liquid medium to be treated inside the photoreactor.

4 Strategies for Efficient Scale-Up

A few important key points about strategies to overcome the current limitations for the scale-up of multiphase photocatalytic reactors is discussed below.

I. Since small volume reactors perform better compared to larger volumes, it might be a good idea to connect several small reactors in parallel to get the desired outcome for the continuous flow process.
II. It is advised to use direct solar light for catalyst activation to reduce the operating cost of the operation. To utilize the solar light to activate the photocatalyst, it is necessary to modify the catalyst in most of the cases. Also, concentrating the solar light into the catalyst surface will also help to enhance the performance. In an industrial scale reactor, the uniform irradiation of the light is difficult, which can be overcome by using several light sources or using multiple small reactors instead of one.
III. In a large-scale multiphase reactor, homogeneous mixing is another issue. Lack of dead zones and proper mixing can only be achieved by placing multiple gas spargers in appropriate places.
IV. It is also advisable and preferable to use immobilized catalyst instead of slurry. The substrate material for the immobilization needs to be chosen carefully so that it should not affect the photocatalytic degradation efficiency of the pollutants.

5 Future Aspects of Multiphase Photocatalytic Reactors for Dye Wastewater Treatment

The future research must further investigate on exploring the main aspects (high throughput, treatment efficiency, eco-friendliness, energy efficiency, and low operating cost) and obtain reactor concepts (configuration/ efficient catalyst/ solar light source) based on contaminants and discuss the quantum efficiency, photochemical thermodynamic efficiency, and photonic efficiency for different photocatalytic reactors. Future research also should include cost evaluation while fabricating a photocatalytic reactor and optimization of the cost while maintaining the high process quality.

Most of the photocatalytic wastewater treatment process showed capability of complete degradation or removal of contaminants but still there are several aspects that need to be improved for an ideal large-scale photocatalytic operation in future, such as (a) reduction of residence time, (b) reuse of photocatalyst for several times without reducing efficiency, (c) high performance catalyst, (d) affective design of photocatalytic reactor, (f) effective use of solar light, (g) reduce the fabrication cost of photocatalytic reactor, (h) high mass transfer between gas and liquid, (i) continuous flow reactor design, and (j) uniform mixing.

6 Conclusions

In this chapter, different reactor configurations are discussed based on the design and performance. The parameters affecting the dye degradation such as light source, catalyst loading, dye concentration in wastewater, residence time, reactor type, pH, and temperature are also discussed in detail. The difficulties encountered in scale-up of the reactors are also listed. It is observed that the reactor with small volume exhibits the highest yield due to the enhanced mass transfer rate, low resistance, and plug flow behaviour of the photocatalytic reactor. The key points to consider in the design and development of novel photocatalytic multiphase reactors operating at industrial scale are as follows:

- The reactor must be designed for efficient utilization of the abundantly available solar light rather than UV or simulated solar light.
- Consequently, the development of more efficient photocatalysts or boosting the efficiency of popular photocatalysts such as TiO_2 is required.
- Immobilized photocatalytic reactors may be a better choice compared to slurry since the reuse of the catalyst several times is very easy.
- Continuous-flow reactor for the photocatalytic dye degradation is a less explored area that needs much more attention in order to develop an industrial scale photoreactor.
- The use of 3-D printing can be considered in the development of novel reactors.
- The integration of multiphase photocatalytic reactors with other conventional or unconventional treatment technologies, i.e., hybrid technologies should be studied carefully to boost the prospects at industrial scale.
- Besides the evaluation of the net reduction obtained in the total organic carbon (TOC) content, the end-products of the degraded dye wastewaters should be carefully evaluated by analytical techniques such as LC–MS and the degradation pathways enumerated.

References

1. Angel J, Loftus A (2019) With-against-and-beyond the human right to water. Geoforum 98:206–213. https://doi.org/10.1016/j.geoforum.2017.05.002
2. Zhao F, Zhou X, Liu Y et al (2019) Super Moisture-absorbent gels for all-weather atmospheric water harvesting. Adv Mater 31:1806446. https://doi.org/10.1002/adma.201806446
3. Zhang Q, Xu P, Qian H (2020) Groundwater quality assessment using improved Water Quality Index (WQI) and Human Health Risk (HHR) evaluation in a semi-arid region of Northwest China. Expo Heal 12:487–500. https://doi.org/10.1007/s12403-020-00345-w
4. Zhang Z, Shi M, Chen KZ et al (2021) Water scarcity will constrain the formation of a world-class megalopolis in North China. npj Urban Sustain 1:13. https://doi.org/10.1038/s42949-020-00012-8
5. Griffiths JK (2017) Waterborne diseases. In: International encyclopedia of public health. Elsevier, pp 388–401

6. Lellis B, Fávaro-Polonio CZ, Pamphile JA, Polonio JC (2019) Effects of textile dyes on health and the environment and bioremediation potential of living organisms. Biotechnol Res Innov 3:275–290. https://doi.org/10.1016/j.biori.2019.09.001
7. Han Y, Li H, Liu M et al (2016) Purification treatment of dyes wastewater with a novel micro-electrolysis reactor. Sep Purif Technol 170:241–247. https://doi.org/10.1016/j.seppur.2016.06.058
8. Rodriguez Couto S, Rodriguez A, Paterson RRM et al (2006) Laccase activity from the fungus Trametes hirsuta using an air-lift bioreactor. Lett Appl Microbiol 060316073800005. https://doi.org/10.1111/j.1472-765X.2006.01879.x
9. Sridewi N, Lee Y-F, Sudesh K (2011) Simultaneous adsorption and photocatalytic degradation of malachite green using electrospun P(3HB)-TiO_2 nanocomposite fibers and films. Int J Photoenergy 2011:1–11. https://doi.org/10.1155/2011/597854
10. Jo W-K, Park GT, Tayade RJ (2015) Synergetic effect of adsorption on degradation of malachite green dye under blue LED irradiation using spiral-shaped photocatalytic reactor. J Chem Technol Biotechnol 90:2280–2289. https://doi.org/10.1002/jctb.4547
11. Selvakumar S, Manivasagan R, Chinnappan K (2013) Biodegradation and decolourization of textile dye wastewater using Ganoderma lucidum. 3 Biotech 3:71–79. https://doi.org/10.1007/s13205-012-0073-5
12. Rajeshwar K, Osugi ME, Chanmanee W et al (2008) Heterogeneous photocatalytic treatment of organic dyes in air and aqueous media. J Photochem Photobiol C Photochem Rev 9:171–192. https://doi.org/10.1016/j.jphotochemrev.2008.09.001
13. Asghar A, Ramzan N, Jamal BU et al (2020) Low frequency ultrasonic-assisted Fenton oxidation of textile wastewater: process optimization and electrical energy evaluation. Water Environ J 34:523–535. https://doi.org/10.1111/wej.12482
14. Patil S (2019) Synthesis and optical properties of Near-Infrared (NIR) absorbing azo dyes. Curr Trends Fash Technol Text Eng 4. https://doi.org/10.19080/CTFTTE.2019.04.555649
15. Gregory P (2003) Metal complexes as speciality dyes and pigments. In: Comprehensive coordination chemistry II. Elsevier, pp 549–579
16. Marchis T, Avetta P, Bianco-Prevot A et al (2011) Oxidative degradation of Remazol Turquoise Blue G 133 by soybean peroxidase. J Inorg Biochem 105:321–327. https://doi.org/10.1016/j.jinorgbio.2010.11.009
17. Riaz M, Ijaz B, Riaz A, Amjad M (2018) Improvement of waste water quality by application of mixed algal inocula. Bangladesh J Sci Ind Res 53:77–82. https://doi.org/10.3329/bjsir.v53i1.35913
18. Javaid R, Qazi UY (2019) Catalytic oxidation process for the degradation of synthetic dyes: an overview. Int J Environ Res Public Health 16:2066. https://doi.org/10.3390/ijerph16112066
19. Hussain SN, Asghar HMA, Sattar H et al (2015) Free chlorine formation during electrochemical regeneration of a graphite intercalation compound adsorbent used for wastewater treatment. J Appl Electrochem 45:611–621. https://doi.org/10.1007/s10800-015-0814-3
20. Pagga U, Brown D (1986) The degradation of dyestuffs: Part II behaviour of dyestuffs in aerobic biodegradation tests. Chemosphere 15:479–491. https://doi.org/10.1016/0045-6535(86)90542-4
21. Pang YL, Abdullah AZ (2013) Current status of textile industry wastewater management and research progress in Malaysia: a review. Clean Soil Air Water 41:751–764. https://doi.org/10.1002/clen.201000318
22. Senthil Kumar P, Saravanan A (2017) Sustainable wastewater treatments in textile sector. In: Sustainable fibres and textiles. Elsevier, pp 323–346
23. Shang N-C, Chen Y-H, Yang Y-P et al (2006) Ozonation of dyes and textile wastewater in a rotating packed bed. J Environ Sci Heal Part A 41:2299–2310. https://doi.org/10.1080/10934520600873043
24. Vacchi FI, Albuquerque AF, Vendemiatti JA et al (2013) Chlorine disinfection of dye wastewater: implications for a commercial azo dye mixture. Sci Total Environ 442:302–309. https://doi.org/10.1016/j.scitotenv.2012.10.019

25. Abid MF, Zablouk MA, Abid-Alameer AM (2012) Experimental study of dye removal from industrial wastewater by membrane technologies of reverse osmosis and nanofiltration. Iranian J Environ Health Sci Eng 9:17. https://doi.org/10.1186/1735-2746-9-17
26. Hassan MM, Carr CM (2018) A critical review on recent advancements of the removal of reactive dyes from dyehouse effluent by ion-exchange adsorbents. Chemosphere 209:201–219. https://doi.org/10.1016/j.chemosphere.2018.06.043
27. Jiang M, Ye K, Deng J et al (2018) Conventional ultrafiltration as effective strategy for dye/salt fractionation in textile wastewater treatment. Environ Sci Technol 52:10698–10708. https://doi.org/10.1021/acs.est.8b02984
28. Kandisa RV, Saibaba KV N (2016) Dye removal by adsorption: a review. J Bioremediation Biodegrad 07. https://doi.org/10.4172/2155-6199.1000371
29. Buthiyappan A, Abdul Aziz AR, Wan Daud WMA (2016) Recent advances and prospects of catalytic advanced oxidation process in treating textile effluents. Rev Chem Eng 32:1–47. https://doi.org/10.1515/revce-2015-0034
30. Arslan İ, Akmehmet Balcioğlu I, Tuhkanen T (1999) Oxidative treatment of simulated dyehouse effluent by UV and near-UV light assisted Fenton's reagent. Chemosphere 39:2767–2783. https://doi.org/10.1016/S0045-6535(99)00211-8
31. Ghaly MY, Farah JY, Fathy AM (2007) Enhancement of decolorization rate and COD removal from dyes containing wastewater by the addition of hydrogen peroxide under solar photocatalytic oxidation. Desalination 217:74–84. https://doi.org/10.1016/j.desal.2007.01.013
32. Sarto S, Paesal P, Tanyong IB et al (2019) Catalytic degradation of textile wastewater effluent by peroxide oxidation assisted by UV light irradiation. Catalysts 9:509. https://doi.org/10.3390/catal9060509
33. Al-Mamun MR, Kader S, Islam MS, Khan MZH (2019) Photocatalytic activity improvement and application of UV-TiO_2 photocatalysis in textile wastewater treatment: a review. J Environ Chem Eng 7:103248. https://doi.org/10.1016/j.jece.2019.103248
34. Chimupala Y, Phromma C, Yimklan S et al (2020) Dye wastewater treatment enabled by piezo-enhanced photocatalysis of single-component ZnO nanoparticles. RSC Adv 10:28567–28575. https://doi.org/10.1039/D0RA04746E
35. Singh S, Mahalingam H, Singh PK (2013) Polymer-supported titanium dioxide photocatalysts for environmental remediation: a review. Appl Catal A Gen 462–463:178–195. https://doi.org/10.1016/j.apcata.2013.04.039
36. Zhang J, Tian B, Wang L et al (2018) Mechanism of photocatalysis, pp 1–15
37. Ashar A, Bhatti IA, Ashraf M et al (2020) Fe^{3+} @ ZnO/polyester based solar photocatalytic membrane reactor for abatement of RB5 dye. J Clean Prod 246:119010. https://doi.org/10.1016/j.jclepro.2019.119010
38. Pelizzetti E, Serpone N (eds) (1986) Homogeneous and heterogeneous photocatalysis. Springer, Netherlands, Dordrecht
39. Loddo V, Bellardita M, Camera-Roda G et al (2018) Heterogeneous photocatalysis. In: Current trends and future developments on (bio-) membranes. Elsevier, pp 1–43
40. Tan HL, Abdi FF, Ng YH (2019) Heterogeneous photocatalysts: an overview of classic and modern approaches for optical, electronic, and charge dynamics evaluation. Chem Soc Rev 48:1255–1271. https://doi.org/10.1039/C8CS00882E
41. Vaiano V, Sannino D, Sacco O (2020) Heterogeneous photocatalysis. In: Nanomaterials for the detection and removal of wastewater pollutants. Elsevier, pp 285–301
42. Nakata K, Fujishima A (2012) TiO_2 photocatalysis: design and applications. J Photochem Photobiol C Photochem Rev 13:169–189. https://doi.org/10.1016/j.jphotochemrev.2012.06.001
43. Saeed K, Khan I, Gul T, Sadiq M (2017) Efficient photodegradation of methyl violet dye using TiO_2/Pt and TiO_2/Pd photocatalysts. Appl Water Sci 7:3841–3848. https://doi.org/10.1007/s13201-017-0535-3
44. Tran Thi Thuong H, Tran Thi Kim C, Nguyen Quang L, Kosslick H (2019) Highly active brookite TiO_2-assisted photocatalytic degradation of dyes under the simulated solar–UVA radiation. Prog Nat Sci Mater Int 29:641–647. https://doi.org/10.1016/j.pnsc.2019.10.001

45. Molinari R, Lavorato C, Argurio P (2020) Visible-light photocatalysts and their perspectives for building photocatalytic membrane reactors for various liquid phase chemical conversions. Catalysts 10:1334. https://doi.org/10.3390/catal10111334
46. Ohtani B (2011) Photocatalysis by inorganic solid materials, pp 395–430
47. Cates EL (2017) Photocatalytic water treatment: so where are we going with this? Environ Sci Technol 51:757–758. https://doi.org/10.1021/acs.est.6b06035
48. Chong MN, Jin B, Chow CWKK, Saint C (2010) Recent developments in photocatalytic water treatment technology: a review. Water Res 44:2997–3027. https://doi.org/10.1016/j.watres.2010.02.039
49. Ahmed SN, Haider W (2018) Heterogeneous photocatalysis and its potential applications in water and wastewater treatment: a review. Nanotechnology 29:342001. https://doi.org/10.1088/1361-6528/aac6ea
50. Yashni G, Al-Gheethi A, Mohamed R, Al-Sahari M (2021) Reusability performance of green zinc oxide nanoparticles for photocatalysis of bathroom greywater. Water Pract Technol 16:364–376. https://doi.org/10.2166/wpt.2020.118
51. Zhou L, Zhang H, Sun H et al (2016) Recent advances in non-metal modification of graphitic carbon nitride for photocatalysis: a historic review. Catal Sci Technol 6:7002–7023. https://doi.org/10.1039/C6CY01195K
52. Serpone N (1997) Relative photonic efficiencies and quantum yields in heterogeneous photocatalysis. J Photochem Photobiol A Chem 104:1–12. https://doi.org/10.1016/S1010-6030(96)04538-8
53. Krishna R, Sie ST (1994) Strategies for multiphase reactor selection. Chem Eng Sci 49:4029–4065. https://doi.org/10.1016/S0009-2509(05)80005-3
54. Peschel A, Hentschel B, Freund H, Sundmacher K (2012) Design of optimal multiphase reactors exemplified on the hydroformylation of long chain alkenes. Chem Eng J 188:126–141. https://doi.org/10.1016/j.cej.2012.01.123
55. Pangarkar VG (2014) Multiphase reactors. Design of multiphase reactors. Wiley, Hoboken, NJ, pp 47–86
56. Alalm MG, Djellabi R, Meroni D et al (2021) Toward scaling-up photocatalytic process for multiphase environmental applications. Catalysts 11:562. https://doi.org/10.3390/catal11050562
57. Teekateerawej S, Nishino J, Nosaka Y (2006) TiO_2 photocatalytic micro-channel reactors using capillary plates. Adv Mater Res 11–12:303–306. https://doi.org/10.4028/www.scientific.net/AMR.11-12.303
58. Das S, Mahalingam H (2019) Exploring the synergistic interactions of TiO_2, rGO, and g-C_3N_4 catalyst admixtures in a polystyrene nanocomposite photocatalytic film for wastewater treatment: unary, binary and ternary systems. J Environ Chem Eng 7. https://doi.org/10.1016/j.jece.2019.103246
59. Liu L, Liu Z, Bai H, Sun DD (2012) Concurrent filtration and solar photocatalytic disinfection/degradation using high-performance Ag/TiO_2 nanofiber membrane. Water Res 46:1101–1112. https://doi.org/10.1016/j.watres.2011.12.009
60. Miranda-Garcia N, Suarez S, Sanchez B et al (2011) Photocatalytic degradation of emerging contaminants in municipal wastewater treatment plant effluents using immobilized TiO_2 in a solar pilot plant. Appl Catal B Environ 103:294–301. https://doi.org/10.1016/j.apcatb.2011.01.030
61. Ehrampoush MH, Moussavi GR, Ghaneian MT et al (2011) Removal of methylene blue dye from textile simulated sample using tubular reactor and TiO_2/UV-C photocatalytic process. Iran J Environ Heal Sci Eng 8:35–40
62. Rao NN, Chaturvedi V, Li Puma G (2012) Novel pebble bed photocatalytic reactor for solar treatment of textile wastewater. Chem Eng J 184:90–97. https://doi.org/10.1016/j.cej.2012.01.004
63. Baghbani Ghatar S, Allahyari S, Rahemi N, Tasbihi M (2018) Response surface methodology optimization for photodegradation of methylene blue in a ZnO coated flat plate continuous photoreactor. Int J Chem React Eng 16. https://doi.org/10.1515/ijcre-2017-0221

64. Sutisna RM, Wibowo E et al (2017) Novel solar photocatalytic reactor for wastewater treatment. IOP Conf Ser Mater Sci Eng 214:012010. https://doi.org/10.1088/1757-899X/214/1/012010
65. Sacco O, Sannino D, Vaiano V (2019) Packed bed photoreactor for the removal of water pollutants using visible light emitting diodes. Appl Sci 9:472. https://doi.org/10.3390/app9030472
66. Li F, Szeto W, Huang H et al (2017) A Photocatalytic rotating disc reactor with TiO_2 nanowire arrays deposited for industrial wastewater treatment. Molecules 22:337. https://doi.org/10.3390/molecules22020337
67. Zhang Z, Wu H, Yuan Y et al (2012) Development of a novel capillary array photocatalytic reactor and application for degradation of azo dye. Chem Eng J 184:9–15. https://doi.org/10.1016/j.cej.2011.02.057
68. Vaiano V, Sacco O, Pisano D et al (2015) From the design to the development of a continuous fixed bed photoreactor for photocatalytic degradation of organic pollutants in wastewater. Chem Eng Sci 137:152–160. https://doi.org/10.1016/j.ces.2015.06.023
69. Di Capua G, Femia N, Migliaro M et al (2017) Intensification of a flat-plate photocatalytic reactor performances by innovative visible light modulation techniques: A proof of concept. Chem Eng Process Process Intensif 118:117–123. https://doi.org/10.1016/j.cep.2017.05.004
70. Claes T, Dilissen A, Leblebici ME, Van Gerven T (2019) Translucent packed bed structures for high throughput photocatalytic reactors. Chem Eng J 361:725–735. https://doi.org/10.1016/j.cej.2018.12.107
71. Jiang H, Zhang G, Huang T et al (2010) Photocatalytic membrane reactor for degradation of acid red B wastewater. Chem Eng J 156:571–577. https://doi.org/10.1016/j.cej.2009.04.011
72. Sacco O, Vaiano V, Sannino D (2020) Main parameters influencing the design of photocatalytic reactors for wastewater treatment: a mini review. J Chem Technol Biotechnol jctb.6488. https://doi.org/10.1002/jctb.6488
73. Braham RJ, Harris AT (2009) Review of Major Design and scale-up considerations for solar photocatalytic reactors. Ind Eng Chem Res 48:8890–8905. https://doi.org/10.1021/ie900859z
74. Lindstrom H, Wootton R, Iles A (2007) High surface area titania photocatalytic microfluidic reactors. AIChE J 53:695–702. https://doi.org/10.1002/aic.11096
75. Regmi C, Lotfi S, Espíndola JC et al (2020) Comparison of photocatalytic membrane reactor types for the degradation of an organic molecule by TiO_2-coated PES membrane. Catalysts 10:725. https://doi.org/10.3390/catal10070725
76. Ling CM, Mohamed AR, Bhatia S (2004) Performance of photocatalytic reactors using immobilized TiO_2 film for the degradation of phenol and methylene blue dye present in water stream. Chemosphere 57:547–554. https://doi.org/10.1016/j.chemosphere.2004.07.011
77. Manassero A, Satuf ML, Alfano OM (2017) Photocatalytic reactors with suspended and immobilized TiO_2: comparative efficiency evaluation. Chem Eng J 326:29–36. https://doi.org/10.1016/j.cej.2017.05.087
78. Adams M, Campbell I, McCullagh C et al (2013) From ideal reactor concepts to reality: the novel drum reactor for photocatalytic wastewater treatment. Int J Chem React Eng 11:621–632. https://doi.org/10.1515/ijcre-2012-0012
79. Vaiano V, Sacco O, Stoller M et al (2014) Influence of the photoreactor configuration and of different light sources in the photocatalytic treatment of highly polluted wastewater. Int J Chem React Eng 12:63–75. https://doi.org/10.1515/ijcre-2013-0090
80. Boyjoo Y, Ang M, Pareek V (2013) Light intensity distribution in multi-lamp photocatalytic reactors. Chem Eng Sci 93:11–21. https://doi.org/10.1016/j.ces.2012.12.045
81. Pareek V, Chong S, Tadé M, Adesina AA (2008) Light intensity distribution in heterogenous photocatalytic reactors. Asia-Pacific J Chem Eng 3:171–201. https://doi.org/10.1002/apj.129
82. Amano F, Nogami K, Tanaka M, Ohtani B (2010) Correlation between surface area and photocatalytic activity for acetaldehyde decomposition over bismuth tungstate particles with a hierarchical structure. Langmuir 26:7174–7180. https://doi.org/10.1021/la904274c
83. Mazinani B, Masrom AK, Beitollahi A, Luque R (2014) Photocatalytic activity, surface area and phase modification of mesoporous SiO2–TiO_2 prepared by a one-step hydrothermal procedure. Ceram Int 40:11525–11532. https://doi.org/10.1016/j.ceramint.2014.03.071

84. Li Q, Zhang N, Yang Y et al (2014) High Efficiency photocatalysis for pollutant degradation with MoS 2 /C 3 N 4 heterostructures. Langmuir 30:8965–8972. https://doi.org/10.1021/la502033t
85. Lv P, Xu C, Peng B (2020) Design of a silicon photocatalyst for high-efficiency photocatalytic water splitting. ACS Omega 5:6358–6365. https://doi.org/10.1021/acsomega.9b03755
86. Rajamanickam D, Shanthi M (2016) Photocatalytic degradation of an organic pollutant by zinc oxide—solar process. Arab J Chem 9:S1858–S1868. https://doi.org/10.1016/j.arabjc.2012.05.006
87. Santhosh C, Malathi A, Daneshvar E et al (2018) Photocatalytic degradation of toxic aquatic pollutants by novel magnetic 3D-TiO_2@HPGA nanocomposite. Sci Rep 8:15531. https://doi.org/10.1038/s41598-018-33818-9
88. Fathinia M, Khataee AR (2013) Residence time distribution analysis and optimization of photocatalysis of phenazopyridine using immobilized TiO_2 nanoparticles in a rectangular photoreactor. J Ind Eng Chem 19:1525–1534. https://doi.org/10.1016/j.jiec.2013.01.019
89. Visan A, van Ommen JR, Kreutzer MT, Lammertink RGH (2019) Photocatalytic reactor design: guidelines for kinetic investigation. Ind Eng Chem Res 58:5349–5357. https://doi.org/10.1021/acs.iecr.9b00381
90. Moon J, Lee K, Kim S (2015) A study of the temperature dependency for photocatalytic VOC degradation chamber test under UVLED irradiations. Korean Chem Eng Res 53:755–761. https://doi.org/10.9713/kcer.2015.53.6.755
91. Serpone N (2018) Heterogeneous photocatalysis and prospects of TiO_2-based photocatalytic DeNOxing the atmospheric environment. Catalysts 8:553. https://doi.org/10.3390/catal8110553
92. Fanourakis SK, Peña-Bahamonde J, Bandara PC, Rodrigues DF (2020) Nano-based adsorbent and photocatalyst use for pharmaceutical contaminant removal during indirect potable water reuse. npj Clean Water 3:1. https://doi.org/10.1038/s41545-019-0048-8
93. Lavand AB, Malghe YS (2015) Synthesis, characterization, and visible light photocatalytic activity of nanosized carbon doped zinc oxide. Int J Photochem 2015:1–9. https://doi.org/10.1155/2015/790153
94. Cho Y, Yamaguchi A, Uehara R et al (2020) Temperature dependence on bandgap of semiconductor photocatalysts. J Chem Phys 152:231101. https://doi.org/10.1063/5.0012330
95. Meng F, Liu Y, Wang J et al (2018) Temperature dependent photocatalysis of g-C_3N_4, TiO_2 and ZnO: differences in photoactive mechanism. J Colloid Interface Sci 532:321–330. https://doi.org/10.1016/j.jcis.2018.07.131
96. Porley V, Robertson N (2020) Substrate and support materials for photocatalysis. In: Nanostructured photocatalysts. Elsevier, pp 129–171
97. Lopez L, Daoud WA, Dutta D et al (2013) Effect of substrate on surface morphology and photocatalysis of large-scale TiO_2 films. Appl Surf Sci 265:162–168. https://doi.org/10.1016/j.apsusc.2012.10.156
98. Yang Z, Liu M, Wang X (2018) Experiment study and modeling of novel mini-bubble column photocatalytic reactor with multiple micro-bubbles. Chem Eng Process Process Intensif 124:269–281. https://doi.org/10.1016/j.cep.2018.01.019
99. Bessegato GG, Cardoso JC, Zanoni MVB (2015) Enhanced photoelectrocatalytic degradation of an acid dye with boron-doped TiO_2 nanotube anodes. Catal Today 240:100–106. https://doi.org/10.1016/j.cattod.2014.03.073
100. Geng Q, Cui W (2010) Adsorption and photocatalytic degradation of reactive brilliant red K-2BP by TiO_2/AC in bubbling fluidized bed photocatalytic reactor. Ind Eng Chem Res 49(22):11321–11330. https://doi.org/10.1021/ie101533x
101. Das S, Mahalingam H (2019) Dye degradation studies using immobilized pristine and waste polystyrene-TiO_2/rGO/g-C_3N_4 nanocomposite photocatalytic film in a novel airlift reactor under solar light. J Environ Chem Eng 7:103289. https://doi.org/10.1016/j.jece.2019.103289
102. Das S, Mahalingam H (2020) Novel immobilized ternary photocatalytic polymer film based airlift reactor for efficient degradation of complex phthalocyanine dye wastewater. J Hazard Mater 383. https://doi.org/10.1016/j.jhazmat.2019.121219

103. Desa AL, Hairom NHH, Sidik DAB et al (2019) A comparative study of ZnO-PVP and ZnO-PEG nanoparticles activity in membrane photocatalytic reactor (MPR) for industrial dye wastewater treatment under different membranes. J Environ Chem Eng 7:103143. https://doi.org/10.1016/j.jece.2019.103143
104. Esquivel K, Arriaga LG, Rodríguez FJ et al (2009) Development of a TiO_2 modified optical fiber electrode and its incorporation into a photoelectrochemical reactor for wastewater treatment. Water Res 43:3593–3603. https://doi.org/10.1016/j.watres.2009.05.035
105. Kim S, Kim M, Lim SK, Park Y (2017) Titania-coated plastic optical fiber fabrics for remote photocatalytic degradation of aqueous pollutants. J Environ Chem Eng 5:1899–1905. https://doi.org/10.1016/j.jece.2017.03.036
106. Gallo JC, Mariano MB, Lucanas AD et al (2015) Photocatalytic degradation of turquoise blue dye using immobilized AC/TiO_2: optimization of process parameters and pilot plant investigation. J Eng Sci Technol 10:64–73
107. Das S, Mahalingam H (2019) Reusable floating polymer nanocomposite photocatalyst for the efficient treatment of dye wastewaters under scaled-up conditions in batch and recirculation modes. J Chem Technol Biotechnol 94:2597–2608. https://doi.org/10.1002/jctb.6069
108. Meng Z, Zhang X, Qin J (2013) A high efficiency microfluidic-based photocatalytic microreactor using electrospun nanofibrous TiO_2 as a photocatalyst. Nanoscale 5:4687. https://doi.org/10.1039/c3nr00775h
109. He Z, Li Y, Zhang Q, Wang H (2010) Capillary microchannel-based microreactors with highly durable ZnO/TiO_2 nanorod arrays for rapid, high efficiency and continuous-flow photocatalysis. Appl Catal B Environ 93:376–382. https://doi.org/10.1016/j.apcatb.2009.10.011
110. Katayama K, Takeda Y, Kuwabara K, Kuwahara S (2012) A novel photocatalytic microreactor bundle that does not require an electric power source. Chem Commun 48:7368. https://doi.org/10.1039/c2cc33525e
111. Nair VR, Shetty Kodialbail V (2020) Floating bed reactor for visible light induced photocatalytic degradation of Acid Yellow 17 using polyaniline-TiO_2 nanocomposites immobilized on polystyrene cubes. Environ Sci Pollut Res 27:14441–14453. https://doi.org/10.1007/s11356-020-07959-2
112. Neolaka YAB, Ngara ZS, Lawa Y et al (2019) Simple design and preliminary evaluation of continuous submerged solid small-scale laboratory photoreactor (CS4PR) using TiO_2/NO_3-@TC for dye degradation. J Environ Chem Eng 7:103482. https://doi.org/10.1016/j.jece.2019.103482
113. Akram T, Ahmad N, Sheikh I (2016) Photocatalytic degradation of synthetic textile effluent by modified sol-gel, synthesized mobilized and immobilized TiO2, and Ag-doped TiO_2. Polish J Environ Stud 25:1391–1402. https://doi.org/10.15244/pjoes/62102
114. Hamal DB, Haggstrom JA, Marchin GL, et al (2010) A multifunctional biocide/sporocide and photocatalyst based on titanium dioxide (TiO_2) codoped with silver, carbon, and sulfur. Langmuir 26. https://doi.org/10.1021/la902844r
115. Khenniche L, Favier L, Bouzaza A et al (2015) Photocatalytic degradation of bezacryl yellow in batch reactors—feasibility of the combination of photocatalysis and a biological treatment. Environ Technol 36:1–10. https://doi.org/10.1080/09593330.2014.934740
116. Mohammed Redha Z, Abdulla Yusuf H, Amin R, Bououdina M (2020) The study of photocatalytic degradation of a commercial azo reactive dye in a simple design reusable miniaturized reactor with interchangeable TiO_2 nanofilm. Arab J Basic Appl Sci 27:287–298. https://doi.org/10.1080/25765299.2020.1800163
117. Mosleh S, Rahimi MR, Ghaedi M et al (2016) Photocatalytic degradation of binary mixture of toxic dyes by HKUST-1 MOF and HKUST-1-SBA-15 in a rotating packed bed reactor under blue LED illumination: central composite design optimization. RSC Adv 6:17204–17214. https://doi.org/10.1039/C5RA24564H
118. Sauer T, Cesconeto Neto G, José H, Moreira RFP (2002) Kinetics of photocatalytic degradation of reactive dyes in a TiO_2 slurry reactor. J Photochem Photobiol A Chem 149:147–154. https://doi.org/10.1016/S1010-6030(02)00015-1

Enhanced Methylene Blue Degradation onto Fenton-Like Catalysts Based on g-C_3N_4-$MgFe_2O_4$ Composites

Andrei Ivanets, Vladimir Prozorovich, and Valentin Sarkisov

Abstract The development of new effective Fenton-like catalysts is of interest for solving a wide range of problems related to toxic organic pollutants' destruction in aqueous media. In this chapter, an attempt was made to obtain g-C_3N_4-$MgFe_2O_4$ composites of various structures and morphologies, as well as to justify their effectiveness as Fenton-like catalysts. The prepared composites were characterized by XRD, FTIR, and SEM.EDX methods. It was shown that all composites were characterized by the formation of g-C_3N_4 with the s-heptazine structure and a different ratio of g-C_3N_4 and $MgFe_2O_4$ on the surface. The catalytic properties of g-C_3N_4-$MgFe_2O_4$ composites in the degradation reaction of the thiazine dye Methylene Blue under various conditions (dark-, visible-, and UV-driven processes), as well as under multiply catalytic cycles, were studied. The most effective sample of composite I under UV irradiation provided 99% Methylene Blue degradation efficiency for 20 min at four cycles. The mechanism of catalytic destruction of Methylene Blue mainly due to the formation of hydroxyl radicals in the reaction mixture was proposed.

Keywords Carbon nitride · Magnesium ferrite · Nanostructured composites · Heterogeneous fenton catalysts · Advanced oxidation processes · Methylene blue degradation

1 Introduction

AOP (*Advanced Oxidation Processes*) are methods of water purification that allow the mineralization of toxic organic pollutants under the action of reactive oxygen-containing spices *ROS* (HO•, O_2•−, HO_2•, etc.) to the formation of H_2O, CO_2, N_2, and other inorganic compounds [10]. ROS is obtained by combining various conditions: O_3/catalyst, O_3/UV, O_3/H_2O_2, O_3/H_2O_2/UV, H_2O_2/UV, H_2O_2/Fe^{2+}, H_2O_2/Fe^{2+}/UV,

A. Ivanets (✉) · V. Prozorovich · V. Sarkisov
Institute of General and Inorganic Chemistry, National Academy of Sciences of Belarus, st. Surganova 9/1, 220072 Minsk, Belarus
e-mail: Andreiivanets@yandex.ru; ivanets@igic.bas-net.by

S. S. Muthu and A. Khadir (eds.), *Advanced Oxidation Processes in Dye-Containing Wastewater*, Sustainable Textiles: Production, Processing, Manufacturing & Chemistry,
https://doi.org/10.1007/978-981-19-0987-0_11

TiO_2/UV, etc. [3]. As a result of the interaction of H_2O_2 and Fe^{2+}, also known as the Fenton process, extremely active hydroxyl radicals HO• are formed, capable of non-selectively mineralizing various organic pollutants [22].

Despite the high efficiency and simplicity of experimental execution, the use of the Fenton process in water treatment is limited due to the high consumption of H_2O_2 and Fe^{2+}, a narrow pH range of ~3.0 and the formation of $Fe(OH)_3$ precipitate at pH > 3.0. To maintain the pH ~3.0 of the treated waters, their constant acidification with acid solutions (usually H_2SO_4) is required, which leads to the formation of undesirable acidic wastewater [22]. The use of a heterogeneous iron-containing catalyst makes it possible to reduce the release of Fe^{2+} ions into the solution and, as a result, avoid these problems. Thus, the interaction between H_2O_2 and Fe^{2+} will occur not in solution, but on the surface of the catalyst. This modification of the Fenton process was called a heterogeneous Fenton-like process. Fe_3O_4, α-Fe_2O_3, γ-Fe_2O_3, α-FeOOH, β-FeOOH, γ-FeOOH, FeS_2, and other semiconductor composites are usually used as heterogeneous Fenton-like catalysts [19].

Carbon nitride with a graphite-like structure (g-C_3N_4) belongs to n-type semiconductors and, due to its unique electronic structure, attracts special attention for the preparation of photocatalytic systems. With an average band gap (~2.7 eV), carbon nitride can absorb light with a wavelength of more than 450 nm, which accounts for most of the solar spectrum. Unlike conventional organic semiconductor analogs, g-C_3N_4 is characterized by high thermal (up to 600 °C) and chemical stability (aqueous solutions of acids and alkalis, organic solvents, etc.). However, the photocatalytic efficiency of g-C_3N_4 is strongly limited by its low quantum efficiency, low surface area, and high charge carrier recombination rate [11, 16, 21].

Previously, photocatalysts based on g-C_3N_4 were synthesized by thermal condensation of melamine. The effect of pretreatment of melamine with acetic acid on the physicochemical properties of g-C_3N_4 was studied [7]. It was found that the use of melamine pretreated with acetic acid led to a slight shift in the peaks of thermal transformations and had practically no effect on the crystal and chemical structure of g-C_3N_4. At the same time, significant differences in the porous structure of the samples were revealed, which led to the formation of a larger-pored carbon nitride sample. It was shown that the modified g-C_3N_4 sample was characterized by higher activity in the reaction of photocatalytic destruction of Rhodamine B, which was due to its large-pore structure and more efficient adsorption of dye molecules.

The analysis of scientific publications showed that a significant number of works are devoted to the development of high-performance photocatalysts based on g-C_3N_4 doped with nonmetal heteroatoms (P, S, N, etc.), as well as metal nanoparticles (Pt, Co, Cu, etc.) and other semiconductors (TiO_2, CdS, etc.) [4, 13, 14]. Composites based on metal ferrites and carbon nitride are actively studied in the reactions of water splitting and the catalytic destruction of organic pollutants. At the same time, the crystal structure, structure, and morphology of these semiconductor composites largely determine their catalytic activity [1, 5].

The work aimed to establish the regularities of the production of magnesium ferrite and carbon nitride (g-C_3N_4-$MgFe_2O_4$) composites depending on the synthesis conditions for the preparation of heterogeneous Fenton and photo-Fenton catalysts for organic pollutants destruction.

2 Synthesis and Properties of $g\text{-}C_3N_4\text{-}MgFe_2O_4$ Composites

Composite samples were obtained by the following methods: synthesis of magnesium ferrite by a modified sol-gel method [8] and its subsequent mixing with melamine and heat treatment of the mixture at 300 °C for 5 h (composite I), synthesis of carbon nitride was carried out according to the method [7], followed by introduction into the glycine-nitrate mixture at the combustion initiation stage and heat treatment at 300 °C for 5 h (composite II); by mixing melamine and glycine-nitrate mixture with subsequent standard operations of the method for obtaining magnesium ferrite [8] (composite III). The calculated weight ratio of $MgFe_2O_4$:g-C_3N_4 in all composites was 1:1.

Iron nitrate $Fe(NO_3)_3$, magnesium nitrate $Mg(NO_3)_2$, melamine $C_3H_6N_6$, and glycine NH_2CH_2COOH (Five oceans Ltd., Belarus) without additional purification were used for the synthesis of magnesium ferrite and carbon nitride composites.

2.1 XRD

X-ray phase analysis (XRD) was performed on a D8 ADVANCE diffractometer (Bruker, Germany) with CuKα radiation in the range $2\Theta = 10\text{–}70°$. Phase identification was performed using a set of interplanar distances (d) using the ICDD PDF-2 database. Calculation of the unit cell parameter a and estimation of the crystallite sizes by the Scherer Eq. (1):

$$d = \frac{K\lambda}{\beta \cos \Theta}, \quad (1)$$

where d is the average crystal size (nm), K is the Scherrer constant, λ is the X-ray wavelength (nm), β is the half-height reflection width, and Θ is the diffraction angle.

Figure 1a presents X-ray patterns of the g-C_3N_4-$MgFe_2O_4$ composites I obtained by mixing magnesium ferrite nanoparticles with melamine and subsequent heat treatment of this mixture at 300 °C.

The narrow and intense diffraction peaks belong to the individual crystal phases characteristic of magnesium ferrite with a cubic spinel lattice of 28.8° (220), 55.4° (422) [8], and graphite-like carbon nitride of 17.7° (100), and 26.2° (002) [16]. The characteristic peaks of g-C_3N_4 are slightly shifted to the region of smaller angles, which indicates a distortion of the crystal lattice and an increase in the interplane distance due to intercalation by magnesium ferrite particles. The calculated value of the parameter a of the crystal lattice of magnesium ferrite was 8.324 Å, which significantly differs from the reference value of 8.370 Å and indirectly indicates the formation of a composite I based on the crystalline phases g-C_3N_4 and $MgFe_2O_4$.

The X-ray patterns of the g-C_3N_4-$MgFe_2O_4$ composite II (Fig. 1b) obtained by introducing carbon nitride into a glycine-nitrate gel at the stage of magnesium ferrite

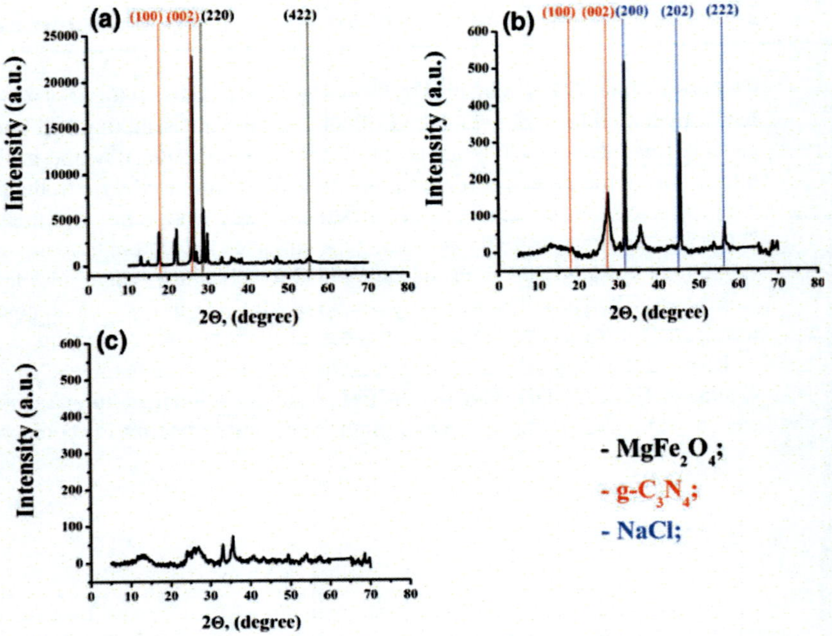

Fig. 1 XRD patterns of **a** composite I, **b** composite II, and **c** composite III g-C_3N_4-$MgFe_2O_4$

synthesis shown peaks characteristic of the graphite-like carbon nitride g-C_3N_4 phase with characteristic peaks for the triclinic crystal lattice of 2Θ 18.1°, 27.4°, and 56.5° belonging to the planes (100), (002), and (222), respectively. The reflexes of NaCl galit with characteristic peaks of 2Θ 31.7° and 45.5° were due to the use as an inert additive and incomplete washing in the preparation of the g-C_3N_4-$MgFe_2O_4$ composites. From the X-ray patterns of the composite II (Fig. 1b), obtained by the introduction of melamine into a glycine-nitrate gel, it can be seen that this composite was X-ray amorphous with a blurred halo in the region of 2Θ 20–40° with the absence of clearly identifiable peaks.

2.2 FTIR

Additional information about the structure of the obtained composites, including X-ray amorphous ones, is provided by the interpretation of the results of the IR spectra (Fig. 2). The IR spectra of the composites were recorded on an IR spectrometer with a Tenzor-27 Fourier converter in the frequency range of 400–4000 cm^{-1}, and the sample weight of 2.0 mg was compressed into tablets with 800 mg KBr. The scanning speed was 10–20 cm^{-1}/min, and the spectral width of the gap in the entire range did not exceed 3 cm^{-1}.

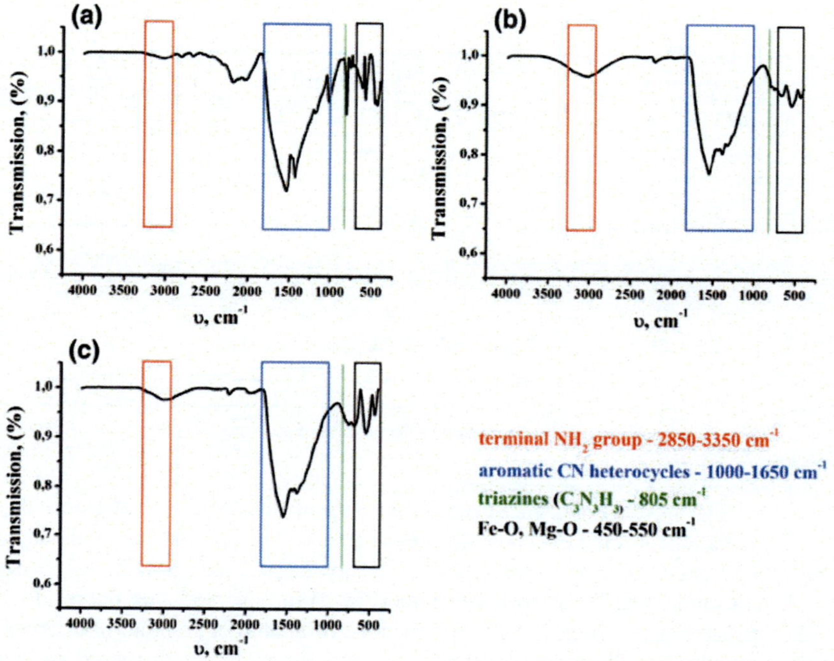

Fig. 2 FTIR spectra of **a** composite I, **b** composite II, and **c** composite III g-C_3N_4-$MgFe_2O_4$

In all the spectra, the most intense absorption bands appeared in the region of 1500–1400 cm^{-1}, which can be attributed to the typical fluctuations of C=N bonds characteristic of the conjugated aromatic structure of s-heptazine. The absorption bands of significantly lower intensity in the region of 800–810 cm^{-1} belong to triazine. This indicates the predominant formation of melem (2) in the process of melamine condensation (1). Also, the peaks at 1238, 1322, and 1402 cm^{-1}, which correspond to the C–N bond in aromatic secondary and tertiary amines, can be attributed to the formation of dimelem (3), which was converted to melon (4) during deammonolysis [9]. The results of IR spectroscopy allowed proposing the following scheme for the formation of graphite-like carbon nitride (Fig. 3).

The absorption band in the region of 3000–3020 cm^{-1} could be attributed to the vibrations of terminal NH_2-groups. At the same time, for the composite I, obtained by mixing magnesium ferrite nanoparticles with melamine and subsequent heat treatment of this mixture at 300 °C. This peak is significantly less pronounced (Fig. 2a) compared to composite II (Fig. 2b) and composite III (Fig. 2b), which indicates a higher degree of condensation of carbon nitride in the g-C_3N_4-$MgFe_2O_4$ composite I. This was due to the condensation of melamine during the combustion of the glycine-nitrate mixture [7].

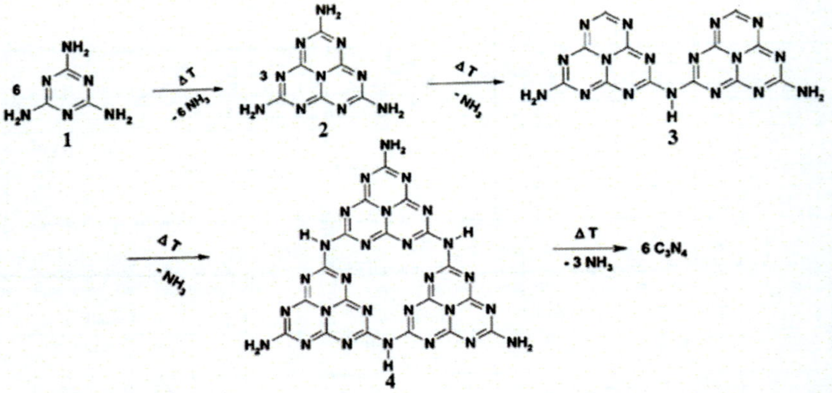

Fig. 3 Scheme of g-C_3N_4 formation in the process of melamine condensation. Adapted from [9]

The presence of intense bands in the range of 560–580 cm^{-1} referred to the valence vibrations of the Fe–O bond located in tetrahedral positions, and the bands at 400–440 cm^{-1} referred to the vibrations of the Mg–O and Fe–O bonds in octahedral positions. The IR spectroscopy data (Fig. 2) fully agreed with the results of X-ray phase analysis (Fig. 1) and confirmed the formation of graphite-like carbon nitride, mainly with a heptazine structure, and magnesium ferrite with a cubic spinel structure in all g-C_3N_4-$MgFe_2O_4$ composites.

2.3 SEM–EDX

The surface morphology and chemical composition of the composites were studied using a scanning electron microscope JSM-5610 LV at an accelerating voltage of 20 kV with an X-ray energy dispersive analysis attachment JED-2201 (JEOL, Japan). Scanning electron microscopy data show that the surface morphology of the g-C_3N_4-$MgFe_2O_4$ composite I was represented by large agglomerates (50–100 microns) covered with fine particles of much smaller size (Fig. 4a). The composite II (Fig. 4b) had a morphology similar to that of magnesium ferrite nanoparticles obtained by the glycine-nitrate method [8], which indicated the formation of ferrite on the surface of carbon nitride. Plate-shaped particles with a lateral size of 10–20 microns covered with fine particles of submicron size were identified on the composite III surface (Fig. 4c).

The chemical composition of the surface of the obtained composites was analyzed using the method of energy-dispersive X-ray spectroscopy (Table 1). Thus, oxygen atoms were not identified on the surface of the g-C_3N_4-$MgFe_2O_4$ composite I, and the content of iron and magnesium atoms was insignificant. At the same time, the highest content of nitrogen atoms was found (70.8 at.%) and carbon (20.0 at.%). This indicates the formation of a carbon nitride shell on the surface of the magnesium ferrite core.

Fig. 4 SEM images of **a** composite I, **b** composite II, and **c** composite III g-C_3N_4-$MgFe_2O_4$

Table 1 Chemical composition of g-C_3N_4-$MgFe_2O_4$ composites

Sample	Element content, at.%				
	C	N	O	Mg	Fe
Composite I	20.0	70.8	–	1.4	7.8
Composite II	29.4	28.8	10.2	3.7	27.9
Composite III	22.0	29.9	15.9	4.7	27.5

The composite II and composite III samples had a similar chemical composition, which was consistent with the similar morphology of their surface (Fig. 4b, c). The higher oxygen atom content for the composite III may be due to the partial evaporation of carbon nitride, which formed during the condensation of melamine during the combustion of the glycine-nitrate mixture. The increased content of iron atoms compared to the content of magnesium atoms on the surface of these composites was characteristic of magnesium ferrite and is in good agreement with the previously obtained data [8].

3 The Catalytic Performance in Methylene Blue Destruction

The development of heterogeneous Fenton catalysts is mainly aimed at finding new catalytic systems characterized by high efficiency when exposed to the visible light range [18]. The ability to absorb visible light is determined by the band gap of the semiconductor and the position of the impurity levels, which can be varied by creating composites [6]. In this regard, the study of the catalytic activity of the developed g-C_3N_4-$MgFe_2O_4$ composites under various conditions (dark-, visible-, and UV-driven processes) was important for substantiating the possibility of the practical application of prepared Fenton-like catalysts.

3.1 *Efficiency of Dark-, Visible-, and UV-Driven Processes*

The catalytic activity of g-C_3N_4-$MgFe_2O_4$ composites was evaluated on model solutions of the thiazine dye Methylene Blue. 25.0 mg of catalyst and 100 μL of hydrogen peroxide solution (20 mmol/L) were added to the 50.0 cm^3 dye aliquot (10.0 mg/L; pH 8.0; 20 °C). To study the kinetic dependences, samples were taken after 5, 10, 20, 30, 40, and 60 min. As a comparative blank experiment, the destruction of Methylene Blue in the presence of hydrogen peroxide was studied without adding a catalyst to the model solution.

The catalytic experiment was carried out using a diode (λ = 650–670 nm) and UV-C lamps (λ = 200–280 nm) as a source of visible and UV radiation, respectively. The dye concentration was measured on a scanning spectrophotometer SP-8001 (Metertech, Taiwan) at the most intense absorption wavelength of 664 nm. All the catalytic experiments were performed in double tests, and the average value was reported. The error of the experiment did not exceed 3%. The efficiency of Methylene Blue catalytic destruction (α%) was calculated by the following Eq. (2):

$$\alpha = (1 - C_0/C_t) \times 100\%, \tag{2}$$

where C_0 and C_t are initial and equilibrium concentrations of Methylene Blue, mg/L; t is reaction time, min.

The values of the apparent rate constant (k', min^{-1}) were determined graphically from the first-order reaction Eq. (3):

$$\ln(C_0/C_t) = k' \times t. \tag{3}$$

To exclude the occurrence of a competing adsorption process, the catalysts were saturated with a solution of Methylene Blue (100 mg/L) for 30 min in the dark.

The kinetic experiment data indicated that the catalysts showed the highest efficiency under UV irradiation (Fig. 5). Thus, for composite I, the efficiency of Methy-

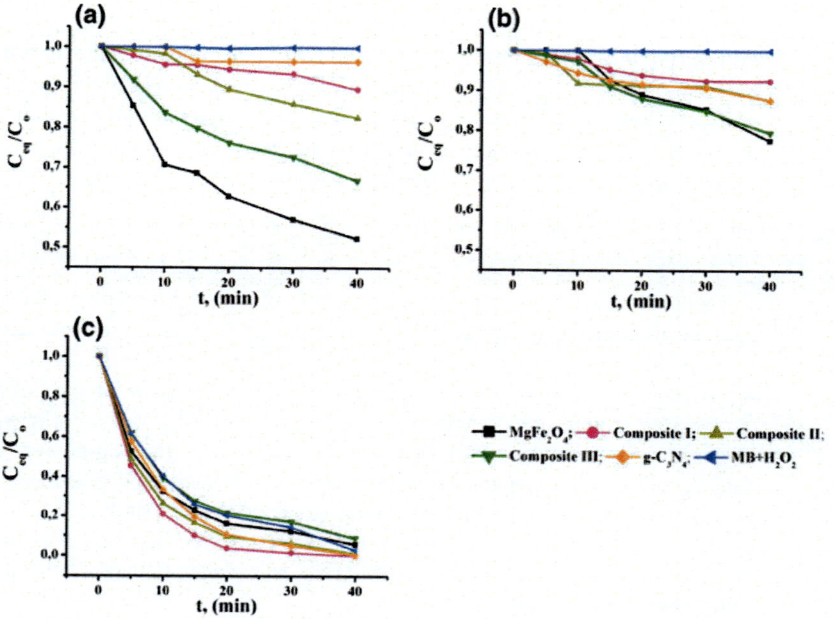

Fig. 5 Kinetics of Methylene Blue degradation on initial g-C_3N_4, $MgFe_2O_4$, and g-C_3N_4-$MgFe_2O_4$ composites under **a** dark-, **b** visible-, **c** UV-driven Fenton-like processes

lene Blue catalytic destruction was ~80% for 10 min and reached 99% in 20 min. Composite II and g-C_3N_4 showed slightly lower efficiency and over similar periods provided the model dye destruction about of 70 and 90%, respectively. It should be noted that the oxidation of Methylene Blue in the presence of hydrogen peroxide was quite high because of their photolysis (Fig. 5c).

Under dark conditions, the activity of the studied catalysts was most clearly differentiated (Fig. 5a). Magnesium ferrite as a common Fenton-like catalyst demonstrated the highest efficiency and the efficiency of Methylene Blue catalytic destruction reached 50% within 40 min of the reaction. Composite III for the same time showed a 35% catalytic destruction, while the other samples were significantly inferior inefficiency (the catalytic efficiency was 5–15%). Thus, in terms of their efficiency under dark conditions, the catalysts were arranged in the following order: $MgFe_2O_4$ < composite III < composite II < composite I < g-C_3N_4, which was due to the different contribution of Fenton and photocatalytic processes to the Methylene Blue destruction in the presence of different catalysts and was completely consistent with their chemical composition.

A somewhat unexpected result was obtained when the experiment was conducted under visible light irradiation conditions (Fig. 5b). Under these conditions, $MgFe_2O_4$ and composite III samples also showed the highest efficiency, but at the same time providing the catalytic destruction of Methylene Blue < 25%. The composite II, composite I, and $g\text{-}C_3N_4$ catalysts demonstrated even lower efficiency and within 40 min the dye destruction did not exceed 10%. This result could be due to the inactivation of hydroxyl radicals formed during the Fenton process, due to the capture of photogenerated electrons (Eq. 4). At the same time, the low photocatalytic activity of $g\text{-}C_3N_4$ could be due to the rapid recombination of the electron–hole pair [15], as well as the low efficiency of visible light absorption [20].

$$OH + e^{-}(h\upsilon) \rightarrow OH^{-} \quad (4)$$

Figure 6 and Table 2 presented the results of kinetics modeling of Methylene Blue destruction using the first-order reaction model. Thus, the data of catalytic experiments were fairly reliably described by the kinetics of the first-order reaction for visible and UV radiation, as well as for dark conditions in the case of $g\text{-}C_3N_4$, composite I, composite II samples, which mainly demonstrated the photocatalytic mechanism of Methylene Blue destruction. This was in good agreement with the data from the study of photocatalytic reactions of various organic pollutants destruction [17, 23]. At the same time, the heterogeneous Fenton catalysts $MgFe_2O_4$ and composite III showed lower confidence and the approximation coefficient R^2 was <0.95. It should be noted that composite III was an amorphous material that it differed from other catalyst samples.

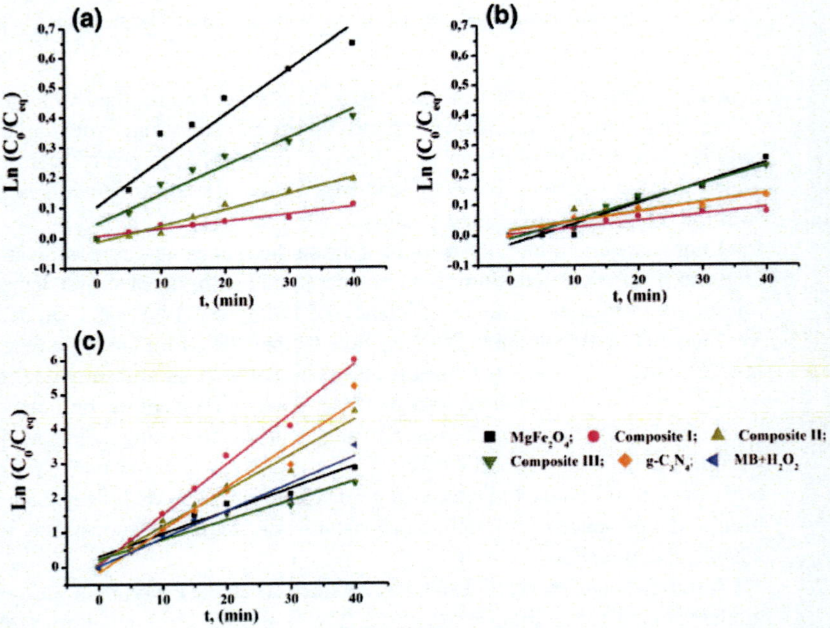

Fig. 6 Linear plots of a first-order kinetic model of Methylene Blue destruction on $g\text{-}C_3N_4$, $MgFe_2O_4$, and $g\text{-}C_3N_4\text{-}MgFe_2O_4$ composites: **a** dark-, **b** visible-, **c** UV-conditions

Table 2 Calculated parameters of a first-order kinetic model for Methylene Blue destruction on $g\text{-}C_3N_4$, $MgFe_2O_4$, and $g\text{-}C_3N_4\text{-}MgFe_2O_4$ composites

Sample	$MgFe_2O_4$	$g\text{-}C_3N_4$	Composite I	Composite II	Composite III
Dark conditions					
k', min^{-1}	0.0152	–	0.0025	0.0054	0.0099
R^2	0.9057	–	0.9703	0.9426	0.9379
Visible light					
k', min^{-1}	0.0067	0.003	0.0022	0.0030	0.0059
R^2	0.9494	0.9268	0.8898	0.7511	0.9731
UV light					
k', min^{-1}	0.0690	0.1297	0.1528	0.1094	0.0592
R^2	0.9626	0.9916	0.9962	0.9969	0.9634

The calculated values of the apparent rate constants k' are presented in Table 2. The highest values of k' were obtained under UV irradiation for $g\text{-}C_3N_4$, composite I, composite II samples, which reached 0.1094–0.1528 min^{-1}. Under the same conditions, the $MgFe_2O_4$ and composite III catalysts showed ~2 times lower activity and the k' values did not exceed 0.0690 min^{-1}. During the visible light experiment, low values of the apparent rate constants were obtained for all samples, which were in the range of 0.0022–0.0069 min^{-1}. Only for the most active $MgFe_2O_4$ and amorphous composite III samples under dark conditions, the calculated values of k' were < 0.010 min^{-1}.

3.2 *Reusability of Fenton-Like Catalysts*

An important characteristic of heterogeneous catalysts is their stability and the preservation of catalytic activity during multiple catalytic cycles. For this purpose, the experiment was performed under the optimal conditions under UV irradiation for four catalytic cycles (Fig. 7).

The composite III and $g\text{-}C_3N_4$ samples showed the lowest stability, while the remaining catalysts showed high reproducibility of the catalytic characteristics at the level of 97–99% under experimental conditions. The obtained data were due to the photodegradation of these samples, as well as the lower stability of the amorphous composite III. At the same time, the formation of $g\text{-}C_3N_4$ (shell) on the surface of the crystalline $MgFe_2O_4$ (core) allowed to increase the stability of the catalytic characteristics of the $g\text{-}C_3N_4$ during multiple catalytic cycles.

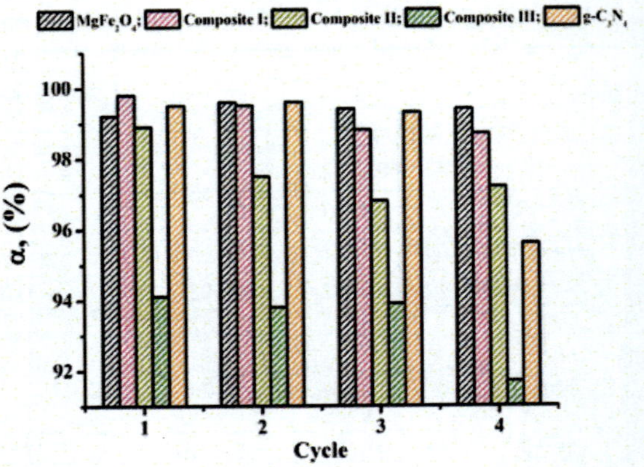

Fig. 7 The catalytic activity of g-C_3N_4, $MgFe_2O_4$, and g-C_3N_4-$MgFe_2O_4$ composites during multiply tests under UV-irradiation

3.3 *Mechanism and Quencher Test*

The study of the mechanism of Methylene Blue catalytic destruction was performed using quencher tests (Table 3). The introduction of certain compounds into the reaction mixture, which is inactivators ("traps") of specified reactive spices, allowed to conduct a qualitative assessment of the contribution of the produced reaction spices based on the analysis of changes in the catalytic reaction kinetics [2, 12].

The analysis of the presented data shows that hydroxyl radicals were the main reaction spices that determined the efficiency of Methylene Blue catalytic destruction in the presence of the studied catalysts. This was evidenced by the lowest values of the apparent rate constants during the reaction in the presence of isopropanol. For $MgFe_2O_4$ and composite III, h^+ also had a significant contribution. For all catalysts, the $\cdot O_2^-$ radical played the least role, which was due to their low concentration and lifetime [2].

Table 3 Calculated parameters for the first-order kinetic model of Methylene Blue destruction on g-C_3N_4, $MgFe_2O_4$, and g-C_3N_4@$MgFe_2O_4$ composites

Sample	Conditions of catalytic test	k', min^{-1}		
		Reactive spices		
		$\cdot OH$	h^+	$\cdot O_2^-$
$MgFe_2O_4$	C_0(MB) 9.0 mg L^{-1}	**0.0397**	0.0629	0.0566
g-C_3N_4	Catalyst dose 1.0 g L^{-1}	**0.0048**	0.0663	0.0545
Composite I	C(H_2O_2) 20.0 mM	**0.0110**	0.1307	0.0677
Composite II	UV-C light (189–254 nm	**0.0182**	0.0906	0.0833
Composite III	t 40 min	**0.0268**	0.0595	0.0621

Quenchers: Isopropanol ($\cdot OH$), ammonium oxalate (h^+), hydroquinone ($\cdot O_2^-$)

4 Conclusions

It was shown that depending on the conditions of synthesis of composites based on $MgFe_2O_4$ and g-C_3N_4, it is possible to change the physical and chemical characteristics of these materials in a directed manner. It was found that by mixing magnesium ferrite with melamine and subsequent heat treatment at 300 °C, g-C_3N_4-$MgFe_2O_4$ composite I with well-identified $MgFe_2O_4$, and g-C_3N_4 crystal phases was obtained. When g-C_3N_4 was introduced into the glycine-nitrate gel at the stage of initiation of the nitrate mixture combustion, a composite II was formed, consisting of a triclinic modification of g-C_3N_4 and an amorphous magnesium ferrite with an admixture of NaCl galit. The X-ray amorphous composite III was obtained by introducing melamine into a glycine-nitrate gel, followed by the initiation of self-combustion high-temperature synthesis. It was found that all composites were characterized by the formation of g-C_3N_4 with the conjugated aromatic structure of s-heptazine. Varying the synthesis conditions made it possible to obtaine composites with a predominant concentration on the surface of g-C_3N_4 or $MgFe_2O_4$.

The catalytic properties of g-C_3N_4-$MgFe_2O_4$ composites in thiazine dye Methylene Blue catalytic destruction under dark-, visible-, and UV-driven processes were investigated. Depending on the conditions, the efficiency of the catalysts decreased in the UV < dark < visible range, which was due to the recombination of the (e^-)–(h^+) pairs. Under UV irradiation for composite I, which was g-C_3N_4 (core)–$MgFe_2O_4$ (shell), the highest efficiency of catalytic Methylene Blue destruction of 99% was achieved within 20 min, and the apparent rate constant k' of 0.01528 min^{-1}. It was shown that the formation of g-C_3N_4 (shell) on the surface of the crystalline $MgFe_2O_4$ (core) increased the stability of g-C_3N_4 catalytic efficiency during multiple catalytic cycles. For all the studied catalysts under UV irradiation, hydroxyl radicals had the greatest contribution to Methylene Blue destruction, and O_{2-} was the smallest influencing. The conducted studies allowed to establish the relationship between the synthesis conditions, the physicochemical properties and the catalytic activity of g-C_3N_4-$MgFe_2O_4$ composites, and to obtain effective heterogeneous Fenton catalysts.

References

1. Aksoy M, Yanalak G, Aslan E, Patir IH, Metin O (2020) Visible light-driven hydrogen evolution by using mesoporous carbon nitride-metal ferrite (MFe_2O_4/mpg-CN; M: Mn, Fe, Co and Ni) nanocomposites as catalysts. Int J Hydrogen Energy 45:16509–16518. https://doi.org/10.1016/j.ijhydene.2020.04.111
2. Collin F (2019) Chemical basis of reactive oxygen species reactivity and involvement in neurodegenerative diseases. Int J Mol Sci 20(10):2407. https://doi.org/10.3390/ijms20102407
3. De Andrade JR, Oliveira MF, Da Silva MGC, Vieira MGA (2018) Adsorption of pharmaceuticals from water and wastewater using nonconventional low-cost materials: a review. Ind Eng Chem Res 57(9):3103–3127. https://doi.org/10.1021/acs.iecr.7b05137

4. Gao G, Zhang L, Chen Q, Fan H, Zheng J, Fang Y, Duan R, Cao X, Hu X (2021) Self-assembly approach toward polymeric carbon nitrides with regulated heptazine structure and surface groups for improving the photocatalytic performance. Chem Eng J 409:127370. https://doi.org/10.1016/j.cej.2020.127370
5. Hassani A, Eghbali P, Ekicibil A, Metin O (2018) Monodisperse cobalt ferrite nanoparticles assembled on mesoporous graphitic carbon nitride ($CoFe_2O_4$/mpg-C_3N_4): a magnetically recoverable nanocomposite for the photocatalytic degradation of organic dyes. J Magn Magn Mater 456:400–412. https://doi.org/10.1016/j.jmmm.2018.02.067
6. Imtiaz F, Rashid J, Xu M (2019) Semiconductor nanocomposites for visible light photocatalysis of water pollutants. In: Rahman MM, Asiri A, Khan A (eds) Concepts of semiconductor photocatalysis. IntechOpen Limited, London. https://doi.org/10.5772/intechopen.86542
7. Ivanets A, Prozorovich V (2020) Effect of melamine acidic treatment on g-C_3N_4 physicochemical properties and catalytic activity. Water Water Purif Technol Sci Tech News 28:26–36. https://doi.org/10.20535/wptstn.v28i3.216100
8. Ivanets A, Prozorovich V, Roshchina M, Kouznetsova T, Budeiko N, Kulbitskaya L, Hosseini-Bandegharaei A, Masindi V, Pankov V (2021) A comparative study on the synthesis of magnesium ferrite for the adsorption of metal ions: insights into the essential role of crystallite size and surface hydroxyl groups. Chem Eng J 411:128523. https://doi.org/10.1016/j.cej.2021.128523
9. Jürgens B, Irran E, Senker J, Kroll P, Müller H, Schnick W (2003) Melem (2,5,8-triamino-tri-s-triazine), an important intermediate during condensation of melamine rings to graphitic carbon nitride: synthesis, structure determination by X-ray powder diffractometry, solid-state NMR, and theoretical studies. J Am Chem Soc 125(34):10288–10300. https://doi.org/10.1021/ja0357689
10. Kanakaraju D, Glass BD, Oelgemöller M (2018) Advanced oxidation process-mediated removal of pharmaceuticals from water: a review. J Environ Manag 219:189–207. https://doi.org/10.1016/j.jenvman.2018.04.103
11. Liang Q, Shao B, Tong S, Liu Z, Tang L, Liu Y, Cheng M, He Q, Wu T, Pan Y, Huang J, Peng Z (2021) A review: recent advances of melamine self-assembled graphitic carbon nitride-based materials: design, synthesis and application in energy and environment. Chem Eng J 405:126951. https://doi.org/10.1016/j.cej.2020.126951
12. Liu W, He T, Wang Y, Ning G, Xu Zh, Chen X, Hu X, Wu Y, Zhao Y (2020) Synergistic adsorption-photocatalytic degradation effect and norfloxacin mechanism of ZnO/ZnS@BC under UV-light irradiation. Sci Rep 10:11903. https://doi.org/10.1038/s41598-020-68517-x
13. Ma S, Zhan S, Jia Y, Shi Q, Zhou Q (2016) Enhanced disinfection application of Ag-modified g-C_3N_4 compositeunder visible light. Appl Catal B 186:77–87. https://doi.org/10.1016/j.apcatb.2015.12.051
14. Maeda K, Wang X, Nishihara Y, Lu D, Antonietti M, Domen K (2009) Photocatalytic activities of graphitic carbon nitride powder for water reduction and oxidation under visible light. J Phys Chem C 113:4940–4947. https://doi.org/10.1021/jp809119m
15. Naseri A, Samadi M, Pourjavadi A, Moshfegh AZ, Ramakrishna S (2017) Graphitic carbon nitride (g-C_3N_4)-based photocatalysts for solar hydrogen generation: recent advances and future development directions. J Mater Chem A 5:23406–23433. https://doi.org/10.1039/c7ta05131j
16. Shorie M, Kaur H, Chadha G, Singh K, Sabherwal P (2019) Graphitic carbon nitride QDs impregnated biocompatible agarose cartridge for removal of heavy metals from contaminated water samples. J Hazard Mater 367:629–638. https://doi.org/10.1016/j.jhazmat.2018.12.115
17. Sidney Santana C, Nicodemos Ramos MD, Vieira Velloso CC, Aguiar A (2019) Kinetic evaluation of dye decolorization by Fenton processes in the presence of 3-hydroxyanthranilic acid. Int J Environ Res Public Health 16(9):1602. https://doi.org/10.3390/ijerph16091602
18. Thomas N, Dionysiou DD, Pillai SC (2021) Heterogeneous Fenton catalysts: a review of recent advances. J Hazard Mater 404(B):124082. https://doi.org/10.1016/j.jhazmat.2020.124082
19. Wang N, Zheng T, Zhang G, Wang P (2016) A review on Fenton-like processes for organic wastewater treatment. J Environ Chem Eng 4(1):762–787. https://doi.org/10.1016/j.jece.2015.12.016

20. Zhang M, Xu J, Zong R, Zhu Y (2014) Enhancement of visible light photocatalytic activities via porous structure of g-C_3N_4. Appl Catal B 147:229–235. https://doi.org/10.1016/j.apcatb.2013.09.002
21. Zhao Z, Sun Y, Dong F (2015) Graphitic carbon nitride based nanocomposites: a review. Nanoscale 7:15–37. https://doi.org/10.1039/c4nr03008g
22. Zhu Y, Zhu R, Xi Y, Zhu J, Zhu G, He H (2019) Strategies for enhancing the heterogeneous Fenton catalytic reactivity: a review. Appl Catal B: Environ 255:117739. https://doi.org/10.1016/j.apcatb.2019.05.041
23. Zulmajdi SLN, Ajak SNFH, Hobley J, Duraman N, Harunsani MH, Yasin HM, Nur M, Usman A (2017) Kinetics of photocatalytic degradation of methylene blue in aqueous dispersions of TiO_2 nanoparticles under UV-LED irradiation. Am J Nanomater 5(1):1–6. https://doi.org/10.12691/ajn-5-1-1

Metal Oxide Heterostructured Nanocomposites for Wastewater Treatment

M. Mondal, M. Ghosh, H. Dutta, and S. K. Pradhan

Abstract Over the past few decades, nanocrystalline metal oxide semiconductor-based photocatalytic technology has drawn significant attention for reducing the recent energy crisis by converting solar energy into potential energies and remediating environmental pollution. Two traditional semiconductors, TiO_2 and ZnO, have been widely investigated as a photocatalyst for water purification among several metal oxide semiconductors. However, the photocatalytic performance of the above-mentioned metal oxide-based photocatalysts has yet to reach the target based on efficiency and stability due to the larger bandgap, inadequate adsorption capacity, insufficient charge carrier generation and transportation, and electron–hole pair recombination, which are some of the significant issues associated with metal oxide photocatalysts. So efforts are given for the synthesis of metal oxide-based heterostructured nanocomposites for wastewater treatment. With this perspective, the present chapter highlights the studies of heterogeneous photocatalysis for better performance. This chapter also deals with the synthesis, structure and microstructure characterization of as-synthesized nanocomposites by analyzing XRD profiles using the Rietveld analysis, TEM and SEM images. Photocatalytic degradation efficiency has been investigated under visible light illumination with RhB as an organic model dye pollutant. The dye degradation kinetics has been correlated with the Urbach energy.

Keywords Metal oxide-based heterostructured nanocomposites · Rietveld analysis · Urbachenegy · Heterogeneous photocatalysis · Wastewater treatment

M. Mondal · M. Ghosh · S. K. Pradhan (✉)
Materials Science Division, Department of Physics, The University of Burdwan, Golapbag, Burdwan, West Bengal 713104, India
e-mail: skpradhan@phys.buruniv.ac.in

H. Dutta
Department of Physics, Vivekananda Mahavidyalaya, Burdwan, West Bengal 713104, India

S. S. Muthu and A. Khadir (eds.), *Advanced Oxidation Processes in Dye-Containing Wastewater*, Sustainable Textiles: Production, Processing, Manufacturing & Chemistry,
https://doi.org/10.1007/978-981-19-0987-0_12

1 Introduction

From the perspective of biology and ecology, water is an essential element for life circulated through a natural hydrologic cycle with self-cleaning capability. In past decades, increasing population and industrial development have become a significant threat to the environment and human health [77]. Advanced water treatment techniques are indispensable to maintain a considerable available water supply and a healthy large scale hydrologic cycle due to tremendous amounts of municipal and industrial wastewater and landfill leachates [15]. According to the World Health Organization (WHO), about 780 million people suffer from affordable drinking water worldwide. Also, there is a substantial limitation to freshwater resources [66]. Another critical step in wastewater remediation technique in removal of organic pollutants causes severe damage to human health and affects marine life. For example, organic azo dyes such as Rhodamine B, Methyl Orange are highly carcinogenic and mutagenic [10, 59]. Hence, advanced wastewater treatment techniques for industrial, agricultural and pharmaceutical wastewater purification have become essential for research communities.

This chapter is focused on the removal of industrial dye pollutants. For this purpose, different strategies had been adopted, such as electrochemical oxidation [14], biodegradation [49], adsorption [1, 2], chlorination [64], ozonation [30], reverse osmosis [32], combined coagulation/flocculation using water treatment plant sludge [51] and so on. However, the heterogeneous photocatalytic process is one of the most effective remediation techniques regarded as a promising green technology to solve environmental concerns and recent energy crises. In this context, metal oxide-mediated heterostructured nanocomposite-based photocatalysis has considerable attention to address environmental issues. Because metal oxides modified with metal, non-metal and semiconductors have the most exciting property such as non-toxicity, solid oxidizing ability, high stability with easy filtration and higher recyclability [57, 58, 60]. This chapter summarizes the water purification mechanisms based on the recent developments in metal oxide modified heterostructured nanocomposites. The various strategies that have been taken on, like heterojunction formation, doping, semiconductor coupling, sensitization, and photocatalytic performance, have been discussed in the chapter.

2 Wastewater Treatment Processes: Photocatalysis

Industrial wastewater treatment is our primary concern. Sources of industrial wastewater include nuclear industry, battery manufacturing, food industry, iron and steel industry, organic chemicals manufacturing, electric power plants, petroleum refining and petrochemicals, oil and gas extraction, paper industry, textile mills, industrial oil contamination, wood preserving and so on [8, 21]. Chemical oxidation processes can be used to destroy the recalcitrant compounds through oxidation and reduction reactions. However, advanced oxidation processes that have been used for the purification of various types of contaminants present in wastewater generated from different industrial processes are investigated in this chapter.

In recent days, the most commonly used chemical treatment method for removing toxic pollutants growing in the wastewater treatment industry is Advanced Oxidation Process (AOP). It is a highly suggested method for remediation of wastewater containing colouring matters, pesticides and other inorganic substances. The primary mechanism of AOPs involves the following two stages [53]:

(i) one of them is the oxidation process, which is conducted through the generation of powerful oxidants (Hydroxyl radicals),
(ii) another stage proceeds through the reaction of these generated oxidants with organic contaminants in wastewater, which in turn transform into less or even non-toxic materials with the advantages of faster reaction rate and non-selective oxidation. It is proved to be the ultimate solution for wastewater treatment. AOPs can be broadly classified into Fenton process, Photo-Fenton process [9], Ozonation [80], ultrasonic treatment [4], gamma radiolysis [55], photocatalysis [3] etc. Among all of them, photocatalytic treatment with UV/visible light irradiation seems to be the most popular technology for wastewater treatment.

2.1 *Photocatalysis: An Advanced Oxidation Process*

A technique employing catalysts activated under the illumination of solar-like light radiation, which is utilized for accelerating chemical reactions, is known to be the photocatalysis process. Photocatalytic science includes phenomena like catalysis of photochemical reactions, photochemical activation of catalytic processes and photoactivation of catalysts [81]. In recent years, semiconductor-based photocatalysis draws more attention due to several advantages, such as being cost-effective and less toxic [61]. The fundamental mechanism of the photocatalytic technique is the photoexcitation of semiconductor photocatalyst with a suitable wavelength. Coordination with the bandgap of the corresponding nanophotocatalyst leads to the generation of the photogenerated electron–hole pair, which again reacts with the organic or inorganic pollutants [82, 83]. The reaction mechanism involving electrons as the main species is a photo-reduction reaction, while the hole involved a photo-oxidation reaction. The significant advantages of the photocatalytic process over existing AOPs are:

(i) Secondary disposal methods are not required.
(ii) Here, ambient oxygen acts as an oxidant. So compared to other AOPs, expensive oxidizing chemicals, hydrogen peroxide/ozone etc., are no longer required.
(iii) Photocatalysts can be recycled, reused and are also self-regenerated.

The photocatalysis technique can be classified as follows:

(a) *Homogeneous Photocatalysis*

In this process, the reactants and the photocatalysts exist in the same phase. For example, Ozonation and Photo-Fenton systems. Mechanism of homogeneous catalysis (Ozonation) can be summarized into the following steps in two ways,

$$O_3 + h\upsilon \rightarrow O_2^{\bullet} + O(1D) \tag{1}$$

$$O(1D) + H_2O \rightarrow OH^{\bullet} + OH^{\bullet} \tag{2}$$

$$O(1D) + H_2O \rightarrow H_2O_2 \tag{3}$$

$$H_2O \rightarrow OH^{\bullet} + OH^{\bullet} \tag{4}$$

In Photo-Fenton processes, additional sources of $OH^{\bullet}$ radicals should be considered through photocatalysis of H_2O_2 and reduction of Fe^{3+} ions under UV light.

$$H_2O_2 + h\upsilon \rightarrow OH^{\bullet} + OH^{\bullet} \tag{5}$$

$$Fe^{3+} + H_2O + h\upsilon \rightarrow Fe^{2+} + H^+ + OH^{\bullet} \tag{6}$$

(b) *Heterogeneous Photocatalysis*

In this process, the photocatalyst and the reactant are in different phases. The following facts can explain a plausible mechanism of heterogeneous photocatalysis. When the photocatalysts absorb a photon of light with energy equal to or greater than the bandgap energy of the respective photocatalyst, then the photogenerated e^--h^+ pairs migrate to the surface of the photocatalysts over rapid recombination and react with the absorbed species following thermal catalytic requirements to generate reactive molecules like H_2O_2 or reactive radicals like hydroxyl radicals ($OH^{\bullet}$) or superoxide ion radicals [22, 23]. Primarily, photocatalysis occurs in the following three steps:

(a) First of all, electron conduction from the valence band (VB) to the conduction band (CB) and creating a hole in the VB.
(b) Then, the excited electron in the CB and hole in the VB migrate towards the surface [70].
(c) Next, according to NHE, as the chemical potential of the electron lies between +0.5 and −1.5 V, it should exhibit a strong reductive potential to react with the electron donor. While the chemical potential of the hole lies between +1.0 and +3 V concerning NHE, it possesses a robust oxidative power and takes part in the reaction with an electron acceptor. The series of oxidation and reduction reactions which can they participate as detailed below [26, 38, 75]:

(i) Hydroxyl radicals are generated due to the oxidation of adsorbed water molecules and hydroxyl ions by the photogenerated holes.

$$H_2O + h^+ \rightarrow OH^\bullet + H^+ \quad (7)$$

$$OH^- + h^+ \rightarrow OH^\bullet \quad (8)$$

(ii) Superoxide anion radicals are produced by reducing dissolved oxygen due to photogenerated electrons resulting in the formation of H_2O_2, which can be summarized through a series of following redox reactions:

$$O_2 + e^- \rightarrow O_2^{\bullet-} \quad (9)$$

$$O_2^{\bullet-} + H^+ \rightarrow HO_2^\bullet \quad (10)$$

$$O_2^{\bullet-} + HO_2^\bullet \rightarrow O_2 + HO_2 \quad (11)$$

$$HO_2^- + H^+ \rightarrow H_2O_2 \quad (12)$$

$$2HO_2^\bullet \rightarrow H_2O_2 + O_2 \quad (13)$$

$$2OH^\bullet \rightarrow H_2O_2$$

(iii) Hydroxyl radicals are yielding on the further decomposition of the photogenerated H_2O_2.

$$H_2O_2 \rightarrow 2OH^\bullet \quad (14)$$

$$H_2O_2 + O_2^{\bullet-} \rightarrow {}^\bullet OH + {}^-OH + O_2 \quad (15)$$

$$H_2O_2 + e^- \rightarrow {}^\bullet OH + {}^-OH \quad (16)$$

(iv) Direct participation of the holes in the oxidation reactions.
(v) Formation and participation of singlet oxygen species in the oxidation.

A simple heterogeneous photodegradation mechanism is schematically represented below for Bi–Fe–O heterostructured nanocomposites (Scheme 1).

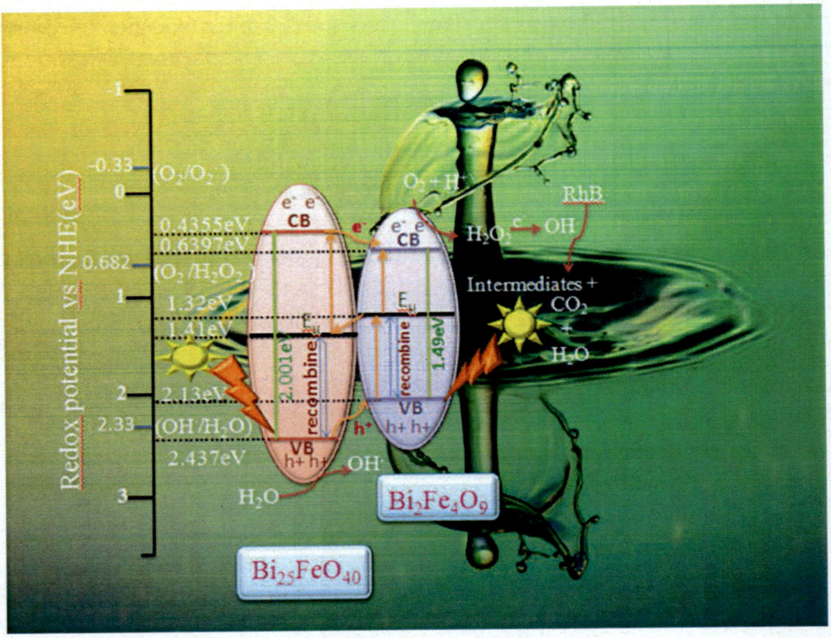

Scheme 1 Simple heterogeneous photodegradation mechanism of a heterostructured nanocomposite

3 A Brief History of Metal Oxide Semiconductors

During the last three decades, in resolving water pollution and the recent energy crisis, semiconductor-based photocatalytic processes have attracted significant attention due to their low-cost synthesis and higher stability [11]. Some traditional metal oxide-based semiconductors such as TiO_2, ZnO, V_2O_5, WO_3, Bi_2O_3, Fe_2O_3 etc., are extensively used worldwide. Structural characteristics and photocatalytic degradation efficiency of a few of these metal oxide semiconductors are epitomized below.

As a single photocatalyst, n-type semiconductor TiO_2 is commonly used in water purification, which posseses a wide bandgap of 3.0–3.2 eV [56, 74]. Despite several advantages, such as high photocatalytic activity, low nano-toxicity, cheap and good chemical and thermal stability [67, 69], there are some limitations of TiO_2 as a visible-light-driven photocatalyst. The large bandgap and higher recombination rate of photogenerated electron–hole pair restrict its catalytic effect in the UV regime. Some strategies were endorsed to enhance its photocatalytic activity by morphology modification, such as spheres zero dimension [12], one-dimensional fibres, rods and tubes [19], 2D sheets [5] and 3D interconnected architectures [27]. However, the efficiency of TiO_2 did not meet the expectation under visible light irradiation without a photosensitizer. Therefore to broaden the range of solar absorption, different metals or ions were incorporated into TiO_2 to maximize its photocatalyst efficiency [39, 78].

There is another n-type semiconductor, ZnO, with a wide bandgap of 3.2 eV is widely used as a photocatalyst due to several advantages, such as high photosensitivity, biocompatibility, low cost, high transparency, non-toxic nature and excellent chemical stability [28]. According to a report, in the destruction of organic pollutants, ZnO has higher activity and more quanta than TiO_2 [54]. However, several shortcomings of ZnO, such as extreme pH facile dissolution and photocorrosion [16], limit its large-scale applicability as a promising photocatalyst.

Among various photocatalysts, the transition metal oxides such as V_2O_5 and VO_2 have drawn much attention due to some outstanding properties, such as narrow bandgap (of approximately 2.4–2.8 eV), good chemical stability, non-toxicity and high visible light absorption [17, 68]. Further, its morphology can be tuned, such as nanobelt, nanotubes and nanorods for far better photocatalytic activity. Due to these morphological modifications, newly generated physical and chemical properties support its diverse significant applicability [47, 72].

These metal oxide semiconductors have restricted applications as efficient visible-light-driven (VLD) photocatalysts due to their large bandgap, and they are mainly active under UV light irradiation. One of the greatest fallibility of using metal oxide semiconductors is poor solar spectrum absorption due to the previously stated factors. However, one net worthy strategy can overcome this drawback, i.e. developing a narrow bandgap semiconductor as a VLD photocatalyst [29, 33]. Of the variety of narrow-band gap semiconducting materials, bismuth oxide (Bi_2O_3) is a prospective candidate with a bandgap varying from 2.1 to 2.8 eV [7, 24]. According to thermodynamic requirements, the valence band of Bi_2O_3 lies at ~+3.13 eV versus NHE, so it possesses adequately high oxidation power of the valence hole. Bi_2O_3 is the desired material as an efficient visible-light photocatalyst. In other viewpoints, it is non-toxic, environmental-friendly and can utilize the complete solar spectrum. However, its conduction band lies at ~−0.33 eV versus NHE, due to which it is unable to scavenge surface oxygen molecules and results in the recombination of e^--h^+ pair [25, 31]. So photocatalytic degradation efficiency against the organic dyes is not much better as expected. Another cheapest narrow bandgap semiconductor is n-type α-Fe_2O_3 with a bandgap of ~2.1 eV. The crystal arrangement of α-Fe_2O_3 is iron cations arranged along the octahedral basal plane (001) between the hexagonal close-packed planes of oxygen anions. Because of such arrangements, FeO_6 octahedrons are generated from Fe^{3+} ions with two different Fe–O bond lengths [43, 44]. However, the expected photocatalytic efficiency is yet to reach.

Thus to enhance the photodegradation efficiency, various techniques have been espoused like co-doping of metal or non-metal, controlling the porosity and trapped defects, tuning the morphology etc. Nevertheless, forming a heterostructure with metal or non-metal oxides is one of the best solutions for desired photocatalytic degradation requirements like lesser recombination rate, better absorption of solar radiation and active under a wide range of the solar spectrum, cost-effective, non-toxicity etc.

4 Recent Developments on Metal Oxide Heterostructured Nanocomposites

Nanostructuring technology reflects as one of the best tools to unlock the previously unachievable properties of technological importance in photocatalytic degradation application. Covalent and Van der Walls interactions play a crucial role in developing unique properties due to nanostructure. It draws considerable attention in electrochemistry, photochemistry, photophysics, photoelectrochemistry and photocatalysis [26]. Forming heterojunctions of two or more metal oxides of suitable bandgap have received significant attention from research communities worldwide due to attaining desired properties by modifying the morphology of the materials, structural and microstructural properties. Approximately 25,000 research papers reported on this field of photodegradation of organic dyes using metal oxide-based heterostructured nanocomposites. A few of them are condensed below with a brief discussion.

Wang et al. [71] fabricated hierarchical SnO_2/TiO_2 composite nanostructures combining the versatility of electrospinning technique and hydrothermal growth of nanostructures. The results of this fabrication reveal the fact that secondary SnO_2 nanoparticles were uniformly distributed over the TiO_2 nanofibres without aggregation. Further, controlling the fabrication parameter morphology and other properties are modified to enhance the photodegradation efficiency against Rhodamine B molecule under UV light irradiation compared to bare TiO_2 nanofibres. Yang et al. [79] prepared heterostructured TiO_2/WO_3 nanocomposite by ultrasonic spray pyrolysis of an aqueous suspension of Degussa P25 particles with ammonium tungstate with different Ti/W molar ratios. This paper reported that the crystallinity of WO_3 increased with the increasing molar ratio and the nanocomposites with 2 mol% WO_3 showed greater efficiency. However, for the other compositions, photochromism occurs due to the accumulation of electrons in the orthorhombic WO_3. The above observations established the fact that the influence of WO_3 on the photocatalytic activity of TiO_2 was related to the crystal phase and electron accumulation ability of WO_3, which makes it more complicated. Uddin et al. [65] synthesized heterostructured SnO_2/TiO_2 nanocomposites through one-pot polyol method result in mesoporous cassiterite SnO_2 and wurtzite ZnO nanocrystallites. Photocatalytic activity of as-synthesized nanocomposites was investigated through methylene blue under UV light irradiation, and the efficiency was prior to the precursor materials mainly due to the higher stability and enhanced charge separation. Kundu et al. [34] reported uniform cylindrical-shaped ZnO nanorods suitable for the lithium-ion battery via a simple chemical route. The nano grain size of synthesized ZnO nanorods was contracted or expanded, which delivered good charging or discharging abilities during alloying or dealloying processes compared to other ZnO materials. Photocatalytic degradation efficiency was also enhanced due to morphological modification. Kundu et al. [35] reported synthesizing polyaniline intercalated vanadium oxide xerogel hybrid nanocomposites via a simple hydrothermal route with a novel structure. The results reflected a phase transformation of the orthorhombic to monoclinic phase due to intercalation of PANI and water layer in the $V_2O_5 \cdot n\ H_2O$ layers.

As a result, there was a substantial increase in the interplanar spacing of the (001) plane. Due to the synergetic effect of PANI and $V_2O_5 \cdot nH_2O$ xerogel established a promising photocatalytic efficiency for the degradation of organic dyes and non-absorbing colourless phenol and kanamycin molecules. Mondal et al. [46] reported mechanosynthesis of V_2O_5–TiO_2 nanocomposites with different morphology. The efficiency of the as-spun nanocomposites increased with the increasing milling hour, and with the addition of H_2O_2, degradation of Rhodamine B molecule increased up to ~89%. Mandal and Pradhan [45] synthesized heterostructured Bi_2O_3–Bi_2WO_6 nanocomposites with nanoflower or nanorod like morphology through mechanical alloying. Approximately 87% of Rhodamine B degraded after 240 min under solar-like radiation. Mandal and Pradhan [46] reported the morphological changes of MoO_2 metal oxides due to sintering at elevated temperature, among which the sample sintered at 650 °C for 2 h reported to have the best degradation efficiency. Majhi et al. [42] reported a series of α-NiS/Bi_2O_3 nanocomposites synthesized by combustion and bismuth subcarbonate decomposition routes. Following the Z-scheme electron transfer, α-NiS/β-Bi_2O_3 nanocomposite showed the highest degradation of about 94% within 3 h. The hydroxyl radicals and h^+ were mainly responsible for the oxidation of tradomol pollutants. Wu et al. [76] reported synthesizing flower-like g-C_3N_4/BiOBr heterostructured nanocomposites as photocatalysts by a simple hydrothermal process with different compositions. Approximately 94% Bisphenol A degraded within 120 min by the 4:1 g-C_3N_4/BiOBr photocatalyst and possessed greater stability against Methyl Orange degradation. Wen et al. [73] reported the fabrication of visible-light-driven p–n junction photocatalyst BiOI/CeO_2 using a facile in situ method. BiOI/CeO_2 nanocomposites with a 1:1 molar ratio reflected the highest degradation for the refractory pollutant Bisphenol A and Methyl Orange with greater stability. This context also established that the introduction of BiOI into the CeO_2 region broadens the visible light capability; hence improved the transfer rate of electron–hole pair and proved to be an efficient photocatalyst for degradation of refractory water pollutant wastewater. Sang et al. [63] also reported the synthesis of flower-like BiOI/CeO_2 nanocomposites by hydrothermal route with different molar ratios of Bi/Ce using polyvinyl pyrrolidone as a surfactant. About 15% CeO_2 incorporated BiOI nanocomposites showed the highest photocatalytic activity of Rhodamine B and Methyl orange degradation and proved to be the best photocatalyst for environmental remediation due to synergistic effect between BiOI and CeO_2.

In this context, synthesis and different physiochemical properties of three metal oxide heterostructured nanocomposites V_2O_5–TiO_2 [45] and Bi–Fe–O [46] have been discussed briefly and also their potentiality as an efficient photocatalyst has been thoroughly investigated. Many synthesis techniques were utilized to synthesize V_2O_5–TiO_2 nanocomposites but less reported about mechanical alloying of V_2O_5–TiO_2 nanocomposites. In this chapter, the mechanosynthesis of V_2O_5–TiO_2 nanocomposites within a short duration and detailed study reveals the correlation between structure and properties.

5 Experimental Section

5.1 *Synthesis*

5.1.1 Mechanical Alloying

In the field of solid-state powder processing, one of the modern age techniques is mechanical alloying, which is a fast and effective process of reducing powder materials size up to nano dimension.

Mechanical alloying or also called high energy ball milling can be carried out in different ball mills such as vibratory ball mill, attrition mill, and planetary ball mill, among which planetary ball mill is conveniently used, as shown in Fig. 1. Each vial is fitted in a rotating disc inside the ball mill following the planetary laws of motion; experience a high impact force between the balls and the vials' walls due to centrifugal force. The precursor powder materials are mixed and ground, yielding a powder mixture or new compounds or new composites with new phases different from precursor materials [36].

Bi–Fe–O [46] and V_2O_5–TiO_2 [46] nanocomposites were synthesized by mechanical alloying.

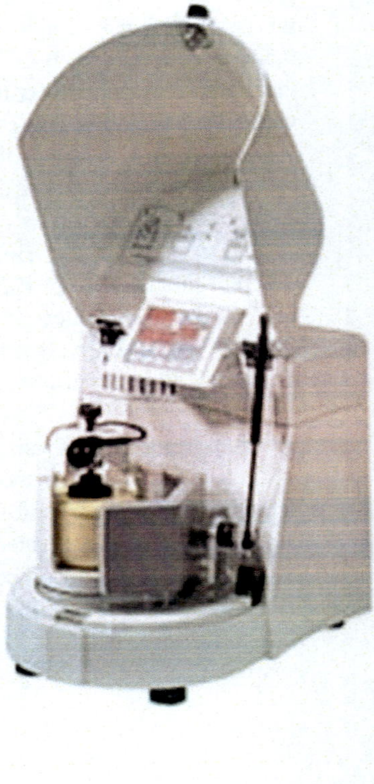

Fig. 1 Image of planetary ball mill (P6)

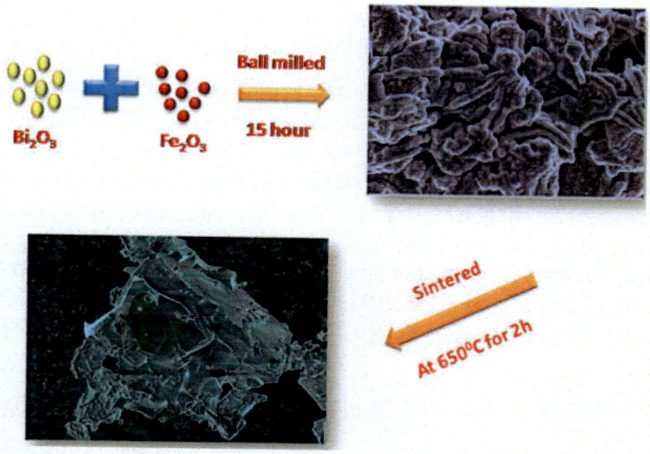

Fig. 2 Schematic diagram of synthesis of Bi–Fe–O based nanocomposites

Synthesis of Bi–Fe–O-based nanocomposites

For obtaining Bi–Fe–O nanocomposites, Bismuth oxide (Bi_2O_3) (SRL, 99% pure) and ferric oxide (Fe_2O_3) (LobaChemie, 98.5% pure) were taken as precursor materials without further purification. A stoichiometric equimolar mixture of Bi_2O_3 and Fe_2O_3 was weighed in the open air, and a homogeneous mixture was prepared using an agate mortar and then placed in an 80 mL vial with 30 chrome steel balls of 10 mm diameter, taken with 40:1 ball-to-powder mass ratio (BPMR). Then the powder mixtures were milled continuously for 5 h, 10 h and 15 h with a successive pause of 15 min to avoid overheating effect in a high energy planetary ball mill (P6, M/S Fritsch, GmbH, Germany) at 200 rpm and at room temperature. Finally, the as-prepared milled powders were sintered at 850 °C for 2 h, as shown schematically in Fig. 2. The as-obtained samples were 5 h, 10 h, 15 h milled and 5 h, 10 h, 15 h milled + sintered.

Synthesis of V_2O_5–TiO_2 nanocomposites

To prepare V_2O_5–TiO_2 nanocomposites, V_2O_5 and anatase TiO_2 elemental powders were taken with 1:1 molar ratio as precursor materials. The homogeneous powder mixture was sealed in an 80 mL vial with 30 chrome steel balls with 10 mm diameter and milled in the open air. Ball milling was carried out with the ball to mass ratio (BPMR) 10:1 in a high energy planetary ball mill (Model-P5, M/s FRITSCH, GmbH, Germany). The as-prepared samples were 0 h, 1 h, 3 h and 5 h milled nanocomposites.

5.2 *Structural Characterization*

5.2.1 Microstructural Characterization Using XRD Data and Rietveld Analysis

The high-quality XRD data of all samples were recorded in step scan mode with a slow scan rate using an X-ray powder diffractometer (Bruker AXS, D8 Advance, Da Vinci) equipped with Ni filtered Kα radiation (λ = 1.5418 Å), within the angular range 20°–80° 2θ with step size 0.02° 2θ and scan time 2 s/step. After the phase identification, all recorded XRD patterns were analyzed by the Rietveld refinement using the Maud software for detailed structural and microstructural characterizations [40].

Microstructural characterization of Bi–Fe–O-based nanocomposites

Figure 3a represents the XRD profiles of all as-synthesized unsintered samples. The peak search-match method is adopted to identify all reflections, and it is found that along with the precursor phase α-Fe_2O_3 (ICSD #9015964), two new phases, namely, $Bi_2Fe_4O_9$ (ICSD #9008148) and $Bi_{25}FeO_{40}$ (ICSD #4030661) have formed after 5 h of milling. At around 30° and 50° 2θ, the peaks are broadened and amorphous-like background intensity is noticed in Fig. 3a. The newly generated $Bi_2Fe_4O_9$ phase has contributed a significant part among all the as-synthesized milled unsintered powders. The relative abundances of these phases decrease with the progress of milling hour, and gradually, the $Bi_{25}FeO_{40}$ phase becomes significant. The relative content of this phase increases and peak broadening with amorphous like background intensity increases with the increase of milling hour up to 15 h. Figure 3b represents the XRD profiles of sintered and milled nanocomposites. It reveals the reflects the facts that after sintering the as-obtained nanocomposites become more crystalline and amorphous like background completely vanishes. In addition to the precursor materials, two new phases as mentioned above, $BiFeO_3$ and β-Bi_2O_3 generate due to sintering. According to the respective ICSD/JCPDS files, all these reflections are identified and indexed accordingly. It is also evident from the figure that the relative intensity distributions of all these phases are varying in different samples. It seems that due to the high energy impact of mechanical alloying, peak-broadening of these overlapping reflections of the milled samples increases enormously; as most of the intense reflections in these regions are composed of both $Bi_{25}FeO_{40}$ and $Bi_2Fe_4O_9$ phases, which results in the formation of amorphous-like background intensity in the XRD patterns of the milled samples. The peak broadening and peak overlapping of successive reflections of these phases increase with increasing milling time. During mechanical alloying, lattice imperfections generated in the milled samples due to the influence of sintering temperature have been reduced extensively, and consequently, peak-broadening also reduces to a large extent with the increasing milling hour. Thus, in Fig. 3b, all these overlapping reflections appear as isolated and distinguishable reflections.

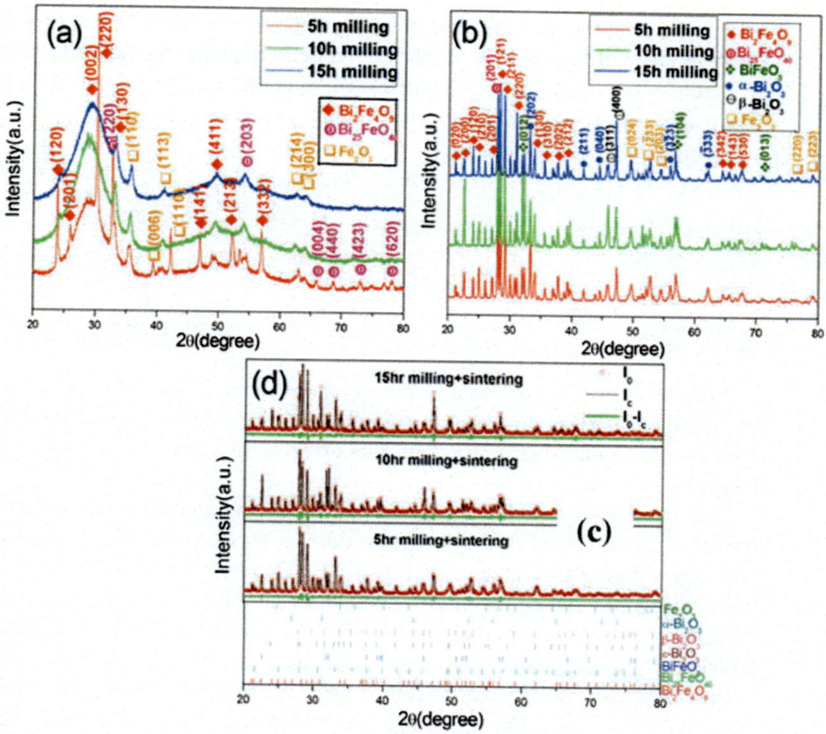

Fig. 3 XRD patterns of **a** unsintered samples, **b** sintered samples properly indexed with JCPDS files, **c** typical Rietveld analysis output of all the sintered samples where the red circles represent the experimental data, the solid black line represents the refined simulated data, and the green line under particular pattern signifies the residual intensity (I_O–I_C), I_O and I_C are the observed and calculated intensities, respectively

For estimating the individual parameters quantitatively, the Rietveld refinement method has been employed, considering all reflections of all pertinent phases in the XRD patterns. The theoretical (calculated) XRD pattern (I_C) is simulated with the contributions of α-Fe_2O_3 (ICSD #9015964; Sp. Gr.- R3c:H; a = 5.0346 Å, c = 13.7473 Å), α-Bi_2O_3 (ICSD #9009850; Sp. Gr.- Fm3m; a = 5.6549 Å), $Bi_{25}FeO_{40}$ (ICSD #4030661; Sp. Gr.- I23; a = 10.21 Å), $Bi_2Fe_4O_9$ (ICSD #9008148; Sp. Gr.- Pbam; a = 7.905 Å, b = 8.428 Å, c = 6.005 Å), β-Bi_2O_3 (ICSD #9007723; Sp. Gr.- P42_1c; a = 7.739 Å, c = 5.636 Å) and $BiFeO_3$ (ICSD #4336775; Sp. Gr.- R3c:H; a = 5.5797 Å, c = 13.87201 Å) phases. By refining several structural and microstructural parameters, experimental (observed) XRD profiles (I_O) of all milled sintered and unsintered samples are fitted to the respective simulated XRD patterns (I_C) (Fig. 3c). Due to the small crystallite size and lattice strain of the nanocomposites, peak broadening arises, and it has been taken care of by considering the pseudo-Voigt (pV) analytical function [41, 62]. A standard Si sample has been utilized to estimate the instrumental broadening, and these parameters were kept unaltered during the entire refinement process [35]. The Marquart Least Squares method has been assigned to conduct the refinement and 'goodness of fit' (GoF) monitored the entire iterative process by converging its values toward unity, which is defined as

$$GoF = R_{wp}/R_{exp}, \tag{17}$$

where,

$$R_e = \frac{(N - P)}{\sum W_i(I_o)^2} \quad \text{and} \quad R_{wp} = \frac{W_i(I_o - I_C)^2}{\sum W_i(I_o)^2} \tag{18}$$

In these reported cases, the GoF values vary between 1.19 and 1.49, which seems to be a good refinement. After completing the refinement process, some of the essential structural and microstructural parameters obtained from the Rietveld refinement are tabulated in Table 1.

It is evident from Table 1 that the relative phase abundance of the crystalline phases of all the milled sintered and unsintered samples varies with milling hour and elevated sintering temperature such as $Bi_2Fe_4O_9$ phase becomes significant in 5 h milled sample, and then the relative amount of this phase decreases from ~80–60% on further milling. Again the phase $Bi_{25}FeO_{40}$ appears in 5 h milled sample, and its relative amount increases up to 10 h and then on further milling, the $Bi_{25}FeO_{40}$ phase disappears and transforms into $BiFeO_3$ phase after 15 h milling. There is no such contribution of α-Fe_2O_3 phase in the milled sample as its relative contribution remains invariant for milling hour. After sintering, significant changes occurred in the crystalline phase abundances. Again $Bi_2Fe_4O_9$ phase becomes a primary contributing phase, and among the sintered nanocomposites, its relative amount increases with the increasing milling hour. It is evident that with the increasing milling hour, the stoichiometric mixture of $Bi_{25}FeO_{40}$ and $BiFeO_3$ disappears and reappears as the $Bi_2Fe_4O_9$ phase. It is also evident from the above table that the crystallite size and the lattice strain change significantly with the increasing milling time and due to sintering. It is noticed that the particle size of all pertinent phases of the milled sample increases, in general, with increasing milling time, establishes the fact of agglomeration of nanoparticles, and the rate of agglomeration increases with the increasing milling time. Due to sintering, there is a significant grain growth noticed in all sintered samples, particularly in the 15 h milled sample. For the $Bi_2Fe_4O_9$ phase, a high grain growth rate is evidenced after 15 h milling followed by sintering it at 850 °C for 3 h. Lattice strain reduces continuously in the course of milling due to agglomeration except for the $BiFeO_3$ phase as it appears due to transformation of the two phases with a very high strain but reduces significantly after sintering along with the other two phases. Lattice strain reduces up to 100 folds of all the pertinent phases due to sintering. So it is evident that after critical size-strain refinement analysis, milling and sintering at elevated temperature favour the growth of heterostructure

Table 1 Refined structural and microstructural parameters of ball-milled and sintered Bi_2O_3-Fe_2O_3 compounds revealed from the Rietveld refinement

Sample specification	Structure and micro-structure parameters	$Bi_2Fe_4O_9$	$Bi_{25}FeO_{40}$	$BiFeO_3$	β-Bi_2O_3	α-Bi_2O_3	ϵ-Bi_2O_3 (orthorhombic)	α-Fe_2O_3 (trigonal)
5 h milling	a (Å)	10.7462	10.7553					5.0133
	b (Å)	3.9755						
	c (Å)	7.3704						13.8921
	D (nm)	122.25	201.02					210.63
	Microstrain	0.250	0.00531					0.0056
	Phase content (Vol%)	79.90	3.27					16.83
5 h milling + sintered	a (Å)	7.965	10.1560	5.5820				
	b (Å)	8.437						
	c (Å)	5.997		13.875				
	D (nm)	384.35	52.18	339.31				
	Microstrain	0.0009	0.0012	0.0000				
	Phase content (Vol%)	61.42	27.31	11.27				
10 h milling	a (Å)	10.7663	10.4667					5.0386
	b (Å)	4.0783						
	c (Å)	7.4873						13.7558
	D (nm)	141.02	2.20					218.39
	Microstrain	0.061	0.0323					0.0055
	Phase content (Vol%)	58.81	29.2					11.9

(continued)

Table 1 (continued)

Sample specification	Structure and micro-structure parameters	$Bi_2Fe_4O_9$	$Bi_{25}FeO_{40}$	$BiFeO_3$	β-Bi_2O_3	α-Bi_2O_3	ϵ-Bi_2O_3 (orthorhombic)	α-Fe_2O_3 (trigonal)
10 h milling + sintered	a (Å)	7.968	10.144	5.577			11.478	4.997
	b (Å)	8.437					6.604	14.182
	c (Å)	5.998		13.864			4.797	
	D (nm)	393.04	311.35	725.22			219.48	282.92
	Microstrain	0.0006	0.0007	0.000			0.0007	0.0003
	Phase content (Vol%)	54.08	26.78	17.76			0.63	0.24
15 h milling	a (Å)	7.4484		6.027				5.0486
	b (Å)	8.9549						
	c (Å)	6.2535		14.4097				13.7587
	D (nm)	192.96		7.38				373.34
	Microstrain	0.043		0.084				0.032
	Phase content (Vol%)	51.88		38.75				10.28
15 h milling + sintered	a (Å)	7.9686	10.1426	5.5769	7.8217	5.7455	11.3131	5.0114
	b (Å)	8.4376		13.8648	5.6670		5.8780	13.5057
	c (Å)	5.9977					5.2578	
	D (nm)	1350.01	700.07	692.98	143.98	101.08	423.05	99.01
	Microstrain	0.0003	0.0005	0.0000	0.0000	0.0006	0.0001	0.0000
	Phase content (Vol%)	81.70	7.46	1.55	0.64	6.87	0.671	1.102

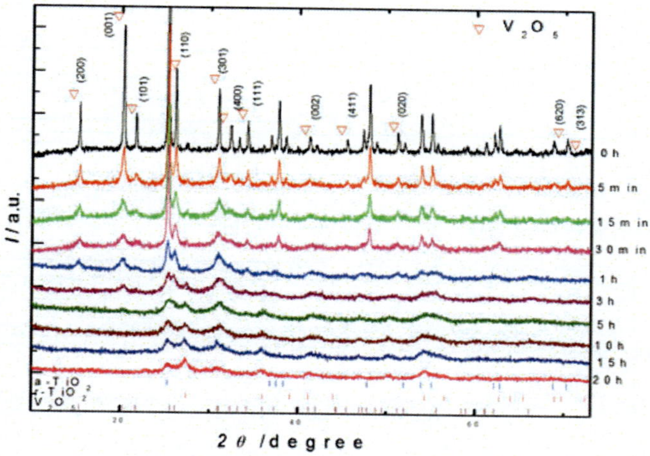

Fig. 4 XRD patterns of all ball-milled V_2O_5–TiO_2 powder mixtures

nanoparticles containing insignificant lattice strain, which drastically reduces peak broadening of all contributing phases.

Microstructure characterization of XRD patterns of V_2O_5–TiO_2 nanocomposites by Rietveld method

Figure 4 represents the XRD profiles of the unmilled and milled powder mixtures of V_2O_5–TiO_2 nanocomposites. The reflections of all these phases are represented by small markers at the bottom of the XRD patterns. Following the reported intensity values, XRD pattern of the unmilled powder sample contains only the reflections of V_2O_5 and anatase (a)—TiO_2 phases. Due to milling, peak broadening appears due to the overlapping of peaks among the corresponding XRD patterns. As most of the peaks are overlapped, the Rietveld refinement method has been employed for quantitative phase estimation and to determine individual phases' structural and microstructural parameters. Figure 5 shows a few of the selected observed XRD patterns (I_O) with the respective calculated best-fitted patterns (I_C). Quality of fitting is judged by the GoF, as defined in Eq. (17) and continues until it approaches 1.0. From Fig. 6a, it is evident that the amount of TiO_2 phase decreases with increasing milling time due to solid-state diffusion of TiO_2 solute into the V_2O_5 matrix, which suggests that the solid solution phase grows within this milling time. Figure 6a represents that the mol fraction value of a-TiO_2 becomes nil during the milling of 5–10 h. After 5 h of the milling, the phase quantity of V_2O_5 decreases with a subsequent increase of the rutile (r)-TiO_2 phase. It seems that the V_2O_5-based solid solution phase decomposes after 5 h of milling, as some quantity of it has transformed to r-TiO_2 phase.

Figure 6b, c illustrates that the increasing milling time crystallite size and r.m.s. lattice strain vary for different phases obtained from Rietveld analysis. From Fig. 6b, it is. evident that the crystallite size of V_2O_5 is found to be anisotropic, and considering the highest intensity ratios of the respective crystal planes of V_2O_5, this anisotropy

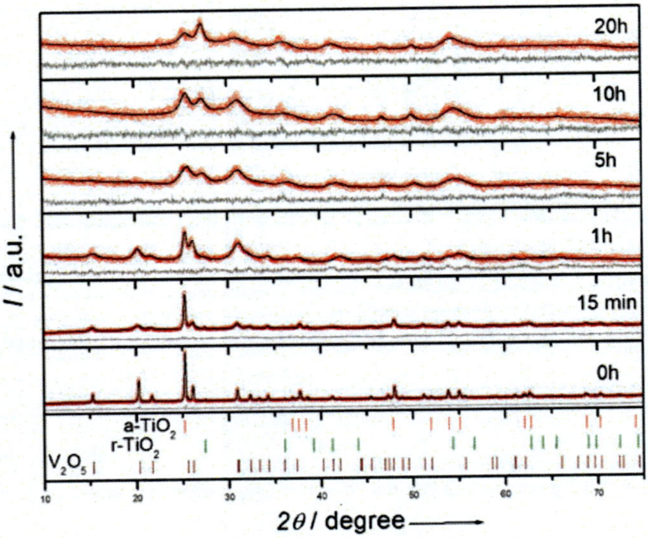

Fig. 5 Experimental (observed) XRD patterns (dots) of some selected ball-milled V_2O_5–TiO_2 powder mixture with refined simulated patterns (continuous lines). The difference between experimental data (I_o) and fitted simulated pattern (I_c) is shown as a continuous line (I_o–I_c) under each XRD pattern

is explained in three particular crystal plane directions. However, the crystallite size of all other phases in the V_2O_5–TiO_2 nanocomposite is found to be isotropic. The substitutional V_2O_5–TiO_2 solid solution phase gradually develops at the early stage of milling, and then the crystallite size of V_2O_5 tends to be isotropic. The almost isotropic crystallite size of V_2O_5 is obtained after 3 h of milling acquired the maximum growth of V_2O_5-based solid solution phase. After the 3 h of milling, crystallite sizes of both the precursor phases a—TiO_2 and V_2O_5 reduce to an average size of 10 nm, though the reduction rate is much faster for V_2O_5. Crystallite size of V_2O_5 becomes anisotropic after 5 h of milling because solid solution phase quantity decreases with a nominal preferential growth along <301>. After 10 h of milling, crystallite sizes of all the phases attained saturation value and became almost invariant till 20 h of milling. R.m.s. lattice strains of as-obtained nanocomposites change accordingly, as shown in Fig. 6c.

5.2.2 FESEM Analysis

FESEM images of the sample surface are produced using a field emission cathode in the electron gun of a scanning electron microscope, which provides narrower probing beams at low and high electron energy to improve the spatial resolution to minimize the sample charging and damage. FESEM (Carl Zeiss; Model: Sigma 300) images

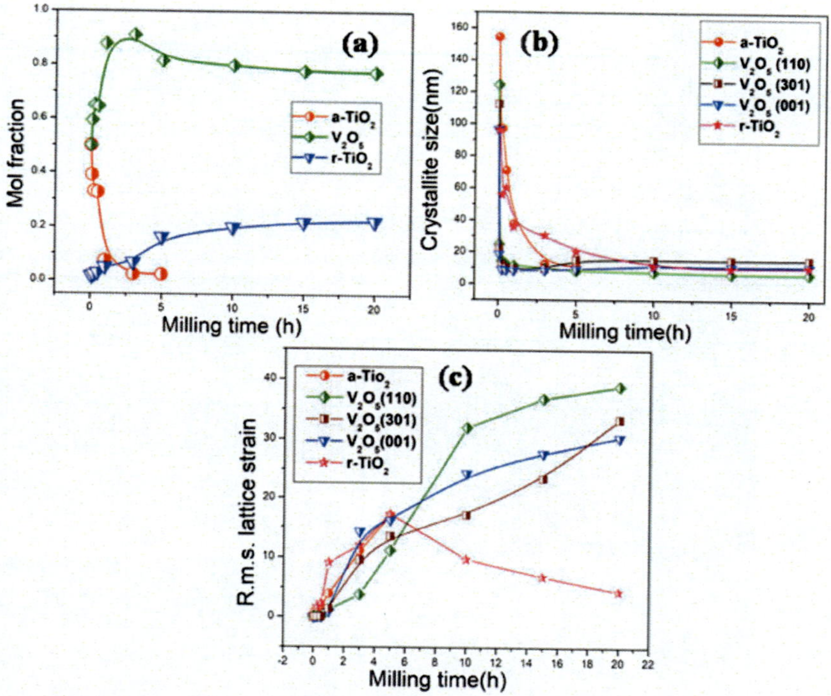

Fig. 6 **a** Mol fraction variations of phases produced in ball-milled V_2O_5–TiO_2 powder mixtures with increasing milling time, **b** crystallite size and **c** r.m.s strain of different phases produced in ball-milled V_2O_5–TiO_2 powder mixtures with increasing milling time

have been used for understanding the morphology of the synthesized products with an accelerating voltage of 5 kV.

FESEM analysis of Bi–Fe–O nanocomposites

Figure 7 represents the surface morphology through FESEM analysis of the as-synthesized Bi–Fe–O nanocomposites ensured the controlled morphology. From Fig. 7a, c, e, it is evident that all the milled samples formed an amorphous like structure with finger-like projections, which become more prominent with the increasing milling hour, supporting the Rietveld refinement results. However, morphology changes spontaneously with a nano brick-like structure due to sintering, which is represented in Fig. 7b, d, f. After 15 h of milling, symmetric agglomeration transforms into a multi-layered exfoliated sheet-like morphology, which enhances the surface area, improving the optical properties of the as-spun nanocomposites.

FESEM analysis of V_2O_5–TiO_2 nanocomposites

Figure 8a represents the FESEM images of the as-obtained 3 h milled sample, which shows a clear and bright pineapple leaf or flower-like morphology. With the

Fig. 7 FESEM images of BiFeO samples, **a** 5 **h** milled sample, **b** 5 **h** milled and sintered sample, **c** 10 **h** milled sample, **d** 10 **h** milled and sintered sample, **e** 15 **h** milled sample and **f** 15 h milled and sintered sample

increasing milling hour, the amount of leaf-like structure increases and after 5 h milling, the synthesized nanocomposites transform into a broccoli or broken leaf-like structure due to the high impact of milling. However, the 3 h milled sample with pineapple morphology shows the greater surface area, enhancing visible light photocatalytic degradation efficiency.

5.2.3 TEM Analysis

A beam of electrons is transmitted through an ultra-thin specimen in a transmission electron microscope (TEM), interacting with the specimen as it passes through it.

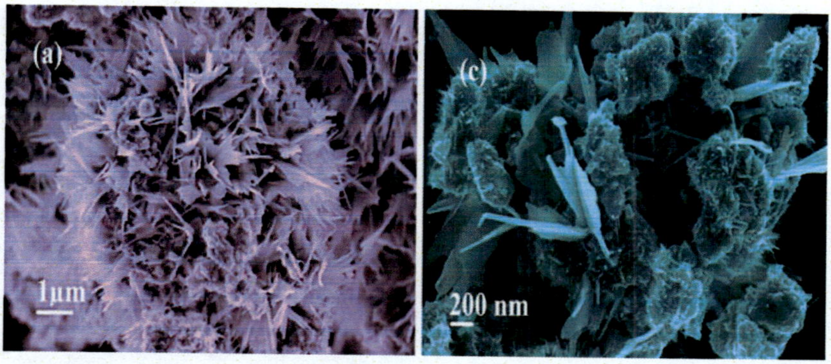

Fig. 8 **a** FESEM images of 3 h ball-milled V_2O_5–TiO_2 nanocomposite sample, pineapple leaf-like morphology, **b** fragmented leaf-like morphology of 5 h ball-milled nanocomposite

As electrons are transmitted through the specimen, a magnified image is formed and focused by an objective lens and appears on an imaging screen. TEM images can be considered as the most direct proof for the formation of nanocrystals.

TEM analysis of Bi–Fe–O nanocomposites.

Figure 9a, d, e represents the TEM images from different places of the subjective grid.

The as-synthesized 15 h milled and sintered sample, which elucidates a cluster type sheet-like images, resembles the FESEM images. The size distribution of some isolated nanoparticles measured from different images is plotted in Fig. 9c, which reveals a wide range of crystallite sizes varying from 100 to 900 nm and most of the particles have crystallite size within 300–500 nm, quite comparable with the Rietveld refinement results. Figure 9b represents the powder diffraction pattern with well-resolved electron diffraction spots, which establishes the fact that the sample is composed of some well-grown tiny single crystals (diffraction spots) as well as few nanocrystalline particles (diffraction rings). Most of the intense and resolved spots and diffraction rings are indexed accordingly, reflecting that the nanocomposite consists of $Bi_{25}FeO_{40}$, $Bi_2Fe_4O_9$, $BiFeO_3$, α-Fe_2O_3, α-Bi_2O_3 and ε-Bi_2O_3 phases. The absence of any other spots establishes that there is no impurity in the sample, and the amorphous-like background intensity disappears after sintering the sample.

TEM analysis of V_2O_5–TiO_2 nanocomposites

Particle size distribution plot of as-spun 3 h milled V_2O_5–TiO_2 nanocomposites possess a Gaussian type distribution, as shown in Fig. 10a, which elucidates that the average particle size is ~25 nm, Resemble well with the Rietveld refinement results. Figure 10b represents the indexed electron diffraction pattern of the as-synthesized 3 h milled sample, which elucidates the formation of V_2O_5–TiO_2 nanocomposite with a trace amount of precursor a-TiO_2 phase and r-TiO_2 phase. According to the JCPDF files, diffraction rings are indexed, and rutile-TiO_2, anatase-TiO_2 and V_2O_5 phases are

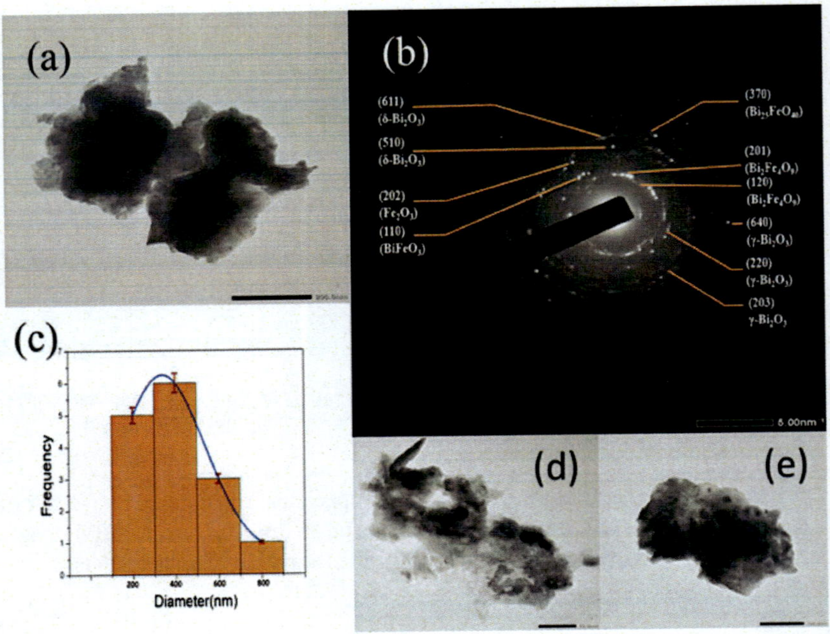

Fig. 9 **a**, **b** and **c** TEM images of as-synthesized 15 h milled and sintered (850 °C) sample, **b** indexed SAED pattern and **c** particle size distribution

noticed to present in the nanocomposites. Figure 10c represents the indexed electron diffraction patterns of the as-obtained 5 h milled sample, establishes the presence of both V_2O_5 and TiO_2 phases and the absence of any other impurity phase. The particle size distribution of the 5 h milled V_2O_5–TiO_2 nanocomposite represented in Fig. 10d reveals that the average particle size is ~12.5 nm, significantly less than the 3 h milled sample.

5.3 *Optical Properties*

5.3.1 Bandgap and Urbach Energy Analysis of Bi–Fe–O Nanocomposites

Figure 11a, b represents the diffuse reflectance spectra of the as-synthesized milled unsintered and sintered nanocomposites reveal a noticeable variation in the region between 600 and 800 nm, which exhibit the d d electronic transition of the Fe element. By applying the Kubelka-Monk function and using the relation [20]

$$F(R) = (1 - R)^2/2R \tag{19}$$

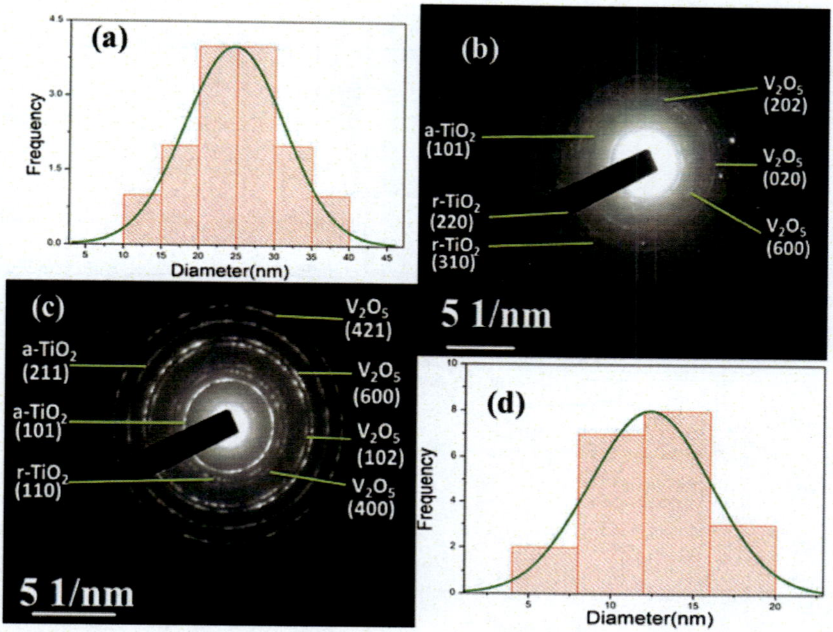

Fig. 10 TEM micrographs showing particle size distributions of as-milled **a** 3 **h** and **d** 5 h milled V_2O_5–TiO_2 nanocomposite; indexed SAED patterns of **b** 3 **h** and **c** 5 h milled V_2O_5–TiO_2 nanocomposite

the bandgap energy of the as-synthesized nanocomposites is calculated, where R is the experimental reflectance. In this present study, due to the absence of a sharp absorption edge, the Tauc plot is employed to determine the bandgap of the as-spun samples following the relation [37, 50],

$$F(R)h\upsilon = A\left(h\upsilon - E_g\right)^n \tag{20}$$

where F(R) is the diffuse reflectance, h is the Plank's constant, ν is the frequency of photons, A is the proportionality constant, n is the exponential constant, which is determined whether the nanocomposites are direct or indirect bandgap semiconductors. For all milled and unsintered samples, there is only one absorption band reflects in the above curve shown in Fig. 11a, which corresponds to the significant phase of the milled nanocomposites, i.e. $Bi_2Fe_4O_9$ and its bandgap sharply decreases from 3.79 to 1.98 eV, with the increasing milling time up to 10 h milling and then almost remain invariant. However, there are two such bands for the sintered samples shown in Fig. 11b corresponding to the sintered nanocomposites' major contributing phases, i.e. for $Bi_2Fe_4O_9$ $Bi_{25}FeO_{40}$. The bandgap of $Bi_{25}FeO_{40}$ only varies from 1.58 eV to 1.49 eV, and the other phase remains almost constant with the increasing milling time. From Fig. 11c, it can be noticed that, unusually, with the increasing milling hour,

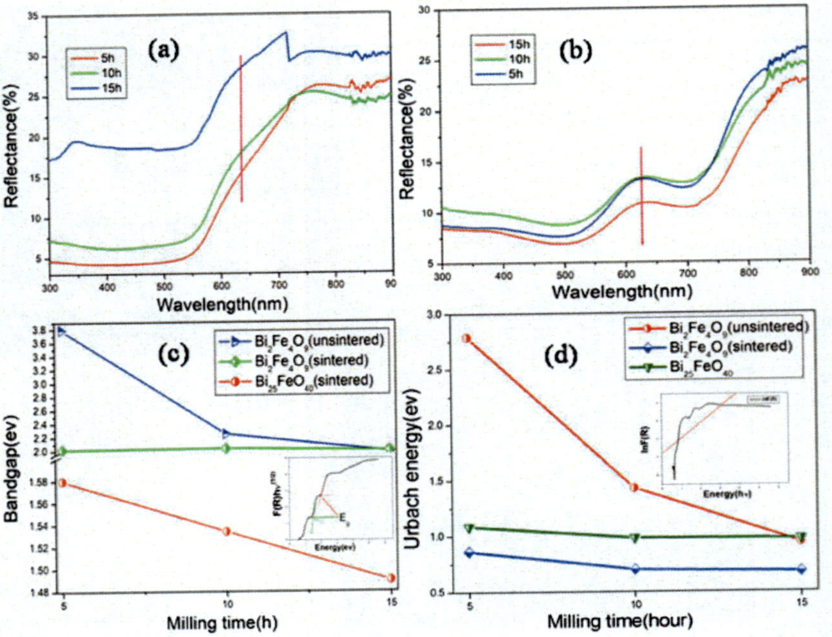

Fig. 11 Reflectance spectra of **a** unsintered and **b** sintered samples, **c** bandgap energy of the unsintered and sintered samples. **d** Variation of Urbach energy of the sintered and unsintered samples

the bandgap of the as-spun nanocomposites decreases due to an increase in particle size due to agglomeration effect and also due to surface effect and formation of new phases, i.e. redshift occurs that enhance the absorption efficiency of the samples. Thus from the above result, it can be concluded that the bandgap energy of the as-obtained nanocomposites can be tuned as a desire for a specific application only by regulating the milling hour and sintering temperature. For further investigation due to the introduction of defect levels, point defects, disorders, or strain between the conduction band (CB) and valance band (VB) band tailing effect is analyzed, and Urbach energy of the nanocomposites is calculated by using a different Kubelka–Monk function [48]

$$\ln F(R) = h\upsilon / E_U, \quad (21)$$

where E_U is the inverse slope of the ln F(R) versus hυ graph. Urbach energy for all the unsintered samples decreases from 2.79 to 0.98 eV with the increasing milling hour, and for all sintered samples, it remains almost invariant ~0.0 eV as shown in Fig. 11d. Urbach energy analysis is confined to the skin layer since most defects are usually trapped in the surface layer. So, whenever there is a change in Urbach energy or bandgap energy, the morphology changes or defects states trapped at the skin layer, which may well modify the optical responses of material, which has been noticed in the case of Bi–Fe–O nanocomposites.

UV–Visible absorption spectra analysis and optical band gap determination of V_2O_5–TiO_2 nanocomposites

Within the wavelength range 200–800 nm, the UV–visible absorbance spectra of the as-synthesized nanocomposites are represented in Fig. 12a. It is evident from Fig. 12a that there are two characteristic peaks arise for 5 h milled sample, respectively, at 265.33 and 425.33 nm, which attribute to the V_2O_5 and TiO_2 phases [35]. However, for 3 and 1 h milled samples, the characteristic peak remains almost the same for the TiO_2 phase but shifts from 260.70 to 256.22 nm for the V_2O_5 phase due to the incorporation of Ti^{4+} ions into the V_2O_5 lattice. For the unmilled sample, there is only one characteristic peak that appears at 253.97 nm, which is assigned to the V_2O_5 phase.

Since there is a sharp absorption peak, the bandgap energy of the as-synthesized samples is calculated using the following relation [18]

$$(\alpha h\upsilon)^{1/n} = A(h\upsilon - E_g) \tag{22}$$

where α is the absorption coefficient, h = plank's constant, ν = frequency of the photon, Eg is the bandgap energy, A = proportionality constant, n = exponent depends upon direct or indirect bandgap semiconductor. In the present study, TiO_2 is considered as an indirect bandgap semiconductor (n = 2), and V_2O_5 is considered as a direct bandgap semiconductor (n = 1/2).

The bandgap energy of both TiO_2 and V_2O_5 decreases with increasing milling time revealed in Fig. 12b, i.e., decreasing the crystallite size, which is attributed to the fact that surface pressure increases as crystallite size decreases reduce in lattice strain and also the diminishing of the bandgap energy. So the observed redshift in

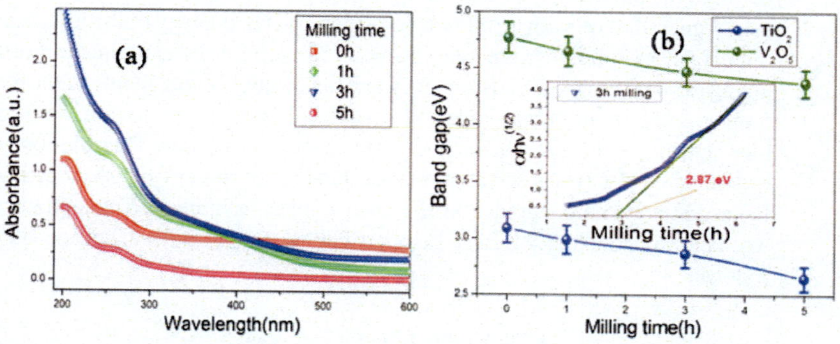

Fig. 12 **a** UV–visible absorbance spectra of unmilled (0 h) and ball-milled V_2O_5–TiO_2 nanocomposites, **b** variation of bandgap energy of V_2O_5–TiO_2 nanocomposites with increasing milling time. Inset: Tauc plot of 3 h milled V_2O_5–TiO_2 powder mixture

Table 2 Urbach energy of the unmilled and ball-milled TiO_2–V_2O_5 samples

Milling hour (h)	Urbach energy (eV)	
	TiO_2	V_2O_5
0	1.302	1.55
1	1.164	1.49
3	1.043	1.061
5	0.788	0.898

the band edge is primarily due to the quantum size effect and surface effect, which enhance the photocatalytic activity of the synthesized nanocomposites.

Table 2 represents the Urbach energy of the as-spun nanocomposites' defect state level using the Eq. (21) mentioned above. The band tailing effect arises due to the introduction of induced defects due to the formation of heterojunction. Urbach energy of the as-synthesized nanocomposites decreases with the increasing milling hour, i.e. with the decreasing bandgap. With the increasing Urbach energy value, more and more defect states are generated between the conduction band and valence band, which effectively enhanced the photodegradation efficiency.

5.3.2 Photocatalytic Activity

The photocatalytic experiment was carried out using Rhodamine B as a model organic pollutant irradiation with solar-like light. For preventing any thermal catalytic effect during the photocatalytic degradation reaction, the whole experimental setup was placed on an ice bath to maintain the room temperature. Previously, a 5 ppm stock solution of Rhodamine B was prepared, and then the photocatalysts were added with a 10 ppm concentration to the 150 mL of the stock solution. Then the solution stirred vigorously at first in dark conditions to attain the adsorption–desorption equilibrium between the photocatalysts and the Rhodamine B molecule and then simulated with solar light radiation with a 200 W halogen lamp. The intensity of the light source and the pH of the solution were kept invariant throughout the experiment. For investigating the photodegradation efficiency and the reaction's kinetic rate of the reactions, 5 mL sample solution was collected periodically after 30 min from the beaker and then centrifuged to separate the photocatalysts for further use. The solution was then examined by UV–Vis absorption spectra about the characteristic peak of Rhodamine B ~ at 553 nm. Photodegradation efficiency of the as-synthesized nanophotocatalysts was determined using the Beer-Lambert law related to molecular concentration as follows,

$$\text{Degradation efficiency} = (C_0 - C)/C_0 \tag{23}$$

For evaluating the kinetic rate constant of the as-obtained photocatalysts, first-order kinetic rate equation has been employed as,

$$\ln C/C_O = kt \tag{24}$$

where k is the kinetic rate constant, C is the instantaneous concentration, and C_0 is the initial concentration.

Photocatalytic activity of Bi–Fe–O nanocomposites

Figure 13 represents the UV–visible spectra of Rhodamine B solution in the presence of 15 h milled and sintered samples, collected at different reaction times. It also reflects that the intensity of the characteristic curve or concentration of Rhodamine B gradually decreases with the increasing illumination time by monitoring the characteristic peak at 553 nm. It is evident from Fig. 13 that Rhodamine B molecule almost degraded ~100% after 360 min.

Figure 14a represents the relative concentration of Rhodamine B molecule in the presence of the as-obtained nanocomposites. It elucidates that the degradation

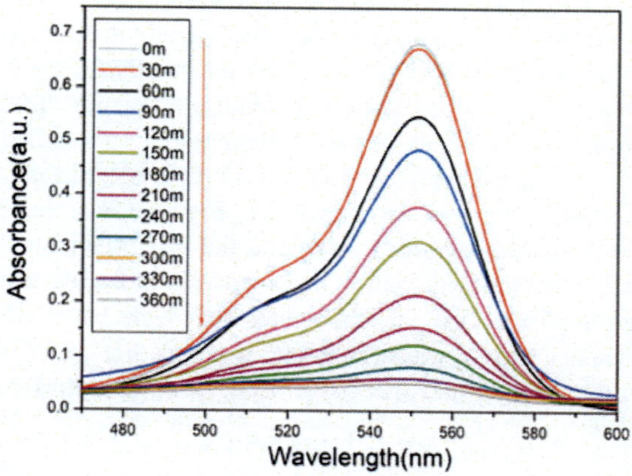

Fig. 13 UV–visible spectra of Rhodamine B molecule in presence of 15 h milled and sintered

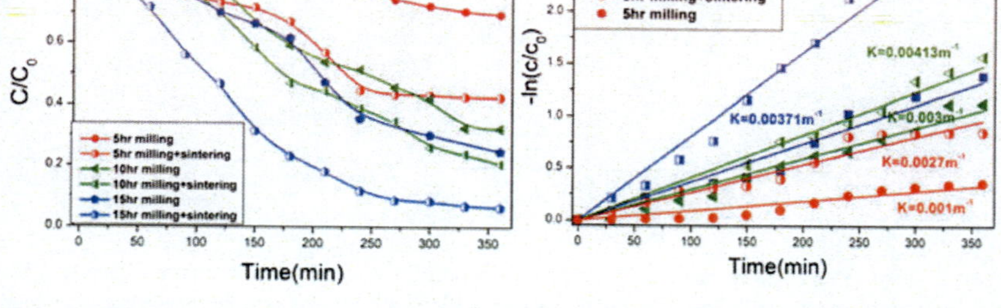

Fig. 14 **a** C/C_0 versus time variation of the RhB degradation, **b** kinetic rate constant of RhBphotodegradation reaction

of the Rhodamine B molecule increases with the increasing milling hour followed by sintering, i.e. with the increasing crystallite size and the highest degradation acquired in the presence of 15 h milled nanophotocatalyst sintered at 850 °C, which is unusual as the photocatalytic experiment is a surface phenomenon. However, with the introduction of lattice imperfection and the skin effect viewpoint, this reverse phenomenon can be explained. Figure 14b represents the kinetic rate constants of the as-spun nanocomposites, and it increases from 0.001 to 0.0081 min^{-1}. Nevertheless, crystallite size is a useful parameter that could explain the light absorption efficiency. Some other facets, such as can explain it with the decreasing particle size. Some defects or local strain creating on the surface may help in further recombination of photogenerated electron–hole pairs [6]. By trapping the charges inside the sample, these inner defects limit the reaction path of the electron and restrict their transportations to the surface. Again the bandgap energy reduces due to some extra levels created by the defects, which effectively reduces the bandgap energy. The other probable reasons may be, (a) skin layer which is the trapping centre for charge carriers [52], (b) a standard feature of the oxide samples, i.e. oxygen vacancy or point defects [13], (c) by enhancing the surface barriers due to formation of heterojunctions by forming the mixed-phase, which eventually promote the charge separation that increases the photocatalytic degradation efficiency. To further investigate the photodegradation efficiency of the as-synthesized 15 h milled and sintered sample, degradation of a non-absorbing molecule, i.e. colourless kanamycin antibiotic mainly found in wastewater, was also performed. Figure 15a represents the degradation of relative intensities concerning time, and the inset elucidates the kinetic rate, K = 00.075 min^{-1} of the degradation of kanamycin molecule taken for a shield photosensitization effect. Also, the RhB molecule's degradation efficiency is investigated in the presence of H_2O_2 molecule, and it seems that the degradation rate becomes faster for the previous and almost 100% degradation occur within 240 min, as shown in Fig. 15b.

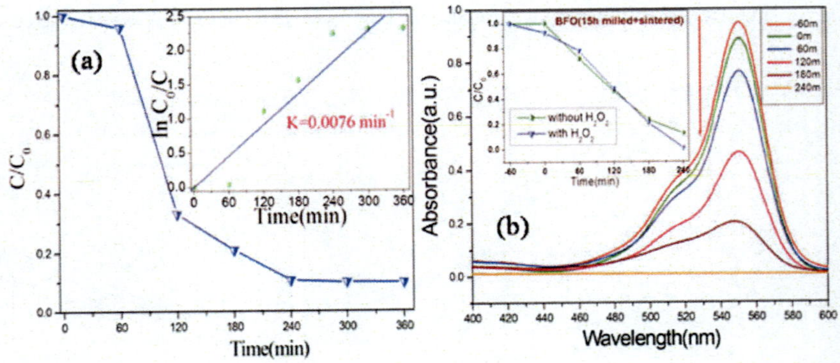

Fig. 15 **a** C/C_0 versus time graph of 15 h milled and sintered samples in the presence of Kanamycin antibiotic (inset: kinetic rate curve for 15 h milled and sintered sample for Kanamycin degradation), **b** RhB degradation in the presence of H_2O_2

Photocatalytic activity of V_2O_5–TiO_2 nanocomposites

Figure 16 represents the UV–visible spectra of Rhodamine B molecule in presence of as-synthesized unmilled and milled V_2O_5–TiO_2 nanocomposites, and the intensity of the curves decreases with increasing irradiation time, which is analyzed by monitoring the shifting of the characteristic peak at ~553 nm. It is shown from the figure that the rate of decreasing the intensity curve is different for different nanocomposites, and it is highest, i.e. 94% within 300 min for 3 h milled photocatalyst resemble with the conclusion of the Rietveld refinement result.

Figure 17a represents the variation of relative intensities of Rhodamine B molecule, i.e. C/C_0 ratio with the irradiation time, which reflects the fact that the degradation efficiency increases with the increasing milling hour up to 3 h milling and then again decreases for 5 h milled nanocomposites. The inset in Fig. 16a elucidates the percentage degradation of the Rhodamine B molecule. To understand the degradation mechanism, the kinetic rate constant is determined by using the Eq. (24).

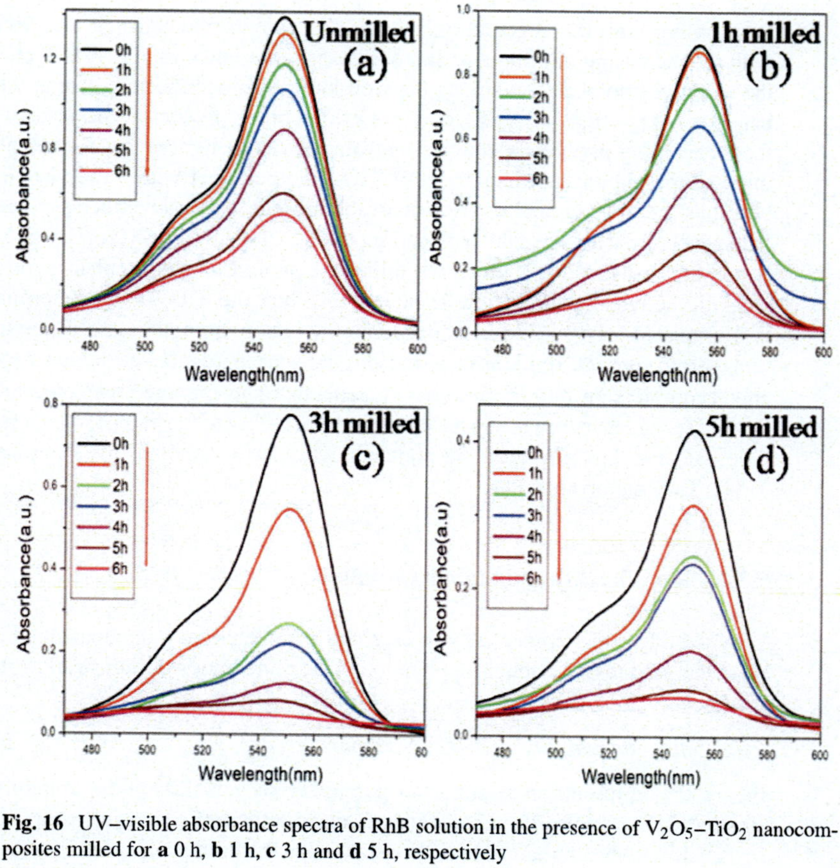

Fig. 16 UV–visible absorbance spectra of RhB solution in the presence of V_2O_5–TiO_2 nanocomposites milled for **a** 0 h, **b** 1 h, **c** 3 h and **d** 5 h, respectively

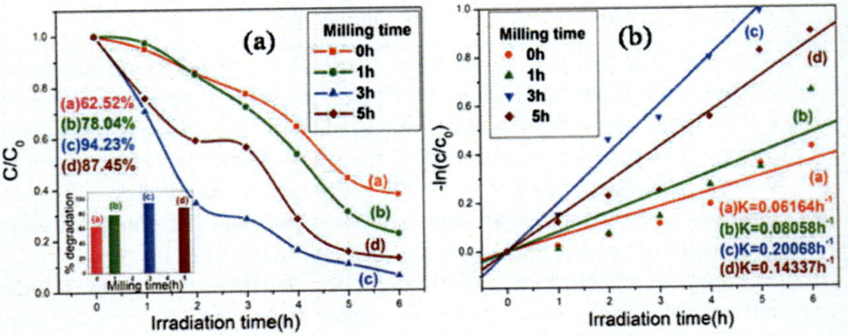

Fig. 17 **a** Variation of reaction rate (C/C_0) with V_2O_5–TiO_2 nanocomposite sample milled for (a) 0 h, (b) 1 h, (c) 3 h and (d) 5 h. Inset: percentage of degradation of RhB with respective milling time. **b** Comparison of the rate constant of RhB degradation with V_2O_5–TiO_2 nanocomposite milled for (a) 0 h, (b) 1 h, (c) 3 h and (d) 5 h, respectively

Resembling with the degradation spectra, the kinetic rate constant also increases with the increasing milling hour and decreases represented in Fig. 17b. TiO_2, being the effective photocatalyst among the well-known semiconductors, due to the wide bandgap energy (E_g ~ 4.78 eV), does not exhibit better photocatalytic activity due to the incapability in photons' absorption from the visible light range. So the enhancement of the photocatalytic activity of TiO_2 photocatalyst within the visible range becomes the most crucial motivation of the present work. Nevertheless, creating a heterojunction with a narrow bandgap semiconductor such as V_2O_5 (E_g ~ 2.4 eV) is one of the best strategies, which can utilize the photons and thus enhance the photocatalytic activity in the visible light range. When the TiO_2–V_2O_5 heterojunction is illuminated with visible light irradiation, electron–hole pairs are generated and transported through the interface and possibly the oxidation–reduction capability and recombination rate of these newly generated and migrated electron and holes decreases, which in turn improves the separation ability of the photoexcited electron–hole pair and thus improves the photocatalytic performance of the heterojunction V_2O_5–TiO_2 nanocomposite.

5.3.3 Recyclability and Stability Analysis

Recyclability and stability are two essential parameters in a photocatalytic experiment that ensures its practical application in the remediation of industrial wastewater treatment.

Reusability and stability of Bi–Fe–O photocatalysts

Figure 18a represents an essential key parameter, recyclability of the 15 h milled and sintered BFO sample. The photocatalytic degradation efficiency decreases after the first cycle and then remains almost constant up to the third cycle, primarily due to the

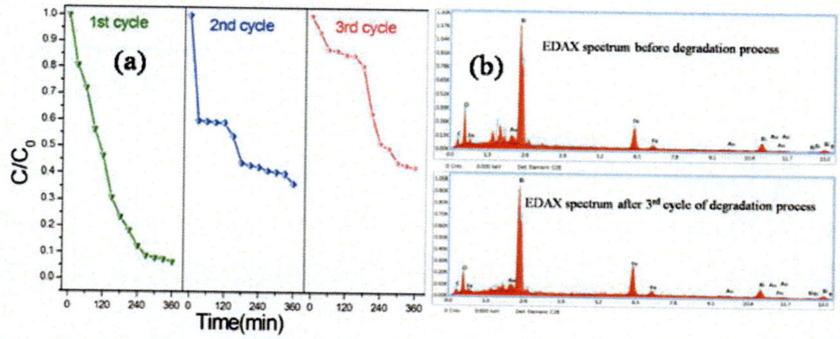

Fig. 18 **a** Recyclability curves for 15 **h** milled and sintered sample, **b** EDX spectra of 15 h milled and sintered BFO samples before and after the degradation process

loss of photocatalyst during filtration and collection. Figure 18b represents the EDX spectra of the BFO sample before the degradation process and after the third cycle of the degradation process, which ensures that the elemental compositions of the BFO catalyst remain unaltered before and after the photocatalytic reactions. Thus, the BFO catalyst can be reused for further photocatalytic reactions.

Reusability and stability of Bi–Fe–O photocatalysts

Figure 19 represents the EDX spectra of 3 h milled V_2O_5–TiO_2 sample before and after the degradation process. No new peak ensures the reusability ability of the as-synthesized nanocomposites.

6 Conclusion

This chapter summarizes the development of metal oxide heterostructured nanocomposites for industrial wastewater treatment using simulated solar radiation. The photocatalytic activities of some metal oxide semiconductors have been reported, and their limitations are also listed. For overcoming these limitations of the metal oxide semiconductors such as poor absorption ability, faster recombination rate etc., synthesis of heterostructured nanocomposite becomes utmost necessary. Along with some previous work, synthesis and characterization of three nanocomposites have been discussed in detail in this chapter. Metal oxide-based heterostructured nanocomposites are synthesized through mechanical alloying. Structural and microstructural characterizations are revealed by employing the Rietveld refinement method to the corresponding XRD profiles using MAUD software. Detailed surface morphology and modification is explored through FESEM image analysis.

Further, the newly generated phases have been verified using TEM images, and the electron diffraction images with no other extra diffraction spot ensure the purity of the phase formation of the corresponding nanocomposites. The semiconducting nature

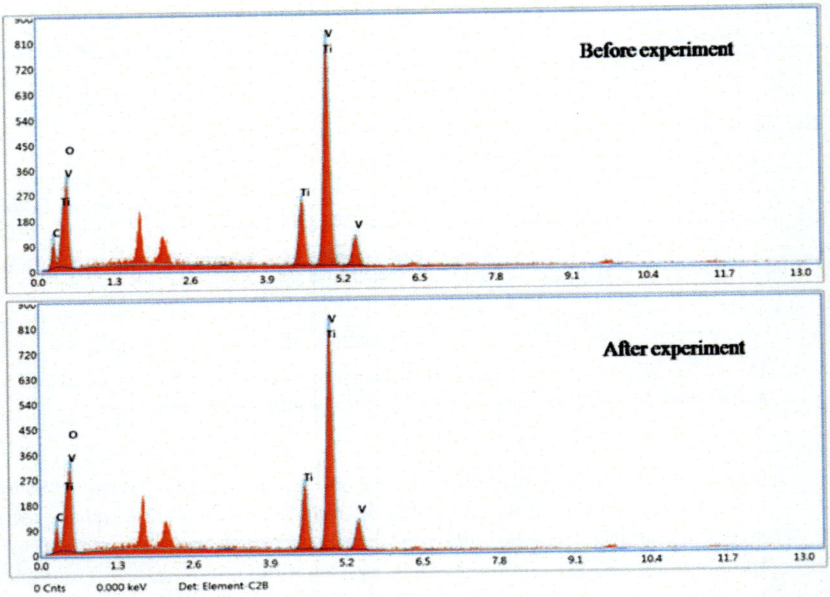

Fig. 19 EDX spectra of 3 h milled V_2O_5–TiO_2 sample before and after the degradation process

of the as-obtained nanocomposites is studied by observing their optical properties examined through diffused reflectance UV–visible spectroscopy and determined the co-relation between Urbach energy values and their corresponding optical properties. Finally, the capabilities of the as spun nanophotocatalysts in industrial wastewater remediation are ensured by investigating the photocatalytic degradation efficiency using primarily Rhodamine B as a model organic pollutant, and further investigation photodegradation of other organic dyes is also studied. Practical application of the as-synthesized nanocomposites in wastewater treatment is also ensured by recyclability and stability analysis.

Conclusive remarks on Bi–Fe–O nanocomposites

- Heterostructured Bi–Fe–O nanocomposites are mechanosynthesized by ball milling at different milling hours followed by sintering at 850 °C. Different physicochemical properties are examined. However, in contrast to the conventional result, particle size increases with the increasing milling hour, probably due to agglomeration or reduced lattice strain of about 100 times.
- After sintering, the XRD pattern changes significantly and proves to be more crystalline due to sintering, which is also reflected in the FESEM images. Surface morphology conveniently changes from cluster type structure to microbrick type structure.
- Again, in contrast to the traditional theory, the photocatalytic activity increases with the increasing particle size, i.e., the reducing surface area though it is a

surface phenomenon. This can be explained by introducing the trapping defect states between the conduction band and valence band, which increases the ability to separate the photogenerated electron–hole pair and migrate them towards the surface.

- The 15 h milled sample shows the highest photodegradation efficiency ~94% within 360 min, and with the addition of a small amount of H_2O_2, efficiency increases up to 100% within 240 min.
- Recycle capability up to third cycle, and prominent stability analysis of the Bi–Fe–O nanocomposites proved to be an efficient photocatalyst.

Conclusive remarks on V_2O_5–TiO_2 nanocomposites

- V_2O_5–TiO_2 nanocomposites are synthesized through mechanical alloying at different milling hours.
- The 3 h milled sample shows the highest photocatalytic activity ~100% within 300 min with a pineapple-like structure.
- Photocatalytic efficiency increases with the decreasing particle size, i.e. with the decreasing band gap and Urbach energy.

Acknowledgements Authors are thankful to the University Grants Commission (UGC) and Dept. of Science and Technology (DST), India, for their financial support in creating sophisticated research facilities in the Dept. of Physics of the University of Burdwan through CAS-I, CAS-II and FIST programs, respectively. M. Mondal wishes to thank the Govt. of West Bengal for financial support through the Vivekananda Merit-Cum-Means fellowship for non-NET fellows.

References

1. Agarwal R, Sanghi A, Ashima S (2011) Rietveld analysis, dielectric and magnetic properties of Sr and Ti codoped $BiFeO_3$ multiferroic. J Appl Phys 110:073909–073915. https://doi.org/10.1063/1.3646557
2. Ali I, Asim M, Khan TA (2012) Low cost adsorbents for the removal of organic pollutants from wastewater. J Environ Manag 113:170–183. https://doi.org/10.1016/j.jenvman.2012.08.028
3. Andreozzi R, Caprio V, Insola A, Marotta R (1999) Advanced oxidation processes (AOP) for water purification and recovery. Cat Today 53:51–59
4. Antonopoulou M, Evgenidou E, Lambropoulou D, Konstantinou I (2014) A review on advanced oxidation processes for the removal of taste and odor compounds from aqueous media. Water Res 53:215–234. https://doi.org/10.1016/j.watres.2014.01.028
5. Aoyama Y, Oaki Y, Ise R, Imai H (2012) Mesocrystalnanosheet of rutile TiO_2 and its reaction selectivity as a photocatalyst. CrystalEngComm 14:1405–1411. https://doi.org/10.1039/B202140D
6. Bai X, Wei J, Tian B, Liu Y, Reiss T, Guiblin N, Gemeiner P, Dkhil B, Ingrid CI (2016) Size effect on optical and photocatalytic properties in $BiFeO_3$ nanoparticles. J Phys Chem C 120:3595–3601. https://doi.org/10.1021/acs.jpcc.5b09945
7. Bessekhouad Y, Robert D, Weber JV (2005) Photocatalytic activity of Cu_2O/TiO_2, Bi_2O_3/TiO_2 and $ZnMn_2O_4/TiO_2$ heterojunctions. Catal Today 101:315–321. https://doi.org/10.1016/j.cattod.2005.03.038

8. Beychok MR (1971) Wastewater treatment. In: Hydrocarbon processing, pp 109–112. ISSN 0818-8190
9. Bokare AD, Choi W (2014) Review of iron-free Fenton-like systems for activating H_2O_2 in advanced oxidation processes. J Hazard Mat 275:121–135. https://doi.org/10.1016/j.jhazmat.2014.04.054
10. Brown MA, Vito DSC (1993) Predicting azo dye toxicity. Crit Rev Environ Sci Technol 23:249–324. https://doi.org/10.1080/10643389309388453
11. Carroll JP, Myles A, Quilty B, McCormack DE, Fagan R, Hinder SJ, Dionysiou DD, Pillai SC (2017) Antibacterial properties of F-doped ZnO visible light photocatalyst. J Hazard Mater 324:39–47
12. Chen JS, Chen C, Liu J, Xu R, Qiao SZ, Lou XW (2011) Ellipsoidal hollow nanostructure assembled from TiO_2 nanosheets as magnetically separable photocatalyst. Chem Comm 47:2631–2633. https://doi.org/10.1039/C0CC04471G
13. Chen S, Wang LW (2012) Thermodynamic oxidation and reduction potentials of photocatalytic semiconductors in aqueous solution. Chem Mater 24:3659−3666. https://arxiv.org/abs/1203.1970
14. Comninellis C (1994) Electrocatalysis in the electrochemical conversion/combustion of organic pollutants for waste water treatment. Electrochim Acta 39:1857–1862
15. Deng W, Zhao H, Pan F, Feng X, Jung B, Abdel-Wahab A, Batchelor B, Li Y (2017) Visible-light-driven photocatalytic degradation of organic water pollutants promoted by sulfite addition. Environ Sci Technol 51:13372–13379. https://doi.org/10.1021/acs.est.7b04206
16. Domenech J, Prieto A (1986) Stability of ZnO particles in aqueous suspensions under UV illumination. J Phys Chem 90:1123–1126
17. Dong F, Sun Y, Fu M (2012) Enhanced visible light photocatalytic activity of V_2O_5 cluster modified N-doped TiO_2 for degradation of toluene in air. J Photoenergy 2012:1–10
18. Fabbiyola S, John Kennedy L, Dakhel AA, Bououdina M, Judith Vijaya J, Ratnaji T (2016) Structural, microstructural, optical and magnetic properties of Mn doped ZnO nanostructures. J Mol Struct 1109:89–96
19. Fan J, Zhao L, Yu J, Liu G (2012) The effect of calcinations temperature on the microstructure and photocatalytic activity of TiO_2 based composite nanotube prepared by an in situ template dissolution method. Nanoscale 4:6597–6603
20. Fei L, Yuan J, Hu Y, Wu C, Wang J, Wang Y (2011) Visible light responsive perovskite $BiFeO_3$ pills and rods with dominant {111}c facets. Cryst Growth Des 11:1049–1053
21. Frank M, Sparrow B, Zoshi J, Low M (2017) Lowering cost and waste in flue gas desulfurization wastewater treatment. Power Mag Electric Power
22. Fu H, Pan C, Yao W, Zhu Y (2005) Visible-light-induced degradation of rhodamine B by nanosized Bi_2WO_6. J Phys Chem B 109:22432–22439
23. Gaya UI, Abdullah AH (2008) Heterogeneous photocatalytic degradation of organic contaminants over TiO_2. J Photochem Photobiol C Photochem Rev 9:1–12
24. Hameed A, Montini T, Gombac V, Fornasiero P (2008) Surface phases and photocatalytic activity correlation of Bi_2O_3/Bi_2O_4-x nanocomposite. J Am Chem Soc 130:9658–9689
25. Hameed A, Gombac V, Montini T, Felisari L, Fornasiero P (2009) Photocatalytic activity of zinc modified Bi_2O_3. Phys Chem Lett 483:254–261
26. Hanaor DA, Sorrell CC (2011) Review of the anatase to rutile phase transformation. J Mater Sci 46:855–874. https://doi.org/10.1007/s10853-010-5113-0
27. Hasegawa G, Morisato K, Kanamori K, Nakanishi K (2011) New hierarchically porous titania monoliths for chromatographic separation media. J Sep Sci 34:3004–3010
28. Hassan MS, Amna T, Yang OB, Kim CH, Khil MS (2012) TiO_2 nanofiber doped with rare earth element and their photocatalytic activity. Ceram Int 38:5925–5930
29. Hirai T, Okubo H, Komasawa I (1999) Size-selective incorporation of CdS nanoparticles into mesoporous silica. J Phys Chem B 103:4228–4230
30. Ikehata K, Gamal El-Din M, Snyder SA (2008) Ozonation and advanced oxidation treatment of emerging organic pollutants in water and wastewater. Ozone Sci Eng 30(1):21−26

31. Jiang HY, Cheng K, Lin J (2012) Crystalline metallic Au nanoparticle-loaded α-Bi_2O_3 microrods for improved photocatalysis. Phys Chem Chem Phys 14:12114–12121
32. Kang G, Cao Y (2012) Development of antifouling reverse osmosis membranes for water treatment: a review. Water Res 46:584–600
33. Kim J, Lee C, Choi W (2010) Platinized WO_3 as an environmental photocatalyst that generates OH radicals under visible light. Environ Sci Technol 44:6849–6854
34. Kundu S, Sain S, Yoshio M, Kar T, Gunawardhana N, Pradhan SK (2014) Structural interpretation of chemically synthesized ZnO nanorod and its application in lithium ion battery. Appl Surf Sci 12:152. https://doi.org/10.1016/j.apsusc.2014.12.152
35. Kundu S, Satpati B, Kar T, Pradhan SK (2017) Microstructure characterization of hydrothermally synthesized PANI/V_2O_5·nH_2O heterojunction photocatalyst for visible light induced photodegradation of organic pollutants and non-absorbing colorless molecules. J Hazard Mater 339:161–173. https://doi.org/10.1016/j.jhazmat.2017.06.0340
36. Lala S (2017) Structural, mechanical and biocompatibility studies of some undoped and doped nanocrystal line hydroxyapatites [dissertation], The University of Burdwan
37. Li S, Lin YH, Zhang BP, Wang Y, Nan CW (2010) Controlled fabrication of $BiFeO_3$ uniform microcrystals and their magnetic and photocatalytic behaviours. J Phys Chem C 114:2903–2908
38. Linsebigler AL, Lu G, Jr JTY (1995) Chem Rev 95:735-758
39. Lv KZ, Li J, Qing XX, Li WZ, Chen QY (2011) Synthesis and photo-degradation application of WO_3/TiO_2 hollow spheres. J Hazard Mater 189:329–335
40. Lutterotti L, MAUD version 2.80. http://maud.radiographema.eu
41. Lutterotti L (2010) Total pattern fitting for the combined size–strain–stress–texture determination in thin film diffraction. Nucl Instrum Methods Phys Res Sect B: Beam Interact Mater At 268:334–340
42. Majhi D, Samal PK, Das K, Gouda SK, Bhoi YP, Mishra BG (2019) α-NiS/Bi_2O_3 nanocomposites for enhanced photocatalytic degradation of tramadol. ACS Appl Nano Mater 2:395–407. https://doi.org/10.1021/acsanm.8b01974
43. Maji SK, Mukherjee N, Mondal A, Adhikary B (2012) Synthesis, characterization and photocatalytic activity of α-Fe_2O_3 nanoparticles author links open overlay panel. Polyhedron 33:145–149
44. Mandal RK, Kundu S, Sain S, Pradhan SK (2019) Enhanced photocatalytic performance of V_2O_5–TiO_2 nanocomposites synthesized by mechanical alloying with morphological hierarchy. New J Chem 43:2804–2816. https://doi.org/10.1039/C8NJ05576A
45. Mandal RK, Pradhan SK (2021) Superior photocatalytic performance of mechanosynthesized Bi_2O_3-Bi_2WO_6 nanocomposite in wastewater treatment. Solid State Sci 115:106587
46. Mandal RK, Pradhan SK (2020) Optimized enhanced photodegradation activity of sintered molybdenum oxide: a morphological hierarchy in wastewater treatment. Mater Res Bull 124:110760. https://doi.org/10.1016/j.materresbull.2019.110760
47. Martha S, Das DP, Biswal N, Parida KM (2012) Facile synthesis of visible light responsive V_2O_5/N, S–TiO_2 composite photocatalyst: enhanced hydrogen production and phenol degradation. J Mater Chem 22:10695–10703
48. Martí X, Ferrer P, Herrero-Albillos J, Narvaez J, Holy V, Barrett N, Alexe M, Catalan G (2011) Skin layer of $BiFeO_3$ single crystals. Phys Rev Lett 106:236101
49. Megharaj M, Ramakrishnan B, Venkateswarlu K, Sethunathan N, Naidu R (2011) Bioremediation approaches for organic pollutants: a critical perspective. Environ Int 37(8):1362–1375
50. Mocherla PSV, Karthik C, Ubic R, Rao MSR, Sudakar C (2013) Tunable bandgap in $BiFeO_3$ nanoparticles: the role of microstrain and oxygen defects. Appl Phys Lett 103:022910
51. Moghaddam SS, Moghaddam MRA, Arami M (2010) Coagulation/flocculation process for dye removal using sludge from water treatment plant: optimization through response surface methodology. J Hazard Mater 175:651–657
52. Mousavi M, Habibi-Yangjeh A (2016) Magnetically separable ternary g-C_3N_4/Fe_3O_4/BiOI nanocomposites: novel visible-light-driven photocatalysts based on graphitic carbon nitride. J Colloid Interface Sci 465:83–92

53. Munter R (2001) Advanced oxidation processes – current status and prospects. Proc Estonian Acad Sci Chem 50:59–80
54. Muruganandham M, Wu JJ (2008) Synthesis, characterization and catalytic activity of easily recyclable ZnO nanobundles. Appl Catal B Environ 80:32–41
55. Neyens E, Baeyens J (2013) A review of classic Fenton's peroxidation as an advanced oxidation technique. J Hazar Mater B 98:33–58
56. Nolan NT, Seery MK, Pillai SC (2009) Spectroscopic investigation of the anatase to rutile transformation of sol-gel synthesisized TiO_2 photocatalyst. J Phys Chem C 113:16151–16157
57. Pare B, Jonnalagadd SB, Tomar HS, Singh P, Bhagwat VW (2010) Photodegradation of Safranine-O dye using visible irradiation and aqueous suspension of ZnO in a slurry batch reactor. J Ind Chem Soc 87:1359–1367
58. Paola AD, Lopez EG, Marci G, Palmisano L (2012) Survey of photocatalytic material for environmental remediation. J Hazard Matter 4:211–212
59. Puvaneswari N, Muthukrishnan J, Gunasekaran P (2006) Toxicity assessment and microbial degradation of azo dyes. Indian J Exp Biol 44(8):618–626
60. Rajeshwar K (1995) Photochemistry and the environment. J Appl Electrochem 25:1067
61. Richards R (2006) Surface and nanomolecular catalysis. CRC/Taylor & Francis Boca Raton, FL. ISBN 9780367390815
62. Rietveld HM (1967) Line profiles of neutron powder-diffraction peaks for structure refinement. Acta Crystallogr 22:151
63. Song H, Wu R, Yang J, Dong J, Ji G (2017) Fabrication of CeO_2 nanoparticles decorated three-dimensional flower-like BiOI composites to build p-n heterojunction with highly enhanced visible-light photocatalytic performance. J Colloid Interface Sci. https://doi.org/10.1016/j.jcis.2017.10.080
64. Slokar YM, Le Marechal AM (1998) Methods of decoloration of textile wastewaters. Dyes Pigm 37(4):335–356
65. Uddin MT, Hoqueand ME, Bhoumick MC (2020) Facile one-pot synthesis of heterostructure SnO_2/ZnO photocatalyst for enhanced photocatalytic degradation of organic dye. RSC Adv 10:23554–23565.https://doi.org/10.1039/D0RA03233F
66. UNICEF, Niang, World Health Organization, and UNICEF. WHO (2012) Progress on drinking water and sanitation
67. Ting Z, Yang S, Yunhang Q, Yong L, Rui X, Jing S, Jianhong W (2017) Facial synthesis and photoreaction mechanism of $BiFeO_3$/$Bi_2Fe_4O_9$ heterojunction nanofibers. ACS Sustain Chem Eng 5:4630–4636
68. Wang Y, Su YR, Qiao L, Liu LX, Su Q, Zhu CQ, Liu XQ (2011) Synthesis of one dimensional TiO_2/V_2O_5 branched heterostructures and their visible light photocatalytic activity towards Rhodamine B. Nanotechnology 22:225702–225709
69. Wang C, Shao C, Zhang X, Liu Y (2009) SnO_2 nanostructures-TiO_2 nanofibers heterostructures: controlled fabrication and high photocatalytic properties. Inorg Chem 48:7261–7268. https://doi.org/10.1021/ic9005983
70. Wang H, Zhang L, Chen Z, Hu J, Li S, Wang Z, Liu J (2014) Semiconductor heterojunction photocatalyst: design, construction and photocatalytic performance. Chem Soc Rev 43:5234–5244
71. Wang Y, Huang Y, Ho W, Zhang L, Zou Z, Lee S (2009) Biomolecule of controlled hydrothermal synthesis of CaC–NaC, S-tridoped TiO_2 nanocrystalline photocatalyst for NO removal under simulated solar light irradiation. J Hazard Mater 169:77–87
72. Weckhuysen BM, Keller DE (2003) Chemistry, spectroscopy and the role of supported vanadium oxides in heterogeneous catalysis. Catal Today 78:25–46
73. Wen XJ, Niu CJ, Zhang L, Zeng GM (2017) Novel p–n heterojunction BiOI/CeO_2 photocatalyst for wider spectrum visible-light photocatalytic degradation of refractory pollutants. Dalton Trans 46:4982–4993
74. Wisitsoraat A, Tuantranont A, Comini E, Sberveglieri G, Wlodarski W (2009) Characterization of n-type and p-type semiconductor gas sensor based on NiOx doped TiO_2 thin film. Thin Solid Films 517:2775–2780

75. Wu CH, Chang HW, Chern JM (2006) Basic dye decomposition kinetics in a photocatalytic slurry reactor. J Hazard Mater 137:336–343
76. Wu J, Xie Y, Ling Y, Dong Y, Li J, Li S, Zhao J (2019) Synthesis of flower-like g-C_3N_4/BiOBr and enhancement of the activity for the degradation of Bisphenol A under visible light irradiation. Front Chem 7:649. https://doi.org/10.3389/fchem.2019.00649
77. Xiao Ping W, Dian Chao S, Tan Dong Y (2016) Climate change and global cycling of persistent organic pollutants: a critical review. Sci China 59:1899–1911
78. Xie W, Yuan S, Mao X, Hu W, Liao P, Tong M, Alshawabkeh AN (2013) Electrocatalytic activity of Pd-loaded Ti/TiO_2 nanotube cathode for TCE reduction in ground water. Water Res 47:3573–3582
79. Yang J, Zhang X, Liu H, Liu YC (2013) Heterostructured TiO_2/WO_3 porous microspheres: preparation, characterization and photocatalytic properties. Catal Today 201:195–202. https://doi.org/10.1016/j.cattod.2012.03.008
80. Yao H (2013) Application of advanced oxidation processes for treatment of air from livestock buildings and industrial facilities. Department of Engineering, Aarhus University. Denmark, 36 pp. - Technical report BCE -TR-8
81. Zhu H, Jiang R, Xiao L, Chang Y, Guan Y, Li X, Zeng G (2009) Photocatalytic decolorization and degradation of Congo Red on innovative crosslinked chitosan/nano-CdS composite catalyst under visible light irradiation. J Hazard Mater 169:933-940. https://doi.org/10.1016/j.jhazmat.2009.04.037
82. Zhang JZ, Wang ZL, Liu J, Chen S, Liu GY (2003) Self-assembled nanostructures, vol 2. Kluwer Academic Publishers, New York
83. Zhao W, Bai Z, Ren A, Guo B, Wu C (2010) Quality of horizontally aligned single-walled carbon nanotubes: is methane as carbon source better than ethanol? Appl Surf Sci 256:3493–3498

Electrocoagulation Technology for Wastewater Treatment: Mechanism and Applications

Prashant Basavaraj Bhagawati, Forat Yasir AlJaberi, Shaymaa A. Ahmed, Abudukeremu Kadier, Hameed Hussein Alwan, Sata Kathum Ajjam, Chandrashekhar Basayya Shivayogimath, and B. Ramesh Babu

Abstract Electrocoagulation (EC) is an excellent and promising technology in wastewater treatment, as it combines the benefits of coagulation, flotation, and electrochemistry. EC is an efficient process to remove both organic and inorganic pollutants from wastewater. During the last decade, extensive research has focused on the treatment of several types of industrial wastewater using electrocoagulation. EC is a popular technique, which is being applied for the treatment of varieties of wastewater, because of its several advantages like compact, cost-effective, efficient, low sludge production, it's automation conveniences, high efficiency, and eco-friendliness. This chapter highlights the EC working principle, mechanism, factors

P. B. Bhagawati (✉)
Civil Engineering Department, Annasaheb Dange College of Engineering and Technology, Ashta, Maharashtra, India
e-mail: prashantbhagawati@gmail.com

F. Y. AlJaberi (✉)
Chemical Engineering Department, College of Engineering, Al-Muthanna University, Al-Muthanna, Iraq
e-mail: furat_yasir@yahoo.com

S. A. Ahmed
Chemical Engineering Department, College of Engineering, University of Baghdad, Baghdad, Iraq

A. Kadier
Laboratory of Environmental Science and Technology, The Xinjiang Technical Institute of Physics and Chemistry, Key Laboratory of Functional Materials and Devices for Special Environments, Chinese Academy of Sciences, Urumqi, China

H. H. Alwan · S. K. Ajjam
Chemical Engineering Department, University of Babylon, Babil, Iraq

C. B. Shivayogimath
Civil Engineering Department, Amruta Institute of Engineering and Management Sciences, Bengaluru, India

B. R. Babu (✉)
Process Engineering Division, Central Electrochemical Research Institute, Karaikudi, Tamil Nadu, India
e-mail: brbabu@cecri.res.in

S. S. Muthu and A. Khadir (eds.), *Advanced Oxidation Processes in Dye-Containing Wastewater*, Sustainable Textiles: Production, Processing, Manufacturing & Chemistry,
https://doi.org/10.1007/978-981-19-0987-0_13

affecting EC process, design aspects, and its application in various industrial effluents. The studies on treatment of various industrial wastewater by global researchers are critically reviewed. The capability of employing this technology in combined systems with other conventional methods to eliminate contaminants from wastewater is reviewed also. The core findings proved the capability of this technique in wastewater treatment whether it is performed alone or combined with other technologies providing many pros, such as the decrement of the operation cost and the sludge formation.

Keywords Electrocoagulation refractory contaminants · Industrial wastewater · Combined systems

1 Introduction

Water is one of the most important resources on earth; water plays a vital role for all human beings, animals, and plants to thrive. Water is an equally important resource for the manufacturing sector and social development. Therefore, there is an urgent need to preserve and protect our ecosystems from all types of pollution [38, 42, 49]. In India, water is used in the industrial sector is about 34 billion m^3 per year, which is estimated to increase four times by 2050. Rapid industrialization has tremendously increased the volume of wastewater to be disposed-off, while the capacity of receiving water to accept the increasing inorganic and organic loads remains the same [16, 52]. This has resulted in the rapid deterioration of the quality of surface water and forcing the concerned more stringent legislation [24, 57, 58]. Accordingly, industries are looking for low-cost solutions to reduce the excess pollutant load. The situation is further aggravated by the threat posed by constant discharge of pollutants by industries such as sugar, paper, steel manufacturing, pharmaceutical, oil refineries, mining, chemical paints, petrochemical, textile, slaughterhouse, distilleries, electroplating, etc. These industries discharge untreated or partially treated wastewater into nearby drains, rivers, stagnant ponds, lagoons, or lakes [1, 28, 64, 66].

The wastewater generated from leather, distillery, cement manufacturing, food and beverages, textile, pharmaceutical, and dairy industries contains various toxic organics, metals, phenolic contaminants that are complex, in nature, and discharge of such wastewater has many negative effects on the receiving water bodies such as low dissolved oxygen (DO), scum formation, thermal impact, and slime growth [14, 29, 62]. Thus continuous discharge of untreated wastewater may harm the aesthetic and scenic beauty of water environment finally leading to totally disturbed ecosystem [31, 40]. In developing countries, huge quantity of effluents are generated by different industries, out of these, only few industries reuse the treated effluent. The potential use of the treated wastewater varies significantly depending upon the degree of treatment and on public acceptance [22, 39, 51]. Nevertheless, this practice is being carried out in many countries with the potential use of treated water for various purposes, which includes irrigation, recreational areas, and landscapes.

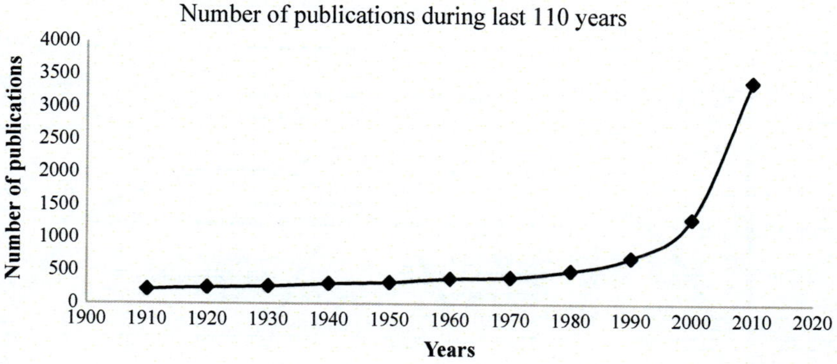

Fig. 1 Number of SCI publications on electrochemical treatment of water and wastewater treatment [70]

In order to trace global research trends in industrial wastewater, the bibliometric analysis is carried out of 110 years as shown in Fig. 1. This figure revealed that many of environmental researchers attracted EC technology due to its inherent advantages—high removal efficiency, environmental compatibility, and clean technology that could be applied in various industrial domains including water and wastewater treatment [70].

Degradation of toxic, refractory contaminants by biological treatment are found to be difficult. Therefore, more effective low-cost wastewater technologies are necessary to fulfill human requirements. Electrochemical treatment technologies have attracted and given attention to their environmental compatibility, high removal efficiency.

Almost all of the electrochemical principles involve the movement of electrons in the conduction media (maybe water or wastewater) and the concept of oxidation and reduction. In recent years, various studies have been focused on the electrocoagulation process, which is an efficient method used to destabilize and remove finely dispersed particles from water and wastewater [43, 47]. Electrocoagulation is a method of treating wastewater containing significant concentrations of organic and inorganic pollutants. Industrial electrochemistry has helped to develop optimal processes and better environmental efficiency.

The EC reactor design cannot be made in isolation as design, flow pattern and operational parameters are dependent on three foundation technologies as shown in Fig. 2. The Venn diagram consists of three lobes, which include electrochemistry, flotation, and coagulation. The intersection of each lobe is highlighted by working principle, characterization method, and tools. In the overall understanding of electrocoagulation, contact pattern kinetics play a key role. The removal path of floatation and settling are represented by vertical arrow. The double-headed arrow (current density) shown in Fig. 2 determines coagulant dosage and bubble production rates. There is a complex interaction between electrochemistry, flotation, and coagulation.

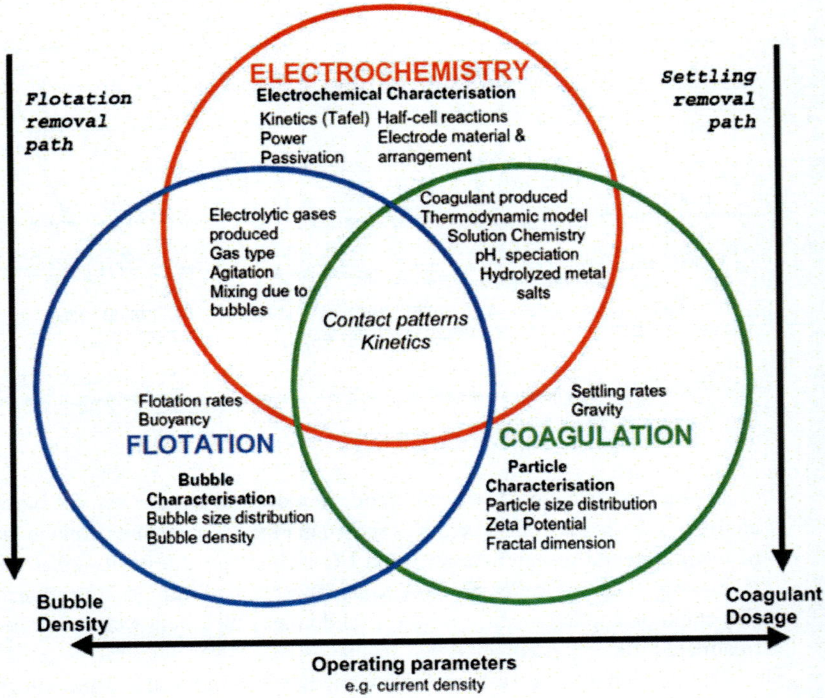

Fig. 2 Conceptual frameworks for electrocoagulation

1.1 Mechanism of Electrocoagulation

EC is a complex separation technique involving a combination of physical and chemical mechanisms. The EC process is a simple and effective method for the treatment of various water and wastewaters. Many investigations in recent years have focused, in particular, on the use of EC due to its advantages such as:

1. A lower amount of coagulant ions is required.
2. A higher rate of pollutant reduction with low treatment time.
3. Simple operation and maintenance, no need for addition of chemicals.
4. The quantity of generated sludge is less and disposal of sludge is simple.
5. Flocs formed settle easily and are readily dewaterable.

Electrocoagulation method has been tested successfully to treat domestic wastewater, dairy, steel manufacturing plant, poultry, olive oil mill, pickle, distillery, textile, dye, pharmaceutical, automobile, photograph processing, electroplating industry wastewater, biodiesel wastewater, petroleum refinery wastewater, paint manufacturing wastewater, tannery wastewater, etc. [3, 5, 64].

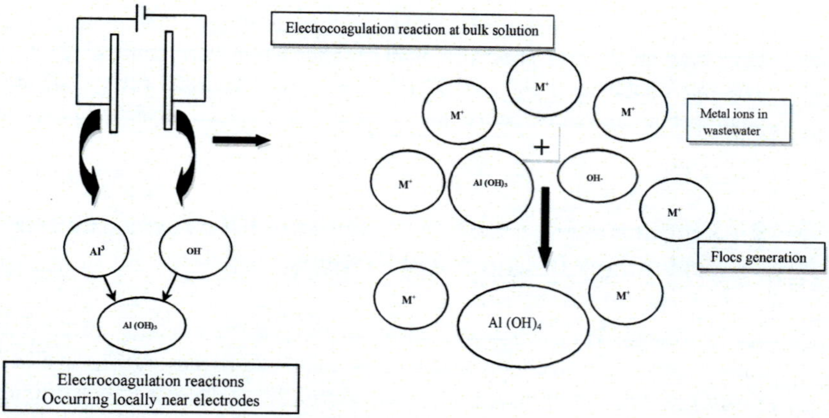

Fig. 3 Electrocoagulation reaction in bulk solution

The simplest EC cell includes a sacrificial anode and a cathode made up of several materials (such as Fe or Al) immersed (partially or fully) in the water or wastewater [60, 67]. An EC system may consist of one or multiple anode–cathode pairs connected either in monopolar or bipolar mode [4, 6]. Three successive stages involved in the EC process are: (i) formation of coagulants by electrolytic oxidation sacrificial anode, (ii) destabilization of pollutants, particulate suspension and breaking of emulsion, and (iii) aggregation of the destabilized phase to form flocs (Fig. 3).

Aluminum is cheap and readily available metal and has an ability to form multivalent ions and various hydrolysis products [44]. Hence, aluminum is the most commonly used anode material. During EC process in a reactor with aluminum electrodes, the important electrochemical reactions that take place are given in Eqs. (1) and (2):

$$\text{At the Anode:} \quad Al_s \rightarrow Al^{3+}_{aq} + 3e^- \tag{1}$$

$$\text{At the Cathode:} \quad 3H_2O^+_{aq} + 3e^- \rightarrow 3/2\,H_2(g) + 3OH^- \tag{2}$$

The Al^{3+} ions generated by the anode and OH^- ions generated by the cathode form various monomeric and polymeric species such as $Al(OH)^{2+}$, $Al_2(OH)^{4-}$, $Al_6(OH)_{15}{}^{9+}$, $Al_8(OH)_{20}{}^{4+}$, $Al_{13}O_4(OH)_{24}{}^{7+}$, and $Al_{13}(OH)_{34}{}^{5+}$ that transform initially into $Al(OH)_{3(s)}$ and finally polymerize to $Al_n(OH)_{3n}$ according to Eqs. (3) and (4) in solution medium:

$$nAl(OH)_3^- \rightarrow Al_n(OH)_{3n} \tag{3}$$

$$Al^{3+} + 3H_2O \rightarrow Al(OH)_{3(s)} + 3H^+ \tag{4}$$

The amphoteric nature of $Al(OH)_{3(s)}$ flocs usually has more surface area that facilitates quick adsorption of soluble matter along with entrapping of colloids. The microbubbles of H_2 and O_2 released at the electrode surface adhere to the agglomerates and carry them to the water surface by electrofloatation [7, 45, 46, 62].

2 Elimination of Contaminants Using Electrocoagulation Technology in Combined Systems

Electrocoagulation technology is also widely performed in integrated systems with other technologies to eliminate numerous pollutants effectively. The hybrid systems could provide several pros, such as low down the sludge generation, minimize the operation cost and consumption of the performed electrodes. The use of this technology in integrated systems has received wide attention from scientists in the last decades. Bilińska et al. [15] combined an electrocoagulation technology and ozonation (O_3) to treat high salinity dye wastewater. They proved the treatability of this hybrid system to obtain 95% dye removal efficacy within 18 min treatment time in comparison with that of 60 min contact time when the ozonation technology was employed alone.

Another study done by Ashraf et al. [10] combined the reverse osmosis process with electrocoagulation to remediate the highly concentrated brine water. They reported 98.9% silica removal within a significant short contact time using iron and/or aluminum electrodes. However, the performance of this system was influenced by generating a thick coating on the used electrodes, for that reason, they performed polarity change to solve this problem. Sharma and Simsek [61] compared the capability of using an integrated system containing of electrocoagulation and electro-oxidation from one side and an electro-peroxidation process on the other side for the remediation of canola oil refinery wastewater. This study found that the integrated system achieved 99% and 95% of COD and dissolved organic carbon (DOC) removal efficiencies, respectively, compared to that obtained using the electro-peroxidation only which removed 77% and 86% only of these pollutants. Hasani et al. [32] investigated the performance of four integrated processes of electrocoagulation flotation system when it was performed for eliminating humic acid from drinking water. The impacts of several technological variables were investigated. They proved that the pulse current-perforated electrode configuration was the most cost-effective in electrode and energy consumption and provided higher contaminants removal efficiency compared to other processes. Tavangar et al. [65] employed an integration system consisting of electrocoagulation and nano-filtration processes to remediate COD and color from real dye wastewater. The cons of each technology were vanished when these technologies were integrated. Another work proposed a combined system containing electrocoagulation and reverse osmosis (RO) technologies. It was documented by Azerrad et al. [11] to enhance the desalination of saline water where higher removal efficiency of carbonate, phosphate, and dissolved organic matter of 98%,

>99%, and (50%), respectively, were achieved using aluminum electrodes, which were larger than in case of iron electrodes. They also concluded that the oxidation of micropollutants was improved by fourfold when this hybrid system was integrated with an advanced oxidation process (AOP), either UVA/TiO_2 or UVC/H_2O_2.

Khani et al. [36] treated wastewater containing olive mill using a hybrid system of electrocoagulation and catalytic sonoperox one processes where the sonoperoxone process consists of the integration of electrooxidation and ozonation addition as well as the ultrasonic process. The impacts of the contact time, current density, and type of electrodes were studied. The (BOD_5/TOC) index was enhanced by 32% in comparison to raw wastewater. Another combination system involved electrocoagulation, membrane distillation, and forward osmosis documented by Sardari et al. [54] for water reuse from high brine wastewater discharged from shale gas extraction process. The findings observed revealed significant removal efficiencies of 78% and 96% of TOC and TSS, respectively, under the selected operational variables. Lalwani et al. [41] performed electrocoagulation and photocatalytic oxidation technologies in an integrated configuration to eliminate TOC from drug wastewater. They found that the elimination of TOC was (33%) using UV-photocatalytic oxidation part compared to 30% of TOC removal in case of natural sunlight-photocatalytic oxidation. Bashir et al. [13] studied the efficiency of a hybrid process containing electrocoagulation and peroxidation technologies to treat wastewater released from the palm oil mill under the impacts of specific operational variables. They concluded that higher removal efficiencies of 100%, 96.8%, and 71.3% were obtained for TSS, color, and COD pollutants, respectively, after 45 min of contact time.

Electrocoagulation and electro-oxidation–reduction steps were investigated individually and together in a combination system by Ghazouani et al. [26] to remediate real wastewater discharged from food processing factories. They found that higher elimination of COD, nitrates, ammonium/ammonia, and phosphates were obtained depending on the configuration set-up employing these two treatment technologies. Shamaei et al. [59] performed a hybrid system of electrochemical coagulation and chemical coagulation to remediate steam-assisted drainage wastewater. A significant removal of 39.8% TOC was achieved after 90 min of contact time using electrodes made of aluminum and connected to the power source in a bipolar mode. Dia et al. [23] studied the ability of an integrated electrocoagulation-biofiltration process to eliminate contaminants from landfill leachates. They obtained 99% and 42% of NH_4 and COD removal. They revealed that the efficient employment of this integration of technologies could remediate different types of wastewater containing numerous kinds of pollutants.

Sardari et al. [53] performed electrocoagulation and ultrafiltration to treat poultry wastewater. They achieved more than 85% removal of pollutants compared to that obtained using the individual methods. The study of the capability of electrocoagulation and membrane distillation was done by Sardari et al. [55] to treat the hydraulic fracturing saline water. A 85% of TSS and organic content removal was achieved and a stable water flux with low fouling had been attained over 434 h experimental period. Al-Malack and Al-Nowaiser [8] tested a hybrid electrocoagulation-membrane bioreactor to eliminate oil and COD and grease from saline oilfield wastewater. They had proved that the removal of these pollutants was decayed with an increment in the concentration of the influent oil. An invented design of the combined system was documented by Sun et al. [63] to remediate humic acid from wastewater, where this cell contains electrocoagulation and electrooxidation integrated with a membrane placed between the electrodes. They had found that this integrated system could be used in wastewater treatment with higher efficacy. Kong et al. [37] documented the ease using of a combined system containing electrocoagulation and electro-peroxone methods, used to remove COD pollutant from shale gas flow-back wastewater. They obtained 35.4% COD removal using 50 mA/cm^2 of current density. Nariyan et al. [50] investigated a hybrid system involving electrocoagulation and chemical precipitation using lime to eliminate sulfate from real wastewater. The obtained findings proved that the hybridization system is useful for eliminating 90% of sulfate pollutants at 25 mA/cm^2 of current density. Zazou et al. [69] had performed hybrid systems consisting of electrocoagulation method and other types of advanced oxidation technologies, involving peroxi-coagulation, anodic oxidation, and the electro-Fenton method to treat wastewater released from textile industries. They had revealed that the integration system of electrocoagulation and electro-Fenton provided the reuse of the treated wastewater belong to WHO specifications.

Barzegar et al. [12] combined electrocoagulation with ozonation methods to improve grey water remediation under the impact of the operational variables of contact time, current density, pH, and the ozone dose was investigated using iron electrodes. They conducted 70% and 85% of TOC and COD removal efficiencies, respectively, after 60 min, 15 mA/cm^2, pH = 7, and 47.4 mg/L of the studied variables. Then the integrated system was combined with UV irradiation where 87% and 95% TOC and COD removal efficiencies were obtained. Jose et al. [34] investigated a hybridization system involving the electrocoagulation and filtration remediation of TDS, TOC, COD, and color from wastewater to attain higher removal. They concluded that the combined system was cost-effective compared to the traditional method. Moradi and Moussavi [48] integrated electrocoagulation technology with UVC/VUV photoreactor to eliminate COD, Cr (III), Cr(VI)) and sulfide from tannery polluted water. They attained higher removal of these pollutants at pH 7 and less and proved the usability of this combined system to treat wastewater released from tannery industries.

A hybrid system containing electrocoagulation and electrooxidation was documented by Costa et al. [21] to remove COD from cashew-nut wastewater discharged from the washing of raw cashew nuts. More than 80% of COD removal efficiency was achieved compared to 51% in the case of using electrocoagulation process alone. GilPavas et al. [27] concluded that the combined system of electrocoagulation-photo-Fenton and electrocoagulation-Fenton followed by adsorption using activated carbon was more valuable to eliminate 76%, 78%, 90%, and 100% TOC, COD, toxicity, and color, respectively, in contrast to that attained using electrocoagulation alone. An integration system of electrocoagulation and sedimentation processes was investigated by Chen et al. [20] to treat polymer containing wastewater under the effects of the operational parameters of current density, flow rate, and tilt angle of electrodes using response surface methodology (RSM) based Box-Behnken method as an experimental design. They achieved higher removal (97%) of oil content at 18.9 mA/cm^2, 5.5 L/h, and (80°) of the applied current density, flow rate, and the tilt angle of electrodes. Keramati and Ayati [35] studied the hybrid system of electrocoagulation and photo-catalytic technologies with ZnO nanoparticles to remediate COD from real oily wastewater. The observed results showed 85% of COD removal under the optimal operating variables. Jiang et al. [33] investigated a hybridization system of electrocoagulation and separation methods to remove oil content and turbidity from a polymer containing wastewater under the impacts of the operating variables of contact time, flow rate, current density, and the tilt angles of electrodes. They attained 96% and 97% of removal efficiencies, respectively, but they minimized with the increase of the tilt angle.

An et al. [9] invented a combined system of electrocoagulation and electro-Fenton for eliminating cyanotoxins and cyanobacteria from wastewater. This efficient system increased the removal of TOC by 30% and minimized the consumption of energy by 92%. Ye et al. [68] performed electrocoagulation and UVA photoelectro-Fenton processes in a hybrid system to eliminate benzophenone-3 from municipal wastewater using aluminum electrodes and boron-doped diamond electrode to achieve total mineralization under the studied conditions. Guan et al. [30] documented an integrated system containing electrocoagulation and electro-Fenton methods to eliminate Cu-EDTA and copper contaminants from wastewater. A higher removal efficacy was observed at pH <7, 72.92 A/m^2 of current density, and 49.4 mM of H_2O_2 concentration, respectively. Bruguera-Casamada et al. [18] constructed a system containing electrocoagulation reactor connected to electro-Fenton or UVA-photoelectro-Fenton cells to treat wastewater discharged from dairy industrial. They attained a 28% TOC removal at pH 3 using UVA-photoelectro-Fenton process and electrocoagulation methods due to the enhancement of the bacterial inactivation by UVA radiation. Another hybrid system consisting of electrocoagulation and forward osmosis methods was proposed by Sardari et al. [56] to eliminate TSS and organic compounds from hydraulic fracturing wastewater. They concluded the treatability of this system for water reuse.

A combined system of photoperoxi-electrocoagulation was performed by Borba et al. [17] to eliminate COD and chromium from tannery wastewater under the influence of the reaction time, pH, the dosage of peroxide, and the current density. They obtained a complete removal of pollutants after 2 h. Adjeroud et al. [2] employed an integrated system containing electrocoagulation and electroflotation processes to remove copper from synthetic wastewater performing aluminum electrodes and opuntia ficus indica (OFI) plant. They achieved a complete removal of copper after 5 min aiding 30 mg OFI/L at pH 7.8 value. Another hybrid system consisting of electrocoagulation and microfiltration was investigated by Changmai et al. [19] for the de-fluorination of contaminated drinking water. This system was applicable to remove 92% of fluoride at pH 7.9 and 15 A/m^2 of current density. Ghanbari et al. [25] tested a combined system involving electrocoagulation, peroxymonosulfate, and electrooxidation to attain higher removal efficiencies of COD, TOC, BOD, and ammonia from landfill leachates.

This summary reveals that effective removal of different pollutants can be attained in the case of combined systems is significantly higher compared to that achieved using individual treatment methods.

3 Conclusion

The electrocoagulation technique is one of the promising technologies to treat different industrial wastewaters. The electrocoagulation process is chosen as an effective method for the treatment of various industrial wastewater, instead of the conventional methods, which have many practical drawbacks. Electrocoagulation cell design is one of the pivotal issues to maintain high mass transfer rates as main reactions take place in electrocoagulation reactor. The selection of appropriate electrode material is also important when designing electrocoagulation cell for maximum removal efficiency. Currently, many of industries use commercially available electrodes made up of BDD, $TiPbO_2$, Ti-SiO_2, graphite, and Zn etc. The cost of these electrodes is high, and they generate toxic elements. This review finds the key achievements and highlights that the major shortcomings in the process are summarized. The fact that EC is a novel technology for water and wastewater treatment that cannot be denied, as it has a number of advantages over traditional methods. However, the technology is still being fine-tuned and refined, and further studies are necessary, to make this technology available all around the world. Due to the overall environmental sustainability, developing membrane technology stands out as the best process to be merged with EC among numerous hybrid EC techniques that have been explored. There must be more research into the optimization of parameters, system design, and economic feasibility to explore the potential of the EC-combined systems.

Acknowledgements This study is a part of research project funded by Science and Engineering Research Board (SERB), Department of Science and Technology (DST) under Teachers Associateship for Research Excellence (TARE) scheme (TAR/2019/000383). The first author sincerely acknowledges and thanks SERB for providing financial support to undertake this study.

References

1. Adeogun AI, Bhagawati PB, Shivayogimath CB (2021) Pollutants removals and energy consumption in electrochemical cell for pulping processes wastewater treatment: artificial neural network, response surface methodology and kinetic studies. J Environ Manag 281. https://doi.org/10.1016/j.jenvman.2020.111897
2. Adjeroud N, Elabbas S, Merzouk B, Hammoui Y, Felkai-Haddache L, Remini H, Madani K (2018) Effect of Opuntiaficusindica mucilage on copper removal from water by electrocoagulation-electroflotation technique. J Electroanal Chem 811:26–36. https://doi.org/10.1016/j.jelechem.2017.12.081

3. Aljaberi FY (2018) Studies of autocatalytic electrocoagulation reactor for lead removal from simulated wastewater. J Environ Chem Eng 6(5):6069–6078. https://doi.org/10.1016/j.jece.2018.09.032
4. AlJaberi FY (2019) Operating cost analysis of a concentric aluminum tubes electrodes electrocoagulation reactor. Heliyon 5(8). https://doi.org/10.1016/j.heliyon.2019.e02307
5. Aljaberi FY (2020) Removal of TOC from oily wastewater by electrocoagulation technology. In: IOP conference series: materials science and engineering, vol 928. IOP Publishing Ltd. https://doi.org/10.1088/1757-899X/928/2/022024
6. AlJaberi FY, Abdulmajeed BA, Hassan AA, Ghadban ML (2020) Assessment of an electrocoagulation reactor for the removal of oil content and turbidity from real oily wastewater using response surface method. Recent Innov Chem Eng (Former Recent Pat Chem Eng) 13:55–71
7. Aljaberi FY, Mohammed WT (2018) Analyzing the removal of lead from synthesis wastewater by electrocoagulation technique using experimental design. Desalin Water Treat 111:286–296
8. Al-Malack MH, Al-Nowaiser WK (2018) Treatment of synthetic hypersaline produced water employing electrocoagulation-membrane bioreactor (EC-MBR) process and halophilic bacteria. J Environ Chem Eng 6(2):2442–2453. https://doi.org/10.1016/j.jece.2018.03.049
9. An J, Li N, Wang S, Liao C, Zhou L, Li T, Feng Y et al (2019) A novel electro-coagulation-Fenton for energy efficient cyanobacteria and cyanotoxins removal without chemical addition. J Hazard Mater 365:650–658. https://doi.org/10.1016/j.jhazmat.2018.11.058
10. Ashraf SN, Rajapakse J, Dawes LA, Millar GJ (2019) Electrocoagulation for the purification of highly concentrated brine produced from reverse osmosis desalination of coal seam gas associated water. J Water Process Eng 28:300–310
11. Azerrad SP, Isaacs M, Dosoretz CG (2019) Integrated treatment of reverse osmosis brines coupling electrocoagulation with advanced oxidation processes. Chem Eng J 356:771–780. https://doi.org/10.1016/j.cej.2018.09.068
12. Barzegar G, Wu J, Ghanbari F (2019) Enhanced treatment of greywater using electrocoagulation/ozonation: investigation of process parameters. Process Saf Environ Prot 121:125–132. https://doi.org/10.1016/j.psep.2018.10.013
13. Bashir MJ, Lim JH, Abu Amr SS, Wong LP, Sim YL (2019) Post treatment of palm oil mill effluent using electro-coagulation-peroxidation (ECP) technique. J Clean Prod 208:716–727. https://doi.org/10.1016/j.jclepro.2018.10.073
14. Bhagawati PB, Shivayogimath CB (2021) Electrochemical technique for paper mill effluent degradation using concentric aluminum tube electrodes (CATE). J Environ Health Sci Eng 19(1):553–564. https://doi.org/10.1007/s40201-021-00627-8
15. Bilińska L, Blus K, Gmurek M, Ledakowicz S (2019) Coupling of electrocoagulation and ozone treatment for textile wastewater reuse. Chem Eng J 358:992–1001. https://doi.org/10.1016/j.cej.2018.10.093
16. Borah SN, Goswami L, Sen S, Sachan D, Sarma H, Montes M, Narayan M et al (2021) Selenite bioreduction and biosynthesis of selenium nanoparticles by Bacillus paramycoides SP3 isolated from coal mine overburden leachate. Environ Pollut 117519
17. Borba FH, Seibert D, Pellenz L, Espinoza-Quiñones FR, Borba CE, Módenes AN, Bergamasco R (2018) Desirability function applied to the optimization of the photoperoxi-electrocoagulation process conditions in the treatment of tannery industrial wastewater. J Water Process Eng 23:207–216. https://doi.org/10.1016/j.jwpe.2018.04.006
18. Bruguera-Casamada C, Araujo RM, Brillas E, Sirés I (2019) Advantages of electro-Fenton over electrocoagulation for disinfection of dairy wastewater. Chem Eng J 376. https://doi.org/10.1016/j.cej.2018.09.136
19. Changmai M, Pasawan M, Purkait MK (2018) A hybrid method for the removal of fluoride from drinking water: parametric study and cost estimation. Sep Purif Technol 206:140–148. https://doi.org/10.1016/j.seppur.2018.05.061
20. Chen Y, Jiang W, Liu Y, Chen M, He Y, Edem MA, Chen J et al (2019) Optimization of an integrated electrocoagulation/sedimentation unit for purification of polymer-flooding sewage. J Electroanal Chem 842:193–202. https://doi.org/10.1016/j.jelechem.2019.04.049

21. da Costa PRF, Emily ECT, Castro SSL, Fajardo AS, Martínez-Huitle CA (2019) A sequential process to treat a cashew-nut effluent: electrocoagulation plus electrochemical oxidation. J Electroanal Chem 834:79–85. https://doi.org/10.1016/j.jelechem.2018.12.035
22. Devi G, Goswami L, Kushwaha A, Sathe SS, Sen B, Sarma HP (2021) Fluoride distribution and groundwater hydrogeochemistry for drinking, domestic and irrigation in an area interfaced near Brahmaputra floodplain of North-Eastern India. Environ Nanotechnol Monitor Manag 100500
23. Dia O, Drogui P, Buelna G, Dubé R (2018) Hybrid process, electrocoagulation-biofiltration for landfill leachate treatment. Waste Manag 75:391–399. https://doi.org/10.1016/j.wasman.2018.02.016
24. Gautam A, Kushwaha A, Rani R (2021) Microbial remediation of hexavalent chromium: an eco-friendly strategy for the remediation of chromium-contaminated wastewater. In: The future of effluent treatment plants. Elsevier, pp 361–384
25. Ghanbari F, Wu J, Khatebasreh M, Ding D, Lin KYA (2020) Efficient treatment for landfill leachate through sequential electrocoagulation, electrooxidation and PMS/UV/$CuFe_2O_4$ process. Sep Purif Technol 242. https://doi.org/10.1016/j.seppur.2020.116828
26. Ghazouani M, Akrout H, Jellali S, Bousselmi L (2019) Comparative study of electrochemical hybrid systems for the treatment of real wastewaters from agri-food activities. Sci Total Environ 647:1651–1664. https://doi.org/10.1016/j.scitotenv.2018.08.023
27. GilPavas E, Dobrosz-Gómez I, Gómez-García MÁ (2019) Optimization and toxicity assessment of a combined electrocoagulation, H_2O_2/Fe^{2+}/UV and activated carbon adsorption for textile wastewater treatment. Sci Total Environ 651:551–560. https://doi.org/10.1016/j.scitotenv.2018.09.125
28. Goswami L, Pakshirajan K, Pugazhenthi G (2020) Biological treatment of biomass gasification wastewater using hydrocarbonoclastic bacterium Rhodococcusopacus in an up-flow packed bed bioreactor with a novel waste-derived nano-biochar based bio-support material. J Clean Prod 256:120253
29. Goswami S, Kushwaha A, Goswami L, Singh N, Bhan U, Daverey A, Hussain CM (2021) Biological treatment, recovery and recycling of metals from waste printed circuit boards. In: Handbook on environmental management of waste electrical and electronic equipment (WEEE). Elsevier, USA. https://doi.org/10.1016/B978-0-12-822474-8.00009-X
30. Guan W, Zhang B, Tian S, Zhao X (2018) The synergism between electro-Fenton and electrocoagulation process to remove Cu-EDTA. Appl Catal B 227:252–257. https://doi.org/10.1016/j.apcatb.2017.12.036
31. Gupt CB, Kushwaha A, Prakash A, Chandra A, Goswami L, Sekharan S (2021) Mitigation of groundwater pollution: heavy metal retention characteristics of fly ash based liner materials. In: Fate and transport of subsurface pollutants. Springer, Singapore, pp 79–104
32. Hasani G, Maleki A, Daraei H, Ghanbari R, Safari M, McKay G, Marzban N, et al (2019) A comparative optimization and performance analysis of four different electrocoagulation-flotation processes for humic acid removal from aqueous solutions. Process Saf Environ Prot. Inst Chem Eng. https://doi.org/10.1016/j.psep.2018.10.025
33. Jiang W, Chen Y, Chen M, Liu X, Liu Y, Wang T, Yang J (2019) Removal of emulsified oil from polymer-flooding sewage by an integrated apparatus including EC and separation process. Sep Purif Technol 211:259–268. https://doi.org/10.1016/j.seppur.2018.09.069
34. Jose S, Mishra L, Debnath S, Pal S, Munda PK, Basu G (2019) Improvement of water quality of remnant from chemical retting of coconut fibre through electrocoagulation and activated carbon treatment. J Clean Prod 210:630–637. https://doi.org/10.1016/j.jclepro.2018.11.011
35. Keramati M, Ayati B (2019) Petroleum wastewater treatment using a combination of electrocoagulation and photocatalytic process with immobilized ZnO nanoparticles on concrete surface. Process Saf Environ Prot 126:356–365. https://doi.org/10.1016/j.psep.2019.04.019
36. Khani MR, Kuhestani H, Kalankesh LR, Kamarehei B, Rodríguez-Couto S, Baneshi MM, Shahamat YD (2019) Rapid and high purification of olive mill wastewater (OMV) with the combination electrocoagulation-catalytic sonoproxone processes. J Taiwan Inst Chem Eng 97:47–53. https://doi.org/10.1016/j.jtice.2019.02.003

37. Kong F, Lin X, Sun G, Chen J, Guo C, Xie YF (2019) Enhanced organic removal for shale gas fracturing flowback water by electrocoagulation and simultaneous electro-peroxone process. Chemosphere 218:252–258. https://doi.org/10.1016/j.chemosphere.2018.11.055
38. Kushwaha A, Goswami L, Lee J, Sonne C, Brown RJC, Kim KH (2021) Selenium in soil-microbe-plant systems: sources, distribution, toxicity, tolerance, and detoxification. Crit Rev Environ Sci Technol. https://doi.org/10.1080/10643389.2021.1883187
39. Kushwaha A, Goswami S, Hans N, Goswami L, Devi G, Deshavath NN, Lall AM (2021) An insight into biological and chemical technologies for micropollutant removal from wastewater. In: Fate and transport of subsurface pollutants. Springer, Singapore, pp 199–226
40. Kushwaha A, Rani R, Agarwal V (2016) Environmental fate and eco-toxicity of engineered nano-particles: current trends and future perspective. In: Advanced nanomaterials for wastewater remediation, pp 387–404
41. Lalwani J, Sangeetha CJ, Thatikonda S, Challapalli S (2019) Sequential treatment of crude drug effluent for the elimination of API by combined electro-assisted coagulation-photocatalytic oxidation. J Water Process Eng 28:195–202. https://doi.org/10.1016/j.jwpe.2019.01.006
42. Lata K, Kushwaha A, Ramanathan G (2021) Bacterial enzymatic degradation and remediation of 2, 4, 6-trinitrotoluene. Microb Natl Macromol 623–659
43. Martínez-Huitle CA, Panizza M (2018) Electrochemical oxidation of organic pollutants for wastewater treatment. Curr Opin Electrochem 11:62–71
44. Mechelhoff M, Kelsall GH, Graham NJD (2013) Electrochemical behaviour of aluminium in electrocoagulation processes. Chem Eng Sci 95:301–312
45. Mollah MYA, Schennach R, Parga JR, Cocke DL (2001) Electrocoagulation (EC)- science and applications. J Hazard Mater 84(1):29–41. https://doi.org/10.1016/S0304-3894(01)00176-5
46. Mollah MYA, Morkovsky P, Gomes JAG, Kesmez M, Parga J, Cocke DL (2004) Fundamentals, present and future perspectives of electrocoagulation. J Hazard Mater 114:199–210
47. Moneer AA, El Nemr A (2012) Electro-coagulation for textile dyes removal. In: Non-conventional textile waste water treatment. Nova Science Publishers, Inc., pp 162–204
48. Moradi M, Moussavi G (2019) Enhanced treatment of tannery wastewater using the electrocoagulation process combined with UVC/VUV photoreactor: parametric and mechanistic evaluation. Chem Eng J 358:1038–1046. https://doi.org/10.1016/j.cej.2018.10.069
49. Muralidhar S (2006) The right to water: an overview of the Indian legal regime. Human Right Water 65:65–81
50. Nariyan E, Wolkersdorfer C, Sillanpää M (2018) Sulfate removal from acid mine water from the deepest active European mine by precipitation and various electrocoagulation configurations. J Environ Manag 227:162–171. https://doi.org/10.1016/j.jenvman.2018.08.095
51. Ranade VV, Bhandari VM (2014) Industrial wastewater treatment, recycling, and reuse-past, present and future. In: Industrial wastewater treatment, recycling and reuse. Elsevier Inc., pp 521–535. https://doi.org/10.1016/B978-0-08-099968-5.00014-3
52. Rizvi M, Goswami L, Gupta SK (2020) A holistic approach for Melanoidin removal via Fe-impregnated activated carbon prepared from Mangifera indica leaves biomass. Bioresour Technol Rep 12:100591
53. Sardari K, Askegaard J, Chiao YH, Darvishmanesh S, Kamaz M, Wickramasinghe SR (2018) Electrocoagulation followed by ultrafiltration for treating poultry processing wastewater. J Environ Chem Eng 6(4):4937–4944. https://doi.org/10.1016/j.jece.2018.07.022
54. Sardari K, Fyfe P, RanilWickramasinghe S (2019) Integrated electrocoagulation – forward osmosis – membrane distillation for sustainable water recovery from hydraulic fracturing produced water. J Membr Sci 574:325–337. https://doi.org/10.1016/j.memsci.2018.12.075
55. Sardari K, Fyfe P, Lincicome D, RanilWickramasinghe S (2018) Combined electrocoagulation and membrane distillation for treating high salinity produced waters. J Membr Sci 564:82–96. https://doi.org/10.1016/j.memsci.2018.06.041
56. Sardari K, Fyfe P, Lincicome D, Wickramasinghe SR (2018) Aluminum electrocoagulation followed by forward osmosis for treating hydraulic fracturing produced waters. Desalination 428:172–181. https://doi.org/10.1016/j.desal.2017.11.030

57. Sathe SS, Goswami L, Mahanta C (2021) Arsenic reduction and mobilization cycle via microbial activities prevailing in the Holocene aquifers of Brahmaputra flood plain. Groundw Sustain Dev 13:100578. https://doi.org/10.1016/j.gsd.2021.100578
58. Sathe SS, Goswami L, Mahanta C, Devi LM (2020) Integrated factors controlling arsenic mobilization in an alluvial floodplain. Environ Technol Innov 17:100525
59. Shamaei L, Khorshidi B, Perdicakis B, Sadrzadeh M (2018) Treatment of oil sands produced water using combined electrocoagulation and chemical coagulation techniques. Sci Total Environ 645:560–572. https://doi.org/10.1016/j.scitotenv.2018.06.387
60. Sharma D, Chaudhari PK, Dubey S, Prajapati AK (2020) Electrocoagulation treatment of electroplating wastewater: a review. J Environ Eng 146:03120009
61. Sharma S, Simsek H (2019) Treatment of canola-oil refinery effluent using electrochemical methods: a comparison between combined electrocoagulation + electrooxidation and electrochemical peroxidation methods. Chemosphere 221:630–639. https://doi.org/10.1016/j.chemosphere.2019.01.066
62. Shivayogimath CB, Bhagawati PB (2017) Separation of pollutants from pulp mill wastewater by electrocoagulation. Int J Energy Technol Policy 13(1/2):166. https://doi.org/10.1504/ijetp.2017.10000614
63. Sun J, Hu C, Zhao K, Li M, Qu J, Liu H (2018) Enhanced membrane fouling mitigation by modulating cake layer porosity and hydrophilicity in an electro-coagulation/oxidation membrane reactor (ECOMR). J Membr Sci 550:72–79. https://doi.org/10.1016/j.memsci.2017.12.073
64. SyamBabu D, Anantha Singh TS, Nidheesh PV, Suresh Kumar M (2020) Industrial wastewater treatment by electrocoagulation process. Sep Sci Technol (Philadelphia) 55:3195–3227
65. Tavangar T, Jalali K, AlaeiShahmirzadi MA, Karimi M (2019) Toward real textile wastewater treatment: membrane fouling control and effective fractionation of dyes/inorganic salts using a hybrid electrocoagulation – nanofiltration process. Sep Purif Technol 216:115–125. https://doi.org/10.1016/j.seppur.2019.01.070
66. Yadav APS, Dwivedi V, Kumar S, Kushwaha A, Goswami L, Reddy BS (2021) Cyanobacterial extracellular polymeric substances for heavy metal removal: a mini review. J Compos Sci 5(1):1
67. Yang Y (2020) Recent advances in the electrochemical oxidation water treatment: spotlight on byproduct control. Front Environ Sci Eng 14
68. Ye Z, Steter JR, Centellas F, Cabot PL, Brillas E, Sirés I (2019) Photoelectro-Fenton as post-treatment for electrocoagulated benzophenone-3-loaded synthetic and urban wastewater. J Clean Prod 208:1393–1402. https://doi.org/10.1016/j.jclepro.2018.10.181
69. Zazou H, Afanga H, Akhouairi S, Ouchtak H, Addi AA, Akbour RA, Hamdani M et al (2019) Treatment of textile industry wastewater by electrocoagulation coupled with electrochemical advanced oxidation process. J Water Process Eng 28:214–221. https://doi.org/10.1016/j.jwpe.2019.02.006
70. Zheng T, Wang J, Wang Q, Meng H, Wang L (2017) Research trends in electrochemical technology for water and wastewater treatment. Appl Water Sci 7(1):13–30. https://doi.org/10.1007/s13201-015-0280-4

Carbonaceous-TiO_2 Photocatalyst for Treatment of Textile Dye-Contaminated Wastewater

Ayushman Bhattacharya and Ambika Selvaraj

Abstract The rising quantity and toxicity of dye-enriched industrial effluents due to rapid industrialization is a foremost and emerging concern throughout the world. The presence of dyes in the waterbodies possesses a threat to aquatic life and can annihilate the environment and human health. Therefore, before discharge, the treatment of dye-contaminated industrial effluents is of paramount importance. Heterogeneous photocatalysis, an advanced oxidation process (AOP), is a sustainable and effective treatment approach that has a greater potential for the catalytic eradication of dyes by generating highly reactive radicals. In general, titanium dioxide (TiO_2) semiconductor photocatalyst is extensively used for the photocatalytic degradation of dyes. However, owing to various drawbacks associated with TiO_2 photocatalysts such as higher electron_hole pair recombination and larger bandgap, the TiO_2 photocatalyst is modified by coupling with carbonaceous materials for higher photocatalytic activity and visible light-harvesting ability. Therefore, this chapter briefly describes the mechanism involved in the photocatalytic degradation of dye and elucidates the different methods for synthesizing carbonaceous-based TiO_2 composites comprising activated carbon, graphene derivatives, carbon doping, and carbon nanotubes for dye removal. Also, the chapter precisely focuses on the existing and recent studies on dye removal using carbonaceous-based TiO_2 materials. Therefore, this study will be useful for engineers and researchers working in the domain of industrial wastewater treatment.

Keywords Photocatalysis · Dye degradation · TiO_2 photocatalyst · Industrial effluents · Graphene oxide · Carbon nanotube · Activated carbon · Doped carbon

---

A. Bhattacharya
Department of Civil Engineering, Indian Institute of Technology Hyderabad, Hyderabad, India

A. Selvaraj (✉)
Adjunt Faculty, Department of Climate Change, Indian Institute of Technology Hyderabad, Hyderabad, India
e-mail: ambika@ce.iith.ac.in

S. S. Muthu and A. Khadir (eds.), *Advanced Oxidation Processes in Dye-Containing Wastewater*, Sustainable Textiles: Production, Processing, Manufacturing & Chemistry,
https://doi.org/10.1007/978-981-19-0987-0_14

1 Introduction

Dyes are unsaturated organic compound that possesses a complex structure and exhibits recalcitrant, xenobiotic, and carcinogenic nature. It is extensively used in leather tanning, textile, food, and paper production industries and is found in their effluents. The uncontrolled and untreated discharge from these industries to water bodies causes an environmental menace and disrupts aquatic life and thus requires treatment. One of the major factors that control the treatment performance of the dye-containing effluent is the nature of dyes.

Synthetic dyes are categorized into soluble and insoluble dyes. Soluble dyes are categorized into cationic and anionic dyes. Cationic dyes comprise of Rhodamine B, Methylene Blue, Brilliant blue-R, Methyl violet, Brilliant green, Crystal violet whereas Eosin Y, Congo red, Methyl Orange, Rose Bengal, Indigo carmine, Eriochrome Black T and Alizarin red S are under anionic dyes [1]. Furthermore, insoluble dyes are categorized into azo, disperse, sulfur, vat, and solvent dyes. Among these, azo dyes such as Acid Red 2, Direct Black 22, Disperse Yellow 7, Acid Orange 20, Methyl Red, Direct Blue 71, and Trypan Blue are majorly used throughout the world [2]. Generally, azo dyes consist of (–N=N–) functional groups combining two symmetrical/unsymmetrical identical non-azo alkyl and aryl radicals. An increase in these azo bonds leads to a higher molecular weight of the azo dyes that causes a reduction in the degradation rate as well as enhances the distribution of the dye in the aquatic bodies [3]. It is reported that ~60–70% of azo dyes are insusceptible to biodegradability and generate daughter compounds. The formation of aromatic amine by-products from the degradation of azo dyes is more carcinogenic than the parent dye molecules and is detrimental to reproductive and immune systems [4].

In addition, the incomplete fixation of dye to the fabric during textile processing makes 15–20% of dye gets into the environment that leads to the generation of dye-coloured containing effluents. The existence of dyes even at lower concentrations is highly visible in the water. Hence, the presence of dyes in the aquatic ecosystem declines the light penetration ability and thereby impedes photochemical action. It results in instability in the ecological balance and is characterized by chemical and biochemical oxygen demand. Besides, dyes persevere as an environmental contaminant and induce biomagnification [5, 6]. The higher solubility of dyes, increased toxicity, and low degradability of dyes raise the difficulty of its removal by conventional methods and make it an emerging environmental concern [2, 3, 5]. These major challenges associated with the degradation of dyes by conventional options suggest moving towards an advanced oxidation process for its effective removal from an aqueous medium.

Advanced Oxidation Process (AOP) has the potential to oxidize either partly or completely the toxic organic pollutants by generating highly reactive hydroxyl radicals ($OH^{\cdot}$) [7]. Several AOPs exist for the treatment of water and wastewater such as photocatalytic processes, UV/H_2O_2-based processes, Fenton-based processes, and ozone-based processes. Among these, heterogeneous photocatalysis has emerged as an environmental friendly and effective treatment for dye removal as it generates the reactive radicals under ambient conditions without the application of potentially hazardous oxidants (chlorination, ozone) and has no secondary waste disposal issues [8–11]. The application of 2D metal oxide semiconductor materials as photocatalysts has significant potential for the photocatalytic degradation of textile dyes [12]. AOP is an emerging field in wastewater treatment and many researchers are exploring the synthesis and evaluation of novel photocatalysts but a deeper investigation is required for large-scale commercialization.

Among various metal oxide photocatalysts, titanium dioxide (TiO_2) has gained widespread attention and is widely investigated because of its inherent properties such as hydrophilicity, stability against photo-corrosion, cost-effective, water insolubility, and less toxicity. However, because of the larger bandgap (3.4 for rutile and 3.2 $_{eV}$ for anatase), the ultraviolet light that occupies only 3–5% of the solar spectrum is used for the excitation of TiO_2. Apart from a larger bandgap, lower surface area and higher recombination of electrons and holes remain a major drawback that restricts its practical applications. To mitigate the above-said drawbacks and to enhance the photocatalytic performance for dye degradation, TiO_2 is modified by integrating with different support materials to modulate the electronic band structure, and for higher quantum efficiency, reaction rate, and adsorption ability [13]. Due to the broader scope of different forms of carbon-based nanomaterials for dye removal and growing research in this field makes it imperative to timely review the ongoing studies and development. Therefore, this chapter begins with a concise explanation of the mechanism behind the photocatalytic degradation of dyes and specifically focuses and in-depth details on the different synthesis methods for synthesizing carbonaceous-based TiO_2 materials. Furthermore, the chapter outlines the advancement and photocatalytic performance of carbonaceous-based TiO_2 materials comprising activated carbon, carbon dopant, graphene derivatives, and carbon nanotubes for dye degradation.

2 Mechanism of Photocatalysis in Carbonaceous-Based TiO_2 Composite

The photocatalytic reaction begins with photon absorption with higher energy than or equal to the bandgap of the photocatalyst and leads to excitation of electrons to the conduction band (CB) and the creation of holes in the valence band. These generated electrons and holes act as reducing and oxidizing agent, respectively, and take part in the redox reactions after migrating to the photocatalyst surface with various groups adsorbed on the catalyst surface [14], resulting in the generation of hydroxyl ($OH^{\cdot}$), hydroperoxyl $\left(HO_2^{\cdot}\right)$, and superoxide $\left(0_2^{\cdot-}\right)$ radicals that induce oxidation of dye molecules [15]. In general, carbonaceous material possesses strong adsorption properties, and therefore coupling it with photocatalyst fosters higher dye adsorption on the surface of the TiO_2 photocatalyst and ultimately leading to higher dye degradation. Furthermore, the formation of chemical bond (Ti–O–C) between

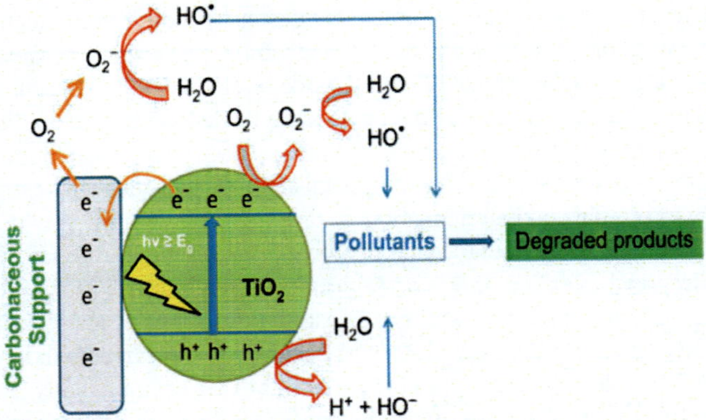

Fig. 1 Photocatalytic degradation of dye molecule as a pollutant [16]

TiO_2 and carbonaceous material extends the light adsorption potential towards the visible spectrum. Besides, the carbonaceous material acts as an electron reservoir and separates the electrons from the electron_hole pairs of TiO_2 photocatalyst, thereby lowering the charge carrier recombination. In summary, the visible light harvesting potential, adsorption of dyes, and separation of charged species can be increased by merging carbonaceous material with the TiO_2, thereby enhancing the photocatalytic activity of TiO_2 photocatalyst [16]. The detailed mechanism of photocatalytic reactions for degradation of dyes molecules under the UV-Visible irradiation in carbonaceous-based TiO_2 composite is shown in Fig. 1.

3 Method of Synthesis for Carbonaceous-Based TiO_2 Composites

3.1 Hydrothermal and Solvothermal Method

The hydrothermal method is used for the synthesis of powdery nanostructures by mixing precursors and suitable agents into the solvent inside sealed Teflon lined high-strength alloy or stainless steel autoclave at specific pressure and temperature [17]. In other words, crystalline nanomaterials directly form from the aqueous solution due to the chemical reaction of substances that occurs in a sealed container and heated aqueous solution at higher temperature (greater than 25 °C) and pressure (greater than 100 kPa). In this method, nucleation of powder and controlled manner of growth results in the formation of crystalline particles with controlled morphology and size. To improve the dissolution of reactants and further nucleation and development of certain types of materials such as carbon nanotube, single crystal of oxides, and diamond, the temperature and pressure can be escalated up to 1000 °C and 500 MPa, respectively. Over a wide range, the particles dimension and morphology characteristics can be varied in the hydrothermal method, and this method is most generally used for the synthesis of ceramics, alloys, silicates, zeolites, oxides, metals, and composites. On the other hand, in the solvothermal method, at an elevated temperature, the synthesis occurs in a non-aqueous solution with organic solvents. The different types of solvents used in solvothermal synthesis are glycols, alcohols, and oleylamine.

For both the hydrothermal and solvothermal methods of synthesis, a higher pressure autoclave or sealed container with alloy or Teflon linings is employed under subcritical and supercritical conditions of the solvent [18]. For example [19] via the solvothermal method, synthesized ternary composite, i.e. TiO_2-rGO-CuO/Fe_2O_3 photocatalyst, and found that TiO_2-rGO-CuO composite exhibits higher photocatalytic performance than TiO_2-rGO-Fe_2O_3 composite as CuO and Fe_2O_3 have different abilities to transfer electrons. Besides, compared to the TiO_2-rGO bandgap, i.e. 3.12 eV, increment in Cu doping up to 0.075 wt% causes lowering of bandgap to 3 eV and thereby leads to higher photocatalytic degradation of Methyl Orange (MO) dye but the further increment in Cu doping up to 0.5 wt% causes increment in bandgap, i.e. 3.15 eV and lowering of photocatalytic performance. This is because of Cu^+/Cu^{2+} introduction into the new energy level of TiO_2. Furthermore, increment in Fe_2O_3 doping up to 1% leads to a reduction in bandgap up to 2.95 $_{eV}$ due to the formation of intra bandgap states because of interaction between 3d and d orbitals of Ti and Fe, respectively. Moreover, the photocatalytic performance of TiO_2-rGO-Fe_2O_3 composite declines with increment in Fe_2O_3 doping due to the amorphous layer formation on the surface of TiO_2 photocatalyst.

3.2 *Sol–Gel Method*

In this method, using sol (colloidal solution) as initial material, a broader range of novel, advanced and functional materials like ceramic, glass, organic–inorganic composites can be prepared, which can have significant applications [20]. The sol–gel method is also referred to as chemical solution deposition. It involves a series of steps beginning from hydrolysis, condensation, drying, and crystallization as shown in Fig. 2a. For the integration of inorganic material into the organic matrix or organic species into the inorganic matrix, lower reaction temperature and mild reaction conditions are suitable [21].

The process initiates with the hydrolysis and partial condensation of alkoxides for the formation of sol that is a stable dispersed colloidal particle or polymer in a solvent. Interaction between sol particles is dominated by hydrogen bonding or Van der Waals forces. Furthermore, a 3D continuous network known as gel forms through polycondensation of precursors that are initially hydrolyzed and possess polymeric substructure. Subsequent drying of gel induces evaporation of solvent leading to a collapse of the porous network thereby forming "xerogel" and this gelation process is an irreversible process in which interaction is a covalent type and after further

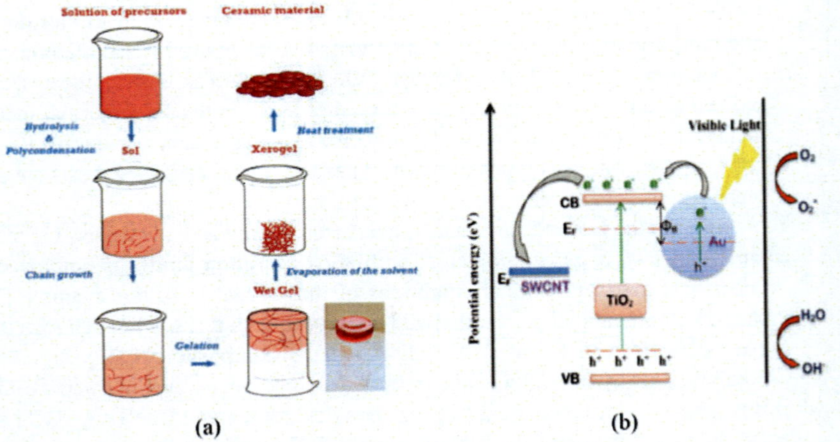

Fig. 2 Schematic diagram of **a** sol–gel process [20] and **b** electron transfer in TiO_2/Au/SWCNT ternary nanohybrid [22]

calcination, mechanically stabilized products, i.e. crystalline substances are obtained. Various factors that influence the sol–gel chemistry include pH, temperature, heat treatment and concentration, and type of additives as well as precursors [23]. The sol–gel method is classified as aqueous and non-aqueous depending upon the types of solvent, i.e. water or inorganic solvent. The sol–gel method is a reliable, economical, and simple method, and nanocomposites synthesized by this method demonstrate even particle size, higher purity, and homogeneity [21].

Through sol–gel technique, Mohammed [22] synthesized ternary TiO_2/Au/SWCNT nanohybrid for the photodegradation of methylene blue (MB) dye. Analysis revealed that under visible light illumination for 30 min, TiO_2/Au/SWCNT ternary nanohybrid demonstrated higher photocatalytic degradation, i.e. 80% as compared with pristine TiO_2 (20%) and TiO_2/Au binary composite (24%) due to smaller bandgap, i.e. 1.95 eV, efficient charge migration from TiO_2 to SWCNT, lower recombination of electrons and holes and higher visible light-harvesting capability. Also, the effect of surface plasma resonance (SPR) of Au contributes to the higher photocatalytic activity of TiO_2/Au/SWCNT ternary nanohybrid. As depicted in Fig. 2b under visible light illumination, the formed electron in Au present in TiO_2 lattice gains sufficient energy to overcome the Schottky barrier and migrate to the conduction band of TiO_2, leaving behind holes in the Au atom that take part in dye degradation. Further, the electron present in the conduction of TiO_2 produces reactive radicals by reacting with oxygen molecules as well as moves to SWCNT resulting in a lowering of recombination rate and increment in the lifetime of charge carriers.

3.3 Ultrasonic-Assisted Method

The ultrasound is generally employed to propel chemical reactions as well as to disperse colloids and nanoparticles, emulsify mixtures, nebulizing solution into fine mixtures. It generates acoustic waves of frequencies greater than 20 kHz [24]. In this method, ultrasound induces high-energy chemistry through acoustic cavitation that fosters structural changes and expedites the chemical reaction as well as breaks the weak non-covalent interactions or disintegrates the aggregated particles. Irradiated by high-intensity ultrasound, the formation, growth, and collapse of bubbles in an aqueous solution are designated as acoustic cavitation. Inside the liquid, ultrasound irradiation creates acoustic waves in alternating expansive and compressive modes that give rise to an oscillating bubble. With an augmentation in size to a certain limit, this oscillating bubble accumulates the ultrasonic energy. Further, these bubbles overgrow and subsequently collapse thereby liberating the stored concentrated energy that leads to high temperature (~5000 K) and pressure (~1000 bar) in the liquid for a short duration [25], and water vapour and oxygen present inside the bubble are dissociated and oxidizing agents such as $OH^{\cdot}$ radicals, O and H_2O_2 and a reducing agent such as $H^{\cdot}$ radicals are formed [26, 27].

The inherent benefits of the ultrasound-assisted method of synthesis are the formation of nanoparticles of uniform size, higher reaction rate, economical, better controllability over reaction conditions [28]. For example, for the Methylene Blue (MB) dye degradation under sunlight, Deshmukh et al. [29] synthesized ultrasonic-assisted rGO/TiO_2 composite as illustrated in Fig. 3. Analysis revealed that with an increment in rGO/TiO_2 photocatalyst loading from 1 to 3 mg/L, MB dye degradation

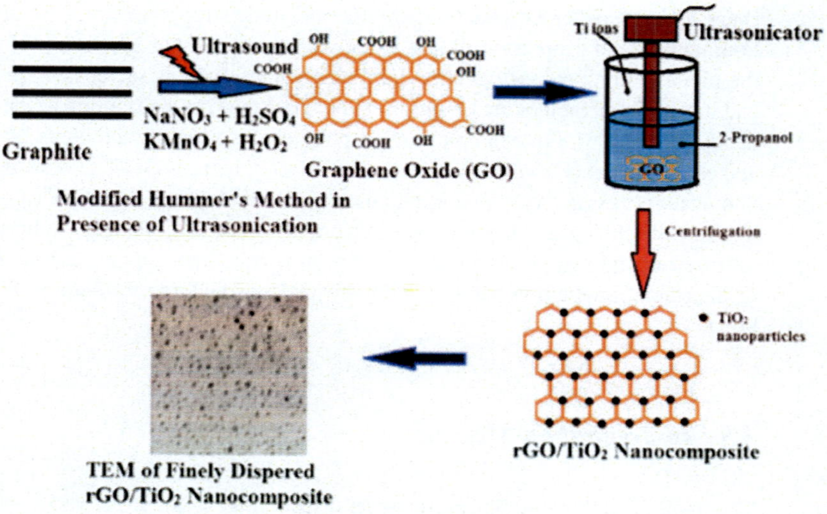

Fig. 3 Schematic representation of rGO/TiO_2 composite synthesis [29]

rises from 18.7 to 52.2% in 30 min as active sites increase. Also, at pH value 13.2, higher MB dye degradation, i.e. 91.3% was observed after 30 min. This is because, at pH value above the point of zero charge (PZC) of the nanocomposite, i.e. 2.62, the rGO/TiO_2 composite surface gets negatively charged and stimulates stronger electrostatic attraction with the MB dye molecule because of its cationic nature thereby leading to a higher degradation of MB dye.

3.4 Co-precipitation

This method is fast, economical, and environment friendly that neither requires higher pressure and temperature nor uses hazardous organic solvents for the synthesis of nanomaterials in an aqueous solution [30]. The sequence of processes involved in the coprecipitation method is simultaneous nucleation, growth, coarsening, and agglomeration. The characteristics of co-precipitation reaction are (i) formation of insoluble products in supersaturation condition, (ii) formation of smaller particles due to nucleation, (iii) post nucleation processes leads to particles aggregation resulting in a change in particle shape, size, morphology, and other properties, (iv) precipitation at the reaction scale induced due to supersaturation. The various stages in the co-precipitation method are (i) formation of nanoparticles from aqueous solutions or by reduction from non-aqueous solutions, electrochemical reduction, and breakdown of metal–organic precursors,(ii) formation of metal chalcogenides due to reactions of the molecular precursors; (iii) coprecipitation at smaller scale facilitated by sonication/microwave and subsequent filtration and drying [31].

The synthesized nanoparticles in the coprecipitation method are pure and their shape and size depend on the temperature, pH, ionic strength, velocity of precipitation, nature, and proportion of salt along with their order of addition [32]. During the precipitation process, various stabilizing agents such as surfactants, polymers, and inorganic molecules are employed to enhance the size distribution of the formed nanoparticles [33]. For example, for the photodegradation of methylene blue (MB) and crystal violet (CV) dye, Ranjith et al. [34] synthesized rGO/TiO_2/Co_3O_4 hybrid photocatalyst using the co-precipitation method and observed that rGO/TiO_2/Co_3O_4 hybrid exhibits higher photocatalytic performance than TiO_2/Co_3O_4 composite due to lower recombination of charge carriers and narrow bandgap, i.e. 2.74 eV because of interfacial interaction between TiO_2/Co_3O_4 and rGO and formation of Ti–O–C bond.

3.5 Impregnation Method

This method is a facile and most general technique to induce a wider variety of heterogeneous reactions by synthesizing supported catalysts. It comprises of three steps commencing with the loading of the solution comprising of metallic precursors on porous support having higher surface area followed by evaporation of solvent at higher temperature and subsequent production of the catalyst due to reduction of metallic precursors under suitable conditions [35]. In this method, the metal precursors at first are contacted with porous support, and the common precursors are inorganic metal salts like chlorides, organic metal complexes, metal sulfates, carbonates, and acetates or nitrates. Owning to the higher solubility of many precursors, water is used as a solvent for inorganic salts whereas, for organometallic precursors, organic solvents are used. In bulk solution, the concentrations of metal precursors should be below supersaturation, to prevent premature deposition. The impregnation method is of two types, namely, the wet impregnation (WI) method and pore volume impregnation (PVI) method.

In the case of WI, an excess quantity of solution is used and after a certain time the separation of solids occurs, and the excess solvent is removed by drying. In the case of PVI, a limited quantity of solution just needed to fill the pore volume of support is used and after the impregnation of the catalyst onto the support, it is then subjected to drying, calcination, or reduction. This process is also referred to as dry impregnation (DI) or incipient wetness impregnation (IWI) because of the dry character of impregnated material at a macroscale [36, 37]. The impregnated support is filtered out in case of WI, as a result, an excess liquid comprising of the precursors that are not retained by the support leaves, and, therefore, this method necessitates the recycling of the excess liquid to curtail the wastage of precursors whereas the DI method eradicates the usage of excess liquid and filtration step. However, lack of filtration in case of DI method can lead to retention of counterions from the metallic precursors in the dried catalyst and therefore further processing is required in case of removal of other substances [38].

The concentration of the precursors and the temperature affect the impregnation method. The temperature affects the solubility of the precursors, viscosity of the solution and as a consequence the wetting time. The precursor's solubility in the solution limits the maximum loading. Furthermore, during impregnation and drying, the condition of mass transfer within the pores determines the concentration profile of impregnated material [36]. For example, through microwave as well as thermal radiation assisted impregnation method with dispersion medium as water, Liu et al. [39] synthesized K doped g-C_3N_4 nanosheets-TiO_2 (KCNN-TiO_2) composite photocatalyst for the photodegradation of Rhodamine B dye and revealed that under visible light illumination the composite with 0.3 TiO_2 to KCNN mass ratio prepared by thermal radiation assisted ($KCNNT_{0.3H}$) and composite with 0.5 TiO_2 to KCNN mass ratio prepared by microwave-assisted impregnation method ($KCNNT_{0.5M}$) shows higher photocatalytic degradation of Rhodamine B dye, i.e. 92.73% and 97.28%, respectively, due to higher carrier separation efficiency. Furthermore, effective heterojunction formation in case $KCNNT_{0.5M}$ than $KCNNT_{0.3H}$ leads to higher dye degradation with rate constant, k (0.01817 min^{-1}) higher than the latter, i.e. 0.01474 min^{-1}.

3.6 Chemical Vapour Deposition (CVD)

In this method, chemical reactions of the precursors in the vapour phase lead to the formation of thin films or crystals due to the deposition of non-volatile solid particles on the substrate [40]. In the CVD process, the substrate is referred to where crystals and thin films are deposited. The precursors are the reactants in the CVD method, and the three reactions involved in the conversion of reactants to products are thermal decomposition, chemical synthesis, and chemical transport. To grow silicon in commercial CVD configuration, vapour precursors (SiH_4 or SiH_2Cl_2) in cohort with a carrier gas (H_2 or Ar) are introduced into the reaction chamber and precise control on the molecules is achieved by varying the flow rate and partial pressure of the precursors.

For deposition of graphene, methane (CH_4) and hydrogen (H_2) are commonly used as precursors and carrier gas, respectively, and Nickel (Ni) or Copper (Cu) is used as substrate. Furthermore, the obtained graphene can be doped with Nitrogen (N) or Phosphorus (P) atoms by introducing ammonia (NH_3) and phosphine (PH_3) gases in the reaction chamber [41]. To grow metallic transitional metal dichalcogenides (TMDCs), a solid phase powder comprising of metal precursors (tungsten oxide, molybdenum oxide) and non-metal precursors (tellurium, sulphur, and selenium) are generally used in a traditional horizontal CVD system and lead to non-uniform and uncontrolled growth results due to mass and time dependence of the sublimation of the solid and spatially non-uniform dynamics of growth whereas the application of gaseous precursors to grow TMDCs in vertical CVD system leads to uniform growth and higher quality product, In general, the application of gaseous precursors in CVD system is better as their flow rate can be accurately controlled over a broader range and size, structure and morphology can be precisely controlled [42]. Furthermore, to activate the gaseous precursors and subsequent gas–solid phase reactions, the CVD process is carried out at higher temperatures [43]. For example, Ahmad et al. [8, 9] prepared the multi-walled carbon nanotubes (MWCNTs) using CVD method and after functionalization with 3-aminopropyltriethoxysilane (APTES), the TiO_2 was loaded to form MWCNTs-APTES-TiO_2 composite and the photocatalytic degradation of methyl orange (MO) dye was investigated. The MWCNTs-APTES-TiO_2 composite with 1:2 w/w of MWCNTs-APTES and TiO_2 demonstrates higher degradation of dye, i.e. 87% as compared to pure TiO_2, i.e. 43% after 180 min due to larger active sites provided by the functionalized MWCNTs.

The obtained product in CVD can significantly vary in terms of size, composition, structure, morphology as the entire growth process depends upon various factors such as gas flow rate, substrate, temperature, growth time, pressure, rate of cooling. The advantages of the CVD method are the formation of a uniform film with lower porosity along with higher purity and stability whereas emission of toxic gases as by-products during chemical reactions and higher cost of the equipment is its demerits [44].

3.7 Photochemical Reduction Method

Under this method, in presence of reducing reagent, it involves ultraviolet irradiation of metal precursor solution for the reduction of metal salts. The nanoparticles are produced at low temperatures and under solid-phase conditions [45]. For example, Koo et al. [46] synthesized Ag-TiO_2-CNT composite using photochemical reduction and found Ag-TiO_2-CNT composite exhibits higher photocatalytic degradation of Methylene Blue (MB) dye as compared with Ag-TiO_2 under artificial light for 180 min because of higher charge separation and larger surface area of Ag-TiO_2-CNT composite with a synergistic effect involving stronger dye adsorption capacity by CNT and photodegradation by Ag-TiO_2.

4 Carbonaceous-Based TiO_2 Composites for Dye Removal

Owning to stronger oxidizing power under ultraviolet irradiation, TiO_2 has been widely used for the removal of contaminants. However, several drawbacks associated with the pristine TiO_2 lower its photocatalytic efficiency and make it unsuitable for practical purposes. Therefore, later sections briefly discuss the various strategies or efforts made to escalate its photocatalytic efficiency by integrating it with suitable dopants, composites or additives, or by using support/carrier for increasing the active surface area of TiO_2.

4.1 Carbon-Doped TiO_2 Composites

For the photocatalytic degradation of Rhodamine B, Zhang et al. [47] prepared a novel SiO_2@C-doped TiO_2 (SCT) hollow spheres as shown in Fig. 4. The steps involved in the synthesis of SCT hollow spheres are the monodisperse cationic polystyrene sphere (CPS) preparation, deposition of inner SiO_2, synthesis of sandwich-like CPS@SiO_2@CPS particles, and then TiO_2 formation at the outer shell. Lastly, the synthesized nanoparticle was subjected to calcination at 450 °C for CPS template removal. The experiment was performed under both ultraviolet (UV) and visible light, and the outcomes were compared with commercial pure P25 (Degussa). Under visible light illumination, after 10 min, SCT hollow spheres degrade 58.5% of Rhodamine B dye whereas it was 11.3% in the case of P25 (Degussa) photocatalyst. Also, after 60 min of prolonging irradiation, SCT hollow spheres degrade dye to 93.6% that is much higher than 48.8% in the case of P25 Degussa thereby inferring higher photocatalytic performance of SCT hollow sphere than P25 (Degussa) in dye degradation. Similarly, under ultraviolet irradiation, after 10 and 60 min, SCT hollow spheres exhibit higher dye degradation, i.e. 39.2% and 61.7%, respectively, than P25 (Degussa). The SCT hollow sphere demonstrates higher photocatalytic activity

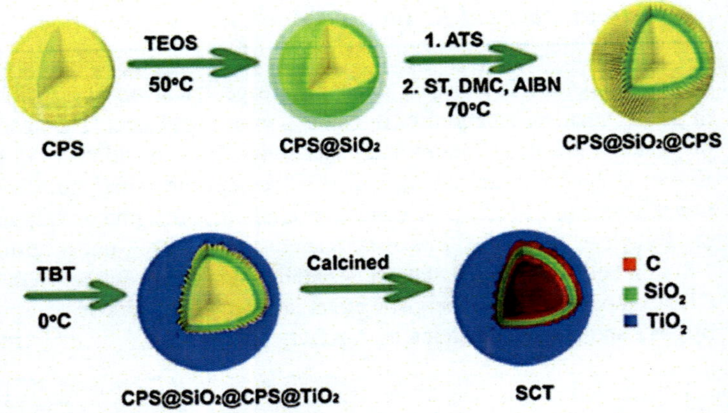

Fig. 4 Schematic diagram of SiO_2@C-doped TiO_2 hollow spheres synthesis [47]

because the presence of carbon lowers the rate of charged species recombination and as hollow SiO_2 acts as a carrier it can provide a better adsorption site and can enhance the dye concentration near the TiO_2 photoactive centre. Furthermore, SCT hollow sphere lowers the diffusion resistance of dyes and the higher surface area of the SCT hollow sphere, i.e. 255.692 m^2/g than P25 (60 m^2/g) enhances the dye accessibility to its surface and thereby leads to faster photodegradation of dyes.

In another study, through microwave dielectric heating for different times, i.e. 0, 2, 4, 6, and 8 min, Luna-Flores et al. [48] synthesized a series of carbon-doped amorphous TiO_2 (C-TiO_2) photocatalyst for the degradation of Rhodamine B. The experimentation was performed for 100 min using a 10 W LED visible light source and a 15 mg photocatalyst was dispersed in 60 mL of Rhodamine B dye solution. Analysis revealed that samples of C-TiO_2 photocatalyst demonstrate 76–96% dye degradation that is higher as compared to P25 (Degussa) titanium oxide. The higher photocatalytic activity of C-TiO_2 is due to the nanodisc amorphous morphology of C-TiO_2 that hastens the charge carrier's migration to the surface due to a shorter migration distance. Also, under visible light irradiation, Ovodok et al. [49] studied the photocatalytic degradation of Rhodamine B dye using sol–gel derived mesoporous carbon-doped TiO_2 photocatalyst and found its photocatalytic activity higher than the P25 (Degussa) because of the higher surface area ensuring a large number of active sites and reduction in bandgap due to carbon doping fosters enhancement of the visible light-harvesting potential. In another study, Ghime et al. [50] explored photocatalytic degradation of eosin yellow dye using carbon-doped TiO_2 (C-TiO_2) photocatalyst. The C-TiO_2 photocatalyst was prepared using the sol–gel method and calcined at two different temperatures, i.e. 200 and 400 °C for 6 h and is designated as C-TiO_2-200 and C-TiO_2-400, respectively. The C-TiO_2-200 manifested 81.88% degradation of eosin yellow dye whereas 75.55% removal was observed in the case of C-TiO_2-400. The C-TiO_2-200 demonstrates higher photocatalytic activity due to higher surface area, i.e. 48.061 m^2/g as compared to C-TiO_2-400 having 10.453 m^2/g.

It was also observed that increment in eosin yellow dye concentration, as well as the higher dosage of H_2O_2 oxidant in TiO_2 photocatalysis, leads to decrement in photocatalytic degradation rate due to non-availability of the required number of free radicals in aqueous solution and formation per-hydroxyl radicals $(OH_2^{\cdot})$ that exhibits lesser reactivity as compared to hydroxyl radicals $(OH^{\cdot})$, respectively.

For sunlight-assisted photocatalytic degradation of Methylene Blue (MB) dye, via calcination, Li et al. [51] synthesized C-doped TiO_2 by doping carbon black into hydrothermally synthesized TiO_2 nanorods. Different weight ratios (0, 0.2, 0.5, and 1.0%) of C-doped TiO_2 composites were used for analysis. Scanning electron microscope analysis reveals that after carbon black doping, the surface of the TiO_2 becomes rough and dispersed with more nanoparticles and TiO_2 nanowires transform into short nanorods. It was observed that carbon doping leads to distortion in the lattice structure of TiO_2 and the formation of the O–Ti–C bond at the carbon black and TiO_2 interface. Also, carbon black doping stimulates the formation of defects and oxygen vacancies, which acts as an electron acceptor and impedes the recombination of charge species, and this phenomenon strengthens with increment in carbon black doping and leads to lowering of TiO_2 bandgap. Therefore, under sunlight, due to smaller bandgap, lower recombination of charge carriers, and lower resistance to charge transfer, a higher photocatalytic degradation rate was found in 1% C-TiO_2 with complete removal of MB dye in 30 min. In summary, the carbon dopant plays a vital role in improving the photocatalytic activity of TiO_2 by narrowing the bandgap and extending the photoresponse towards the visible region.

4.2 *Carbon Nanotubes (CNTs)-TiO_2 Composites*

Owning to excellent optical, electrical, thermal, chemical, and structural properties, carbon nanotubes (CNTs) have gained greater emphasis in photocatalytic applications. It comprises sp^2 hybridized carbon atoms and acts as an electron acceptor and integrating it with TiO_2 photocatalyst leads to higher photocatalytic activity. For the photocatalytic degradation of methylene blue (MB) dye, Azzam et al. [52] using the sol–gel method, synthesized TiO_2@CNTs composite, and silver (Ag) nanoparticles were decorated independently and in the presence of cationic surfactant (C10) and is therefore referred to as TiO_2@CNTs/Ag (without surfactant) and TiO_2@CNTs/Ag/C10 (with surfactant), respectively. The experiment was performed under visible light irradiation, and the effect of various components such as H_2O_2 concentration, photocatalyst dosage, and initial concentration of dye on the photocatalytic performance of TiO_2@CNTs/Ag/C10 was analyzed. It was observed that after 240 min, the CNTs exhibits 32% photocatalytic degradation of MB dye that aroused to 66% after TiO_2 loading onto CNTs (TiO_2@CNTs) because of the larger surface area of the composite as compared to CNTs alone and increment in photoactivity of TiO_2 by loading onto CNTs. Also, as a photosensitizer, CNTs eject electrons to the conduction band of TiO_2, which intensifies the formation of reactive

OH˙ thereby amplifies the MB dye degradation. Furthermore, the presence of well-distributed Ag nanoparticles impedes the recombination of electrons and holes, and therefore TiO_2@CNTs/Ag composite demonstrates 68% photocatalytic degradation of MB dye in 240 min whereas 100% degradation of MB dye was observed in TiO_2@CNTs/Ag/C10 composite in 180 min due to lower bandgap, i.e. 2.25 eV, higher surface area, i.e. 146 m^2/g and presence of C10 surfactant increase the life-time of charge carriers that leads to higher photocatalytic performance. In the case of TiO_2@CNTs/Ag/C10 composite, the increment in H_2O_2 concentration from 0.139 to 0.782 mol/L boosts the photodegradation of MB dye due to the increased formation of reactive OH˙ formed from the H_2O_2 decomposition as well as an increment in the photocatalytic degradation of MB dye was observed from 60 to 100% after 120 min with an increase in photocatalyst dosage from 0.2 to 0.5 g/L but further increase in photocatalyst dosage lowers the photocatalytic degradation MB dye due to agglomeration of nanoparticles. Also, under constant TiO_2@CNTs/Ag/C10 photocatalyst loading, i.e. 0.5 g/L, decrement in photocatalytic degradation of MB dye was observed with an increase in the concentration of dye from 10 to 100 mg/L due to insufficient availability of reactive radicals.

To develop higher performance photocatalyst, Chen et al. [53] prepared CNT/TiO_2 nanocomposite as shown in Fig. 5 at different calcination temperatures, i.e. 350, 450, 550, 650, 750, and 850 °C for 120 min and found photocatalytic degradation of Rhodamine B reached 91.14, 100, 79.36, 69.41, 63.58, and 86.52%, respectively, at 25 min inferring that with an increase in calcination temperature, photocatalytic performance reduces as photoelectron yield capacity of CNT/TiO_2 declines. The CNT/TiO_2 calcined at 450 °C demonstrates higher photocatalytic performance due to lower recombination of charge carriers, higher surface area, i.e. 129.20 m^2/g as

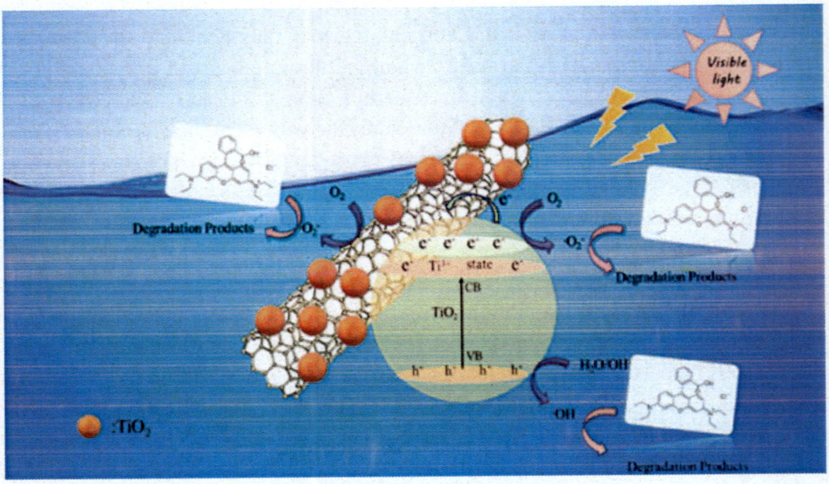

Fig. 5 Diagram depicting photocatalytic degradation of Rhodamine B dye using CNT/TiO_2 nanocomposite [53]

the presence of CNT impedes the agglomeration of TiO_2 particles and higher visible light responsiveness due to lower bandgap, i.e. 2.18 $_{eV}$ as TiO_2 and CNT interaction leads to the formation of Ti_O–C bond that introduces energy state in TiO_2 bandgap. In summary, the specific structure and morphology of CNTs allow them to serve as supporting material to provide higher surface area and to boost the adsorption activity as well as to lower the electron–hole pair recombination for higher photocatalytic activity.

4.3 Graphene-TiO_2 Composite

A 2D allotrope of carbon, i.e. graphene as well as its derivatives manifested the potential to boost the photocatalytic activity of TiO_2. In graphene, the delocalized π conjugated bonds accept the electrons and reduce the recombination of charge species, as well as the oxygenated functionalities on the surface of the graphene, promote the integration of graphene sheets and semiconductors. Through a simple chemical process, Ali et al. [54] synthesized 1 and 3% TiO_2@rGO nanocomposite photocatalyst for the photocatalytic degradation of Rhodamine B and the effect of pH (3–9), photocatalyst dosage (0.1–0.5 g), initial concentration of dye (5–30 mg/L) and different time scales of UV irradiation (15, 30, 45, 60, and 120 min) on photocatalytic degradation were investigated. The analysis at different pH values reveals that at pH 9 and after 60 min of UV irradiation, the 3% TiO_2@rGO demonstrates higher photocatalytic degradation of Rhodamine B dye, i.e. 99.2% as compared with 98.2% for 1% TiO_2@rGO. The higher photocatalytic performance in the alkaline condition is due to the formation of TiO^- species as the point of zero charge for TiO_2 is 6.3 that leads to higher adsorption because of the cationic nature of Rhodamine B dye. Also, the photocatalytic degradation rate of the 1% TiO_2@rGO and 3% TiO_2@rGO nanocomposite escalated with an increase in UV irradiation time and reached the maximum rate after 60 min, i.e. $k = 0.026$ and 0.028 min^{-1}, respectively, as light energy falling on photocatalyst surface increases with prolong irradiation time and induces generation of photoexcited species. Furthermore, a steep increase in photocatalytic degradation rate was observed with an increase in photocatalyst dosage from 0.1 to 0.25 g and become constant with further increment in photocatalyst dosage, i.e. from 0.25 to 0.5 g because increased photocatalyst dosage leads to a collision between activated and ground-state molecules and as a result the activated molecule gets deactivated. Also, the rate of photocatalytic degradation escalates with a rise in initial dye concentration up to 20 mg/L but further increment shows an insignificant change in the photocatalytic degradation rate. Ambika et al. successfully generated exfoliated graphene from industrial graphite that has shown excellent adsorption of five different textile dyes. The tested dyes were royal blue (RB), turquoise blue (TB), black supra (BS), navy blue (NB), and deep red (DR) for various environmental conditions. The order of adsorption at equilibrium was found to follow, DR > TB > BS > NB > RB at circum-neutral pH in the range of 5–25 mg/L of dye, having 0.2 gm of exfoliated graphene [55].

In another study, under visible light illumination, the photocatalytic activity of synthesized TiO_2 composites with graphene oxide (GO-T), reduced graphene oxide (rGO-T), and nitrogen/sulphur (N/S) or boron (B) doped reduced graphene oxide (rGONS-T or rGOB-T) was studied for the degradation of Orange G (OG) dye. Experiment was carried out using 1500 W Xenon lamp as a source of illumination, and it was noticed that after 60 min, photocatalytic degradation of OG dye increased to 99.8, 90, 98.2, and 96.5%, respectively, as compared to 47.6% for TiO_2 inferring higher efficiency of graphene-derived TiO_2 composite as compared to pristine TiO_2. This is attributed to lower bandgap and higher generation of reactive radicals because of reduced recombination of photogenerated charged species due to efficient charge transfer between graphene derivatives and TiO_2. As compared with GO-T composite, the lower photocatalytic performance of rGO-T, rGONS-T, and rGOB-T is due to poor interaction between graphene derivatives and TiO_2 because of the reduced number of oxygen functionalities. The lower bandgap, i.e. 2.98 eV due to the formation of Ti–O–C bonds in synergy with good interfacial coupling and assembly between TiO_2 and GO sheets leads to the higher photocatalytic performance of GO-T composite [56].

To overcome the associated drawback of pristine TiO_2 and for better reusability and recovery, Elshahawy et al. [57] prepared the nanocomposites (PVA/PAAc)-GO and (PVA/PAAc)-rGO by trapping the synthesized GO and rGO in acrylic acid (PAAc) and polyvinyl alcohol (PVA) hydrogels and further TiO_2 was incorporated into the matrix to form (PVA/PAAc)-TiO_2, (PVA/PAAc)-GO-TiO_2 and (PVA/PAAc)-rGO-TiO_2 for the photocatalytic decolorization of Direct Blue71 (DB71) dye under UV irradiation for 180 min. It was found that with the increase in the concentration of DB71 dye from 20 to 40 mg/L, the % decolorization increases due to higher surface area and presence of unfilled active sites on the photocatalyst but % decolorization declines with further increment in concentration of DB71 dye, i.e. beyond 40 mg/L as the intensity of UV illumination reaching the photocatalyst surface reduces as dye molecule occupies the surface area and the active sites of the photocatalyst. The maximum % decolourization was observed in (PVA/PAAc)-rGO-TiO_2 nanocomposite, i.e. 52.7% followed by (PVA/PAAc)-GO-TiO_2 (50.7%) and (PVA/PAAc)-TiO_2 (40.4%). This is because higher generation of reactive radicals and lower recombination of charge carriers in case of rGO-TiO_2 and stronger interaction of GO/rGO with TiO_2 to form chemical bonds, i.e. Ti–C and Ti_O–C bonds foster higher photocatalytic performance. Also, the synthesized nanocomposites exhibit higher % decolorization with increment in H_2O_2 loading up to 2 mL/L but the further increment in H_2O_2 loading leads to lowering of % decolorization due to generation of hydroperoxyl radicals ($^{\cdot}OOH$) having lower oxidation potential as compared to $OH^{\cdot}$.

Rgo et al. [58] synthesized heterogeneous rGO/BiOCl/TiO_2 nanocomposites as shown in Fig. 6 using the hydrothermal method for the photocatalytic degradation of methylene blue (MB), amido black-10B (AB-10B), methyl orange (MO), and rhodamine B (RhB) dye. It was found that the Brunauer–Emmett–Teller (BET) surface area tends to increase initially and then decreases with an increase in TiO_2 loading from 0 to 30%, and rGO/BiOCl/TiO_2-10% exhibits higher surface area,

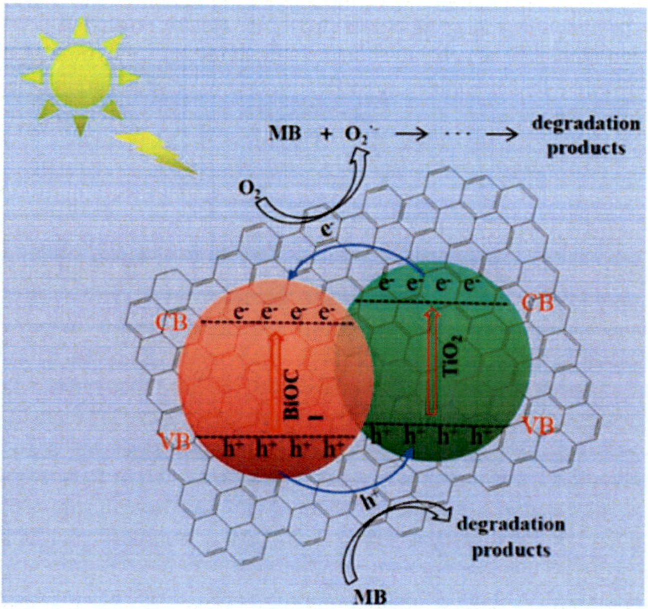

Fig. 6 Schematic illustration of photocatalytic dye degradation by rGO/BiOCl/TiO_2 composites [58]

i.e. 64.21 m^2/g. and in the case of 50% TiO_2 loading, BET surface area raises to 270.59 m^2/g. This is because surface morphologies vary in nanocomposites of different compositions. The rGO/BiOCl/TiO_2-10% nanocomposite exhibits higher adsorption of MB dye, i.e. 13.1 mg/g due to higher specific surface area and larger pore structure whereas in the case of rGO/BiOCl/TiO_2-50%, despite the higher specific surface area, the smaller size of the pore restricts the adsorption of MB dye. Therefore, after 40 min of visible light irradiation, rGO/BiOCl/TiO_2-10% nanocomposite with 1 g/L as photocatalyst dose demonstrates higher photocatalytic degradation of MB dye, i.e. 98.2% with maximum degradation rate constant of 0.070 min^{-1} that is 3.18 times higher than rGO/BiOCl composite. In the case of RhB, AB-10B, and MO dye, with rGO/BiOCl/TiO_2-10% nanocomposite, the degradation rate ranges from 0.239 to 0.313 min^{-1} which is 3.40–4.47 times higher as compared to the degradation rate of MB dye because of variation in structures of dye molecules. Therefore rGO/BiOCl/TiO_2-10% nanocomposite displays 96, 98.3, and 90.5% degradation for AB-10B, MO, and RhB dye after 10, 15, and 5 min, respectively. Furthermore, the increment in MB dye concentration from 5 to 40 mg/L leads to higher adsorption of dye molecules on the rGO/BiOCl/TiO_2-10% nanocomposite but no significant increase in adsorption takes place when adsorption sites reach the saturation limit with a further increase in MB dye concentration. In summary, reduced graphene oxide (rGO) exhibits higher surface area, rapid electron mobility, higher conductivity, and chemical stability as compared to normal graphene oxide (GO) and therefore TiO_2 loaded with rGO as cocatalyst leads to higher photocatalytic degradation of dyes.

4.4 Activated Carbon-TiO_2 Composite

Powdered activated carbon (PAC) is a porous material that is used for the elimination of contaminants from water and wastewater. During the treatment, the contaminants are just adsorbed in the pores and are therefore not referred to as clean processes as contaminants are just transferred from liquid to solid phase. To decompose or degrade the adsorbed contaminant and to regenerate the PAC as well as to boost the photocatalytic performance of the photocatalyst, the PAC is coupled with a photocatalyst. For example, Belayachi et al. [59] prepared TiO_2/GMAC composite by impregnating TiO_2 as a coating agent on powdered grape marc-based activated carbon (GMAC) by chemical activation for photocatalytic degradation of Reactive Black (RB-5) azo dye. The Chemical Oxygen Demand (COD) was determined for estimating the dye degradation. Results revealed that TiO_2/GMAC composite exhibits higher degradation of RB-5 dye as compared to individual TiO_2 and GMAC because of synergistic effect as GMAC induces higher adsorption of RB-5 dye due to larger surface area and lower active sites on the TiO_2 foster quick degradation of RB-5 dye to release active site for other molecules. Therefore, under 60 min of UV irradiation, a higher COD reduction was observed in TiO_2/GMAC composite, i.e. from 153.60 to 4.80 mg/L than TiO_2 and GMAC.

To develop a novel and high-performance photocatalyst for the eradication of organic contaminants using agricultural waste, Moosavi et al. [60] prepared activated carbon (AC) from the coconut shell to synthesize different ratios of Fe_3O_4/AC:TiO_2 composite for the eradication of Methylene Blue (MB) dye under UV irradiation. It was observed that increment in TiO_2 ratio from 1:1 to 1:2 leads to an increase in Fe_3O_4/AC/TiO_2 particle size and further increment from 1:2 to 1:4 lowers the particle size. Also, the weight loss percentage of Fe_3O_4/AC/TiO_2 photocatalyst declines with increment in TiO_2 ratio thereby preventing the loss of carbon from AC that plays a major role as an adsorbent for dye molecules. The higher MB dye removal efficiency, i.e. 98% was observed in Fe_3O_4/AC/TiO_2 composite in ratio of 1:2 after 60 min and due to aggregation of TiO_2 nanoparticle on AC, Fe_3O_4/AC/TiO_2 composite in ratio of 1:4 demonstrates lower dye removal efficiency, i.e. 66%. Furthermore, in the case of (1:2) Fe_3O_4/AC/TiO_2 composite, the increment in pH from 10 to 12 leads to escalation in the MB dye degradation efficiency from 91.4 to 98.6% but further increment to pH 13 leads to lowering of MB dye degradation efficiency to ~84% after 120 min ensuing pH plays a vital role in influencing the dye degradation efficiency.

5 Conclusion

Advanced oxidation process is an effective and promising technique for dye removal. The uncontrolled discharge of untreated industrial effluents comprising of dyes needs proper treatment and a strict regulatory framework to mitigate the environmental menace. The integration of carbonaceous materials with TiO_2 enhances the photocatalytic performance of TiO_2 and leads to a higher degradation of dyes. The development of advanced and novel visible light-responsive TiO_2 composite or potential alternative needs to be focussed to boost the dye removal along with lower energy recovery and higher reusability of the photocatalyst. Furthermore, the research should emphasize the commercialization, advanced design, and upscaling of the photocatalytic reactor to the industrial level for large-scale treatment of dyes.

Funding Sources The corresponding author would like to acknowledge the Science Engineering and Research Board (SERB), India for their support for the computing facility under Start up Research Grant (File Number: SRG/2020/000793) and Seed Grant (letter dated 15.5.2020) from Indian Institute of Technology Hyderabad, India.

References

1. Shanker U, Rani M, Jassal V (2017) Degradation of hazardous organic dyes in water by nanomaterials. Environ Chem Lett 15:623–642. https://doi.org/10.1007/s10311-017-0650-2
2. Benkhaya S, M'rabet B, El Harfi A (2020) A review on classifications, recent synthesis and applications of textile dyes. Inorg Chem Commun 115:107891. https://doi.org/10.1016/j.inoche.2020.107891
3. Benkhaya S, M'rabet S, El Harfi A (2020) Classifications, properties, recent synthesis and applications of azo dyes. Heliyon 6. https://doi.org/10.1016/j.heliyon.2020.e03271
4. Saini RD (2017) Textile organic dyes: polluting effects and elimination methods from textile waste water. Int J Chem Eng Res 9:975–6442
5. Lellis B, Fávaro-Polonio CZ, Pamphile JA, Polonio JC (2019) Effects of textile dyes on health and the environment and bioremediation potential of living organisms. Biotechnol Res Innov 3:275–290. https://doi.org/10.1016/j.biori.2019.09.001
6. Sharma K, Dalai AK, Vyas RK (2017) Removal of synthetic dyes from multicomponent industrial wastewaters. Rev Chem Eng 34:107–134. https://doi.org/10.1515/revce-2016-0042
7. Holkar CR, Jadhav AJ, Pinjari DV, Mahamuni NM, Pandit AB (2016) A critical review on textile wastewater treatments: possible approaches. J Environ Manag 182:351–366. https://doi.org/10.1016/j.jenvman.2016.07.090
8. Ahmad A, Razali MH, Mamat M, Kassim K, Amin KAM (2020a) Physiochemical properties of TiO_2 nanoparticle loaded APTES-functionalized MWCNTs composites and their photocatalytic activity with kinetic study. Arab J Chem 13:2785–2794. https://doi.org/10.1016/j.arabjc.2018.07.009
9. Ahmad K, Ghatak HR, Ahuja SM (2020b) A review on photocatalytic remediation of environmental pollutants and H_2 production through water splitting: a sustainable approach. Environ Technol Innov Elsevier B.V. https://doi.org/10.1016/j.eti.2020.100893
10. Konstantinou IK, Albanis TA (2004) TiO_2-assisted photocatalytic degradation of azo dyes in aqueous solution: kinetic and mechanistic investigations: a review. Appl Catal B Environ 49:1–14. https://doi.org/10.1016/j.apcatb.2003.11.010

11. Zangeneh H, Zinatizadeh AAL, Habibi M, Akia M, Hasnain Isa M (2015) Photocatalytic oxidation of organic dyes and pollutants in wastewater using different modified titanium dioxides: a comparative review. J Ind Eng Chem 26:1–36. https://doi.org/10.1016/j.jiec.2014.10.043
12. Zhang L, Mohamed HH, Dillert R, Bahnemann D (2012) Kinetics and mechanisms of charge transfer processes in photocatalytic systems: a review. J Photochem Photobiol C Photochem Rev 13:263–276. https://doi.org/10.1016/j.jphotochemrev.2012.07.002
13. Mohd Adnan MA, Muhd Julkapli N, Amir MNI, Maamor A (2019) Effect on different TiO_2 photocatalyst supports on photodecolorization of synthetic dyes: a review. Int J Environ Sci Technol 16:547–566. https://doi.org/10.1007/s13762-018-1857-x
14. Subramaniam MN, Goh PS, Lau WJ, Ng BC, Ismail AF (2018) Development of nanomaterial-based photocatalytic membrane for organic pollutants removal. In: Advanced nanomaterials for membrane synthesis and its applications. Elsevier Inc., Amsterdam. https://doi.org/10.1016/B978-0-12-814503-6.00003-3
15. Leong S, Razmjou A, Wang K, Hapgood K, Zhang X, Wang H (2014) TiO_2 based photocatalytic membranes: a review. J Memb Sci 472:167–184. https://doi.org/10.1016/j.memsci.2014.08.016
16. Khalid NR, Majid A, Tahir MB, Niaz NA, Khalid S (2017) Carbonaceous-TiO_2 nanomaterials for photocatalytic degradation of pollutants: a review. Ceram Int 43:14552–14571. https://doi.org/10.1016/j.ceramint.2017.08.143
17. Ng JJ, Leong KH, Sim LC, Oh WD, Dai C, Saravanan P (2020) Environmental remediation using nano-photocatalyst under visible light irradiation: the case of bismuth phosphate. In: Nanomaterials for air remediation. Elsevier Inc., Amsterdam. https://doi.org/10.1016/B978-0-12-818821-7.00010-5
18. Suvaci E, Özel E (2020) Hydrothermal synthesis. In: Ref. Modul. Mater. Sci. Mater. Eng. pp 1–10. https://doi.org/10.1016/b978-0-12-803581-8.12096-x
19. Li D, Liang Z, Zhang W, Dai S, Zhang C (2021) Preparation and photocatalytic performance of TiO_2-RGO-CuO/Fe_2O_3 ternary composite photocatalyst by solvothermal method. Mater Res Express 8. https://doi.org/10.1088/2053-1591/abdc3b
20. Esposito S (2019) "Traditional" sol-gel chemistry as a powerful tool for the preparation of supported metal and metal oxide catalysts. Materials (Basel) 12:1–25. https://doi.org/10.3390/ma12040668
21. Mudhoo A, Paliya S, Goswami P, Singh M, Lofrano G, Carotenuto M, Carraturo F, Libralato G, Guida M, Usman M, Kumar S (2020) Fabrication, functionalization and performance of doped photocatalysts for dye degradation and mineralization: a review. Environ Chem Lett. Springer International Publishing. https://doi.org/10.1007/s10311-020-01045-2
22. Mohammed MKA (2020) Sol-gel synthesis of Au-doped TiO_2 supported SWCNT nanohybrid with visible-light-driven photocatalytic for high degradation performance toward methylene blue dye. Optik (Stuttg) 223:165607. https://doi.org/10.1016/j.ijleo.2020.165607
23. Valverde Aguilar G (2019) Introductory chapter: a brief semblance of the sol-gel method in research. In: Sol-gel method—Des. Synth. New Mater. with Interes. Phys. Chem. Biol. Prop. pp 3–8. https://doi.org/10.5772/intechopen.82487
24. Kharissova OV, Kharisov BI, Valdés JJR, Méndez UO (2011) Ultrasound in nanochemistry: recent advances. Synth React Inorg Met Nano Metal Chem 41:429–448. https://doi.org/10.1080/15533174.2011.568424
25. Bang JH, Suslick KS (2010) Applications of ultrasound to the synthesis of nanostructured materials. Adv Mater 22:1039–1059. https://doi.org/10.1002/adma.200904093
26. Bang JH, Suslick KS, Kharissova OV, Kharisov BI, Valdés JJR, Méndez UO (2010) Ultrasound in nanochemistry: recent advances. Synth React Inorg Met Nano Metal Chem 22:1039–1059. https://doi.org/10.1080/15533174.2011.568424
27. Theoretical and experimental sonochemistry involving inorganic systems. Springer Netherlands, Dordrecht, pp 1–404. https://doi.org/10.1007/978-90-481-3887-6
28. Ashokkumar M, Cavalieri F, Chemat F, Okitsu K, Sambandam A, Yasui K, Zisu B (2016) Handbook of ultrasonics and sonochemistry. Handbook of Ultrasonics and Sonochemistry. https://doi.org/10.1007/978-981-287-278-4

29. Deshmukh SP, Kale DP, Kar S, Shirsath SR, Bhanvase BA, Saharan VK, Sonawane SH (2020) Ultrasound assisted preparation of rGO/TiO_2 nanocomposite for effective photocatalytic degradation of methylene blue under sunlight. Nano Struct Nano Objects 21:100407. https://doi.org/10.1016/j.nanoso.2019.100407
30. Cruz IF, Freire C, Araújo JP, Pereira C, Pereira AM (2018) Multifunctional ferrite nanoparticles: from current trends toward the future, magnetic nanostructured materials: from lab to fab. https://doi.org/10.1016/B978-0-12-813904-2.00003-6
31. Athar T (2014) Smart precursors for smart nanoparticles. In: Emerging nanotechnologies for manufacturing, 2nd edn. Elsevier Inc., Amsterdam. https://doi.org/10.1016/B978-0-323-28990-0.00017-8
32. Varanda LC, De Souza CGS, Perecin CJ, De Moraes DA, De Queiróz DF, Neves HR, Junior JBS, Da Silva MF, Albers RF, Da Silva TL (2019) Inorganic and organic-inorganic composite nanoparticles with potential biomedical applications: synthesis challenges for enhanced performance. In: Materials for biomedical engineering: bioactive materials, properties, and applications. https://doi.org/10.1016/B978-0-12-818431-8.00004-0
33. Huang G, Lu CH, Yang HH (2018) Magnetic nanomaterials for magnetic bioanalysis. In: Novel nanomaterials for biomedical, environmental and energy applications. Elsevier Inc., Amsterdam. https://doi.org/10.1016/B978-0-12-814497-8.00003-5
34. Ranjith R, Renganathan V, Chen SM, Selvan NS, Rajam PS (2019) Green synthesis of reduced graphene oxide supported TiO_2/Co_3O_4 nanocomposite for photocatalytic degradation of methylene blue and crystal violet. Ceram Int 45:12926–12933. https://doi.org/10.1016/j.ceramint.2019.03.219
35. Tsao KC, Yang H (2018) Oxygen reduction catalysts on nanoparticle electrodes. In: Encyclopedia of interfacial chemistry: surface science and electrochemistry. Elsevier, Amsterdam. https://doi.org/10.1016/B978-0-12-409547-2.13334-7
36. Deraz NM (2018) The comparative jurisprudence of catalysts preparation methods: I. Precipitation and impregnation methods. J Ind Environ Chem 2(1):19–21
37. Munnik P, De Jongh PE, De Jong KP (2015) Recent developments in the synthesis of supported catalysts. Chem Rev 115:6687–6718. https://doi.org/10.1021/cr500486u
38. Mehrabadi BAT, Eskandari S, Khan U, White RD, Regalbuto JR (2017) A review of preparation methods for supported metal catalysts. Advances in catalysis, 1st edn. Elsevier Inc., Amsterdam. https://doi.org/10.1016/bs.acat.2017.10.001
39. Liu Y, Tian J, Wei L, Wang Q, Wang C, Yang C (2020) A novel microwave-assisted impregnation method with water as the dispersion medium to synthesize modified g-C_3N_4/TiO_2 heterojunction photocatalysts. Opt Mater (Amst) 107:110128. https://doi.org/10.1016/j.optmat.2020.110128
40. Ruppi S, Larsson A (2001) Chemical vapour deposition of κ-Al_2O_3. Thin Solid Films 388:50–61. https://doi.org/10.1016/S0040-6090(01)00814-8
41. Cai Z, Liu B, Zou X, Cheng HM (2018) Chemical vapor deposition growth and applications of two-dimensional materials and their heterostructures. Chem Rev 118:6091–6133. https://doi.org/10.1021/acs.chemrev.7b00536
42. Tang L, Tan J, Nong H, Liu B, Cheng HM (2020) Growth of two-dimensional compound materials: controllability, material quality, and growth mechanism. arXiv. https://doi.org/10.1021/accountsmr.0c00063
43. Zhang T, Fu L (2018) Controllable chemical vapor deposition growth of two-dimensional heterostructures. Chem 4:671–689. https://doi.org/10.1016/j.chempr.2017.12.006
44. Karfa P, Majhi KC, Madhuri R (2019) Synthesis of two-dimensional nanomaterials. In: Two-dimensional nanostructures for biomedical technology: a bridge between material science and bioengineering. Elsevier B.V., Amsterdam. https://doi.org/10.1016/B978-0-12-817650-4.00002-4
45. Qiao SZ, Liu J, Max Lu GQ (2017) Synthetic chemistry of nanomaterials. In: Modern inorganic synthetic chemistry, 2nd edn. Elsevier B.V., Amsterdam. https://doi.org/10.1016/B978-0-444-63591-4.00021-5

46. Koo Y, Littlejohn G, Collins B, Yun Y, Shanov VN, Schulz M, Pai D, Sankar J (2014) Synthesis and characterization of Ag-TiO_2-CNT nanoparticle composites with high photocatalytic activity under artificial light. Compos Part B Eng 57:105–111. https://doi.org/10.1016/j.compositesb.2013.09.004
47. Zhang Y, Chen J, Hua L, Li S, Zhang X, Sheng W, Cao S (2017) High photocatalytic activity of hierarchical SiO_2@C-doped TiO_2 hollow spheres in UV and visible light towards degradation of rhodamine B. J Hazard Mater 340:309–318. https://doi.org/10.1016/j.jhazmat.2017.07.018
48. Luna-Flores A, Sosa-Sánchez JL, Morales-Sánchez MA, Agustín-Serrano R, Luna-López JA (2017) An easy-made, economical and efficient carbon-doped amorphous TiO_2 photocatalyst obtained bymicrowave assisted synthesis for the degradation of Rhodamine B. Materials (Basel) 10. https://doi.org/10.3390/ma10121447
49. Ovodok E, Maltanava H, Poznyak S, Ivanovskaya M, Kudlash A, Scharnagl N, Tedim J (2018) Sol-gel template synthesis of mesoporous carbon-doped TiO_2 with photocatalytic activity under visible light. Mater Today Proc 5:17422–17430. https://doi.org/10.1016/j.matpr.2018.06.044
50. Ghime D, Mohapatra T, Verma A, Banjare V, Ghosh P (2020) Photodegradation of aqueous eosin yellow dye by carbon-doped TiO_2 photocatalyst. In: IOP Conf. Ser. Earth Environ. Sci., vol 597. https://doi.org/10.1088/1755-1315/597/1/012010
51. Li W, Liang R, Zhou NY, Pan Z (2020) Carbon black-doped anatase TiO_2 nanorods for solar light-induced photocatalytic degradation of methylene blue. ACS Omega 5:10042–10051. https://doi.org/10.1021/acsomega.0c00504
52. Azzam EMS, Fathy NA, El-Khouly SM, Sami RM (2019) Enhancement the photocatalytic degradation of methylene blue dye using fabricated CNTs/TiO_2/AgNPs/surfactant nanocomposites. J Water Process Eng 28:311–321. https://doi.org/10.1016/j.jwpe.2019.02.016
53. Chen Y, Qian J, Wang N, Xing J, Liu L (2020) In-situ synthesis of CNT/TiO_2 heterojunction nanocomposite and its efficient photocatalytic degradation of Rhodamine B dye. Inorg Chem Commun 119:108071. https://doi.org/10.1016/j.inoche.2020.108071
54. Ali MHH, Al-Afify AD, Goher ME (2018) Preparation and characterization of graphene—TiO_2 nanocomposite for enhanced photodegradation of Rhodamine-B dye. Egypt J Aquat Res 44:263–270. https://doi.org/10.1016/j.ejar.2018.11.009
55. Ambika S, Srilekha V (2021) Eco-safe chemicothermal conversion of industrial graphite waste to exfoliated graphene and evaluation as engineered adsorbent to remove toxic textile dyes. Environ Adv 4:100072. https://doi.org/10.1016/j.envadv.2021.100072
56. Pérez-Molina Á, Morales-Torres S, Maldonado-Hódar FJ, Pastrana-Martínez LM (2020) Functionalized graphene derivatives and TiO_2 for high visible light photodegradation of azo dyes. Nanomaterials 10:1–17. https://doi.org/10.3390/nano10061106
57. Elshahawy MF, Mahmoud GA, Raafat AI, Ali AEH, Soliman EsA (2020) Fabrication of TiO_2 reduced graphene oxide based nanocomposites for effective of photocatalytic decolorization of dye effluent. J Inorg Organomet Polym Mater 30:2720–2735. https://doi.org/10.1007/s10904-020-01463-3
58. Rgo O, Tio B, Jing Z, Dai X, Xian X, Zhang Q, Zhong H, Li Y (2020) Enhanced adsorption and visible-light induced
59. Belayachi H, Bestani B, Benderdouche N, Belhakem M (2019) The use of TiO_2 immobilized into grape marc-based activated carbon for RB-5 azo dye photocatalytic degradation. Arab J Chem 12:3018–3027. https://doi.org/10.1016/j.arabjc.2015.06.040
60. Moosavi S, Li RYM, Lai CW, Yusof Y, Gan S, Akbarzadeh O, Chowhury ZZ, Yue XG, Johan MR (2020) Methylene blue dye photocatalytic degradation over synthesised Fe_3O_4/AC/TiO_2 nano-catalyst: degradation and reusability studies. Nanomaterials 10:1–15. https://doi.org/10.3390/nano10122360